BL
13.46

OCEANOGRAPHY: An Introduction to the Marine Environment

OCEANOGRAPHY

AN INTRODUCTION TO THE
MARINE ENVIRONMENT

PETER K. WEYL

Professor of Oceanography
State University of New York at Stony Brook
and
Research Oceanographer
Marine Sciences Research Center
State University of New York

John Wiley & Sons, Inc. New York · London · Sydney · Toronto

To Muriel

Preface

When I started to teach a general introductory course in oceanography, I asked myself why the college student should learn about the ocean. The obvious answer is that an educated person should know about the ocean because it forms a large portion (about 71 percent) of the surface of our planet. We can study the ocean because it is there; however, a more profound justification is because we exist. The evolution and maintenance of complex forms of life, as we know them, are only conceivable on an ocean planet. Although man evolved on land, the vertebrates originated in the sea, and the maintenance of a stable surface environment for life on earth depends critically on the large capacity of the ocean for heat and chemical substances.

In developing the subject matter of oceanography, therefore, I have not been primarily concerned with a description of the ocean. I am more interested in an exploration of how the ocean has stabilized the surface environment of the earth, not as a passive stage for life, but as a total environmental-life happening. I do not want to know how deep the ocean is but why the ocean is deep. I am not primarily interested in the configuration of the sea floor but in the mechanisms that sculpt it. My task is not to prepare a classified inventory of the life in the sea but to understand how the organisms in the sea interact with each other, with seawater and with the atmosphere, and with the floor of the ocean. Not finding a text that approached oceanography from this point of view, I was led to write this volume.

In organizing the material to be included, I was faced by a dilemma. On the one hand, the orderly development of the subject matter demands a separation into the physical, geological, chemical, and biological sub-disciplines of oceanography. The stability of the marine environment, however, primarily depends on complex interactions between physical, geological, chemical, and biological processes. To meet these conflicting requirements, I have divided the text into six parts.

Part I introduces the time-space scale appropriate to a study of the ocean. Part II considers the dynamics of the earth's fluid envelope. The light from the sun drives the atmosphere and the ocean. These motions depend on the properties of the two fluids—air and water—and are modified by the rotation of the earth.

Part III examines the solid earth beneath the fluid envelope. It develops a dynamic picture from the motion of an eroded rock fragment to the slow drift of the continents of this ever-changing planet. Part IV considers the chemical substances dissolved in seawater, with special emphasis on the carbon-dioxide system that forms a bridge between life and the solid, liquid, and gaseous envelopes of the earth. Finally, it reviews the chemical history of seawater, which leads to a discussion of the origin of life.

Part V examines the requirements for life in the ocean and considers the diversity of plant and animal life in the sea. It places particular emphasis on the manner in which the marine organisms interact with each other and with the physical and chemical aspects of the environment. Part VI, the final part, attempts a partial synthesis by illustrating different types of marine environments. It examines a coral reef to show how communities of organisms interact with the geography of tropical shore-lines. It discusses estuaries where man has had the greatest impact on the marine environment. This leads to a consideration of the circulation of mediterranean seas and to an examination of the vertical circulation in the world ocean. The concluding chapter considers how the circulation of the ocean interacts with the climate to produce stability and change.

My choice of material and its development is biased by my scientific style and background. I make no apology for this because, without strong feelings as to how the ocean should be studied, I would not have undertaken this labor. To be honest with the reader, however, I must comment briefly on my personal adventure with the science of the sea. The reader might think that personal prejudices and a personal style are irrelevant in the natural sciences. The answers to scientific questions must be arrived at by objective techniques and thus, are independent of the personality of the scientist. However, the particular questions we ask of nature and the relative importance we place on the answers we obtain involve subjective judgments. Therefore, while the individual scientific

truths are verified by objective means, the organization, style, and structure of a particular science are the result of the historically integrated
personalities of its past and present practitioners.

My scientific style evolved while I studied for a doctorate in experimental, low-energy nuclear physics. Because of this training, I look at the
world from a dynamic and mechanistic, rather than a descriptive, point of
view. I first began to think about the ocean when I worked in the research
laboratory of an oil company. To contain petroleum, a rock must be
porous. Thus I became interested in the porosity of sedimentary rocks and
how that porosity is modified in the transition from loose sediment to
sedimentary rock. This led me to the shores of the ocean where sediments
are being deposited and transformed. I spent many years studying the
solution interaction between carbonate sediments and seawater. Field
observations and laboratory experiments taught me that there is more to
the solution chemistry of carbonates than can be found in textbooks of
physical chemistry and that biological processes are of particular
importance.

Having been a physicist working on geological problems in the chemistry
department of an oil company, I then became a professor of chemical
oceanography. I studied the physical chemistry of seawater and learned
about the oceans from my colleagues. While on the shores of the Pacific
Ocean, I became intrigued by the contrast between the circulation of this
ocean and that of the better-known Atlantic. This led me to a new
hypothesis about the possible reasons for the climatic variations that we
call the ice ages. As a result, I became interested in the deep circulation
of the ocean, and I accepted a position as professor of oceanography
with an emphasis on the physical aspects.

My previous interest in the chemical stability of the ocean, combined
with studies on the density stratification of the ocean, led me to specu-
late about the origin and early development of life in the sea. Currently,
I am continuing research on the deep circulation of the ocean, and I am
embarking on a study of the effect of man on the shallow-water environment. My present concern with environmental problems has pointed out
the need for an interdisciplinary approach in order to cope with these
problems. Although universities have an excellent record in training
specialists, the departmental organization of universities makes it difficult
to train scientists to work in the exciting interdisciplinary areas that are

of prime importance if we are to cope with the environmental problems created by industrial man. I hope that this volume will focus attention on these problems, which seem to me to be at the frontier of a new era of science.

I am surprised at how much of the contents of this text is the result of recent research. I have had to depend heavily on the recent primary scientific literature. Therefore, to assist the serious student in learning more about the subject and to document statements in the text, the bibliography is divided into two parts. At the end of each chapter there is a "Supplementary Reading" list. This furnishes general references that should be available in most libraries. Articles that require little or no scientific background are marked by an asterisk to distinguish them from more technical sources. References to the scientific literature are cited by the name of the author, followed by the year of publication (Smith, 1966). These literature references are compiled alphabetically at the end of the book, starting on page 514.

A summary at the end of each chapter briefly reviews the high points. This is followed by study questions designed to test the student's comprehension of the material. The questions contain both computational problems and essay topics. Appendix 1 is intended for the student who is not familiar with the exponential decimal notation. The units used in this book are exclusively metric. A summary of the metric system, as well as conversion factors to English units, are found in Appendix 2.

Peter K. Weyl

Stony Brook, New York
June 1969

Acknowledgments

Having had no formal training in the science of the sea, I am grateful to many colleagues who patiently contributed to my education in oceanography. Since one never really learns a subject until one has to teach it, I am particularly grateful to my students. Most of the illustrations have been superbly drawn by John Balbalis. The assistance of Susan Wechsler in securing photographs from various sources is greatly appreciated. I thank the many individuals, institutions, publishers, and organizations who have generously permitted the use of illustrations. The specific credits are given in the captions.

I am especially grateful to my friend, Thomas F. Goreau, for supplying the photographs in the chapter on coral reefs and for carefully reviewing the contents of that chapter. I am also grateful for the review of the preliminary edition of this text by: John P. Barlow, Edward Baylor, John V. Byrne, Thomas F. Goreau, M. Grant Gross, Joel W. Hedgpeth, J. Murray Mitchell, Jr., Boyd E. Olson, Allison Palmer, John G. Weihaupt, and George Williams. I thank the Shell Development Company, the National Science Foundation, and the Office of Naval Research for supporting my research in oceanography. Without this support, I would never have learned enough about the ocean to write this book. The editorial help of Gladys Topkis and the production assistance of Kenneth Burke are gratefully acknowledged. Finally, I thank my wife Muriel who, after working my way through graduate school and raising a family while following me on a scientific journey of exploration from the atom to the ocean, has exerted herself, for better or worse, beyond the call of wifely duty, as secretary, editor, critic, and comfort.

P. K. W.

Contents

III   THE EARTH BENEATH THE SEA

PART IV   THE SALT OF THE SEA

PART V   LIFE IN THE SEA

PART VI   THE MARINE ENVIRONMENT

OCEANOGRAPHY: An Introduction to the Marine Environment

Part I
Perspectives

Oceanography, the study of the marine environment, requires a special perspective. As human beings, we are accustomed to man-size scales in time and space. But these human scales are not appropriate to a study of the oceans of the earth. The oceans are much larger than man, and the earth is much older and has evolved at a much slower rate than its human inhabitants. To get a feeling for the mixture of art, adventure, and science that is oceanography, we briefly review the history of man's explorations of the secrets of the sea. A large part of that history is oceanography in the restricted sense—that is, the mapping of the distribution of the oceans and landmasses on the surface of the globe. Oceanography in the larger sense is concerned with the study of the waters of the ocean, the life within the sea, and the solid earth beneath it.

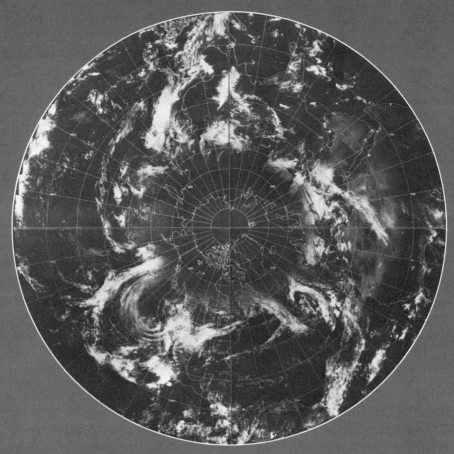

The cloud cover over the northern hemisphere as revealed by the ESSA III satellite on January 6, 1967 (ESSA).

A salinity-temperature-depth sensor being lowered into the Weddell Sea from the icebreaker *Glacier*. (Official U. S. Coast Guard photograph.)

1 Introduction

The nineteenth-century British philosopher and critic George Henry Lewes wrote in his *Seaside Studies* (1858):

The fact is, the sea is a passion. Its fascination, like all true fascination, makes us reckless of consequences. The sea is like a woman; she lures us and we run madly after her; she ill uses us, and we adore her; beautiful, capricious, tender and terrible! There is no satiety in this love; there never is satiety in true affection.

Although many people share Lewes' highly romantic view of the sea, our interest in it has a more practical focus as well. Man is not only fascinated by the oceans; he requires a knowledge of the seas for many useful purposes. The mariner, for example, must know about the geographical distribution of the oceans and their currents and tides in order to bring his ship safely and expeditiously from port to port. The fisherman must know where and how to capture the life that abounds in the sea. The military man must know how to operate on and in the sea, how to hide and find submarines, and when and how to perform amphibious operations. The oil operator and the diamond prospector must know how to find and recover the wealth that lies buried beneath the sea. The meteorologist must know how the sea affects the weather on our globe.

There is also a more profound reason for studying the oceans. So far, there have been two great intellectual revolutions in the natural sciences. The first was the revolution in physics initiated in the sixteenth century by Galileo, who not only removed the earth from the center of the universe but changed the method of approach to scientific truth. He insisted that true answers to the questions of nature could be obtained only by experimentation and not by searching among the writings of the ancient Greek philosophers.

The second great revolution arose three hundred years later, from a combination of developments in geology and biology. The study of fos-

3

sils entombed in rocks led to a new perspective on the history of the earth. The teaching of such geologists as Sir Charles Lyell led the biologists Charles Darwin and Alfred Wallace to the theory of evolution, which transformed the mere enumeration of ancient life forms into a story of gradual and progressive development. It is no coincidence that Darwin and Wallace each arrived independently at the theory of evolution after an extensive ocean voyage to study forms of life on islands separated by the sea.

Later, in the early years of the twentieth century, a controversy arose regarding the antiquity of the earth. The evolutionists insisted that the earth must have had a very long history; the classical physicists, spearheaded by Lord Kelvin (1824–1907), held that the laws of heating and cooling proved that the earth and its sun were relatively young. The controversy was not reconciled until Ernest Rutherford pointed out the significance of the newly discovered phenomenon of radioactivity for the age of the earth. During a lecture to the Royal Institution, Rutherford spotted Lord Kelvin in the audience.

I . . . realized that I was in trouble at the last part of my speech, dealing with the age of the earth, where my views conflicted with him. To my relief, Kelvin fell asleep, but as I came to the important point, I saw the old bird sit up, open an eye and cock a baleful glance at me. Then sudden inspiration came, and I said Lord Kelvin had limited the age of the earth, *provided no new source* [of heat] *was discovered.* That prophetic utterance refers to . . . radium. Behold! The old boy beamed upon me!

Thus Rutherford showed that the laws of physics were consistent with the concept of an ancient earth. The next day the newspapers heralded the longer life expectancy of the cooling earth by the headline: "Doomsday Postponed." Today we tend to be less optimistic regarding the relationship between doomsday and radioactivity.

Developments in biology and particularly the application of the physical sciences to biological problems have solved many of the mysteries of illness and disease and promise to find answers to most of the remaining ones. The application of physics and chemistry to engineering has provided us with a standard of living unknown to the richest men of antiquity. Provided that we can learn to live with one another and avoid a nuclear doomsday, the prospects for the future look bright.

Man has learned how to cure most of the ills of his body and how to shape his immediate environment to satisfy his needs. However, he has not yet learned to live in harmony with the natural environment. We build great cities and are forced to breathe polluted air. We develop great industries and make the water in our rivers unfit for fish and man. The lead we burn in the internal-combustion engines of our automobiles is beginning to collect in the water of the oceans. Mineral resources that have been accumulating for hundreds of millions of years are being exhausted in hundreds of years. Man has been altering the face of the earth at a fantastic rate. Some of the current consequences of these alterations are very apparent, but the future response of the environment to the actions of man cannot be predicted at the present time. Will man learn the limits of the natural life-support system on this planet before he inadvertently exceeds these limits? Are the safety factors of nature adequate, or will the careless use of our resources lead to the extinction of *Homo sapiens* and perhaps most other species? In view of man's technological progress, it is clear that we had better concern ourselves with the stability of the environment.

There are many sciences that deal with our environment, among them meteorology, oceanography, geomorphology, glaciology, and climatology. Each considers a special aspect of the environment and tries to answer problems within the special framework of its discipline. Thus the meteorologist takes the surface conditions on land and on the ocean as given and tries to derive the motions of the atmosphere. The physical oceanographer considers the atmospheric conditions as given and tries to derive the motions of the ocean. The hydrologist tries to predict the water resources in streams, lakes, and the groundwater from measurements of rainfall.

But nature does not recognize the boundaries of the various disciplines. The ways in which she stabilizes the environment as a stage for life transcend the areas of the specialists. It is only as a result of the interactions among the ocean, the atmosphere, and the land that life as we know it has evolved on earth. The development of organisms on earth has not been governed by the physical environment. Rather, the biological and physical aspects of the environment have interacted in such a way as to permit the continuation of life.

As we study the factors that stabilize the environment on earth, it will become clear that the ocean plays a dominant role. Because its size enables the sea to absorb huge quantities of heat and chemical substances without significantly altering conditions on the surface of the earth, the ocean moderates variations and acts as a stabilizing influence. The most important reason for studying the ocean, then, is to investigate the part it plays in the environment. How has the ocean, by interacting with the atmosphere, the land, and the biology on land and sea, been able to stabilize the environment on earth?

The study of the environment is not an experimental science, for it is not feasible to perform controlled experiments that modify the environment on a global scale. However, industrial man is "experimenting" with the environment without conscious design. For example, by burning coal and oil we are significantly increasing the carbon dioxide in the atmosphere, and we have at least doubled the sulphate content in the world's rivers. To learn about the operation of the natural system, we can also turn to the history of the earth, using information from the geologic past to help us understand the mechanics of the environment.

Supplementary Reading

GENERAL REFERENCES ON OCEANOGRAPHY

Dietrich, Günter (1963). *General Oceanography: An Introduction,* translated by Feodor Ostapoff. New York: Interscience Publishers.

Hill, M. N., ed. (1962). *The Sea,* 3 vols. New York: Interscience Publishers.

Murray, J., and J. Hjort (1912). *The Depths of the Ocean.* London: J. Cramer, Weinheim reprint, 1965.

Sverdrup, H. U., Martin W. Johnson, and Richard H. Fleming (1942). *The Oceans: Their Physics, Chemistry, and General Biology.* Englewood Cliffs, N.J.: Prentice-Hall.

2 Man and the Earth

According to the ancient Greeks, the sea was ruled by the whims of the god Poseidon, a brother of Zeus. Today we attempt to understand the behavior of the ocean in terms of physical laws rather than as the result of the volition of manlike deities. The development of science, however, has been conditioned by man's size and lifespan.

Before the invention of the microscope and similar devices, the phenomena examined in the laboratory were perceived by the unaided human senses. Therefore these investigations were limited in size and time. The development of techniques and instruments with which to study the very small and the very fast radically altered the classical sciences. Thus when the microscope and the electron microscope made it possible to study unicellular forms of life, the previous division of these forms into plants and animals and the corresponding division of the sciences into botany and zoology were found to be inadequate. When physicists began to study molecules, atoms, and the parts of atoms, the macrophysics of Newton had to be modified by quantum mechanics. It is now possible, with the aid of technology, to overcome the limitations of our senses to investigate the detailed anatomy of a cell and the kinetics of an explosion.

When we wish to investigate the surface environment of our planet, however, we must operate at the other extremes of time and space. We are concerned with systems that are much larger than man (Fig. 2-1), and the processes of interest generally produce changes over a period that is much longer than a lifetime. When we are studying our environment, technology is of little value in helping us to overcome the limitations of the human body. Time machines which could be used to study the evolution of the landscape exist only in science fiction. With airplanes and satellites, we can survey the surface of the earth from a distance and so encompass a larger fraction of the earth in our view. At

Figure **2-1** Astronaut Edwin E. Aldrin, Jr., working in space with the earth in the background. Picture taken by astronaut James A. Lovell through a window of Gemini XII spacecraft. (Photograph: NASA)

present, however, these observation platforms do not permit a detailed examination of the environment.

To overcome the physical limitations of our bodies, we must use our brains. Thus we must cast aside our usual sense of scale and contemplate the earth from a frame of reference appropriate to its size and age. We must enlarge our horizons to encompass the entire earth, and we must think of history, not in terms of generations, but in terms of millions and billions of years. Only by transcending human scales can we appreciate the environmental drama around us.

Magnitudes

Oceanography is concerned with the properties of microscopic entities, such as the water molecule, and huge systems, such as the Pacific Ocean. It investigates the propagation of sound and electricity through seawater, and the history of the ocean basins. Such everyday terms as "large" and "small," "fast" and "slow," become meaningless in the immense spectrum of magnitudes. "Large" and "small" have meaning when applied to the size of men, but what is a large amount of water? Is it the water needed to fill a swimming pool, Long Island Sound, the

Mediterranean, or the Pacific Ocean? To express magnitudes, therefore, we must use numbers rather than vague adjectives. For example, a water molecule is approximately 0.000,000,03 cm long, and the radius of the earth is 640,000,000 cm. It is more convenient to write these numbers as 3×10^{-8} cm and 6.4×10^{8} cm, respectively. The advantage of *exponential notation,* as this system is called, becomes even greater when we must make computations with very small or very large numbers. (The algebra of exponential numbers is briefly reviewed in Appendix 1.)

To facilitate our handling of numbers, we shall adopt the metric system, which is universally used in science. The use of this system eliminates the awkward conversions of the English system—for example, from inches to feet to miles $(1:12:5280)$. Instead, the conversions from centimeter to meter to kilometer involve only powers of ten $(10^{-2}:1:10^{3})$. (The metric system and its English equivalents are summarized in Appendix 2.)

When a quantity is designated, it is necessary to specify the numerical value as well as the units of measurement of the quantity. The fundamental units are length, mass, and time. Other units are often designated in terms of these fundamental units. Thus speed is the distance traveled in unit time, and the unit of speed is a unit of length divided by a unit of time—for example, centimeters per second or kilometers per hour. These compound units can be written in a number of ways; we can write centimeters per second as cm/sec, $\frac{cm}{sec}$, cm per sec, or cm sec^{-1}. As long as the units are fairly simple, any of these methods is satisfactory. Complex units, however, are preferably designated by positive and negative exponents. Thus the units of the Stefan-Boltzmann constant (p. 74) are cal cm^{-2} sec^{-1} deg^{-4}, which is equivalent to $\frac{cal}{cm^2\ sec\ deg^4}$. For consistency, we shall designate all units in terms of positive and negative exponents.

Unfortunately, there are no decimal units of time. Therefore, in general, we shall limit ourselves to the units seconds and years; one year is approximately 3.16×10^{7} sec. Another conversion factor that will prove useful is that 2^{10} is approximately 10^{3}. Ten generations, or 250 years, ago you had approximately 1000 great great . . . grandparents. In the year zero the number of your ancestors was 10^{24}.

To get an idea of the disparity in magnitude between man and the earth, let us compare their dimensions:

	Man (Maximum)	*The Earth*
Length	Height 2 meters	Radius 6.4×10^6 meters
Age	10^2 years	4.5×10^9 years

Length

Whereas the English unit of length, the foot, is derived from the average length of a part of the human anatomy, the size of the meter is derived from the size of the earth. In 1795, in order to obtain a convenient unit of length, the French National Assembly set up a commission to determine the distance on the earth's surface from the equator to the North Pole along a north-south line (*meridian*) running through Paris. The meter (which is approximately equal to the length of the yard) was then defined as 10^{-7} of that distance. Since the distance represents one-fourth of the circumference of the globe, or $2\pi R/4$, the radius of the earth $R = 4 \times 10^7/2\pi$. And since $\pi = 3.14$, $R = 6.4 \times 10^6$ m.

The meter has had a varied legal history since its adoption by the French National Assembly in December 1799. In France a heavy penalty was exacted for the use of any unit other than the meter, but in England there was a period when it was a crime to use or have in one's possession weights or measures of the metric system. Fortunately this law was abolished in 1897, and the meter has been universally legalized. By 1970, according to a recent act of Parliament, the metric system will be official throughout England, but its adoption by the United States, although probably inevitable, is yet to be realized. It has been easier to measure the earth than for governments to agree on the units in which to express that size.

The earth, to a very good approximation, is a sphere. This fact is more apparent on the level ocean than on the often irregular terrain of the land. We learn that the earth is round so early in our education and take it so much for granted that some eminent scientists have ignored this fact. For convenience, when we draw a cross section of part of the earth's surface, we represent sea level by a straight line. Some geologists inferred from this convention that the downbuckling of a basin would lead to an elongation of the surface and so to tensions. Actually, when we take the

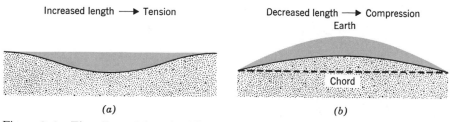

Figure **2-2** The effect of downbuckling on the surface of the earth. (*a*) Conventional presentation. Downbuckling appears to result in increased length and, therefore, tension. (*b*) Actual effect on the spherical earth. Downbuckling results in shortening, hence compression.

curvature of the earth into account, we find that the exact opposite is true. The downbuckling brings the surface closer to the chord; therefore the length shrinks, producing compressions (Fig. 2-2). Thus eminent scientists have been fooled by their ways of representing the earth. Overlooking the obvious can result in a wrong picture of the world.

The curvature of the earth determines the distance from the observer to the horizon. If our eyes are 2 m above the surface, the distance to the horizon is 5 km. This distance increases as the square root of the observer's height above the surface.

This statement can be proved by considering the right triangle formed by our eyes, a point on the horizon, and the center of the earth (Fig. 2-3). Let r be the radius of the earth, d the distance to the horizon, and h the height above the ground. According to the Pythagorean theorem,

$$(r + h)^2 = r^2 + d^2 = r^2 + 2rh + h^2$$

Since the height h is very much smaller than the radius of the earth, we can ignore h^2 relative to $2rh$. Thus we obtain

$$d^2 = 2rh \quad \text{and} \quad d = \sqrt{2rh}$$

Substituting $r = 6.4 \times 10^6$ m and $h = 2$ m gives

$$d^2 = 25 \times 10^6$$
$$d = 5 \times 10^3 \text{ m} = 5 \text{ km}$$

When we are standing on level ground, our vision encompasses an area of about 80 km², or about 1.5×10^{-7} of the surface area of our

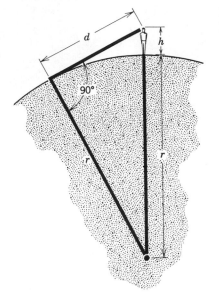

Figure **2-3**   Determining the distance to the horizon (not to scale).

planet. To see a significant fraction of our earth, then, we must contemplate it from a height of several hundred kilometers. It is only since the space age that a few men have seen the earth as it is revealed on a globe.

Time

We can overcome the space limitation of our vision by observing from a great height or by traveling over the surface of the earth. The time limitation, however, is not so easily overcome. The historical record spans only the last 5000 years, and direct observations of a phenomenon by a single investigator seldom exceed a year or two and never span 100 years.

To get a feeling for the time scale appropriate to the earth, consider the human heartbeat. Man's pulse beats about 72 times per minute. During an average lifetime of 60 years, therefore, the heart beats 2.3×10^9 times. If we consider a year, the time the earth takes to rotate once around the sun, as the "heartbeat" of the earth, then the earth has existed for the equivalent of two lifetimes, about 4.5×10^9 years. *Homo sapiens* has existed for only about 2×10^6 years, corresponding to the last 18 days of our earth's two lives.

It is very difficult for us to conceive of the immense period of time that the earth has existed. Perhaps we can form some idea if we assume that every year a sheet of paper is laid down, each piece on top of the last. Let us assume, further, that the paper is so thin that it will take 1000 sheets to produce a thickness of 1 cm. If this process is continued for a time equal to the history of the earth, a tower of paper 45 km high—many times the height of the highest mountain—would be built. In a similar way, processes that would be imperceptible during a man's lifetime can drastically alter the face of the earth.

A study of the ocean involves a wide spectrum of time scales. We measure the depth of the sea with a high-frequency sound wave; for a single ocean wave to pass a fixed point takes from seconds to minutes; the tide rises and falls twice a day; and the seasons recur every 365 days. The geologic period we live in, the Quaternary, is marked by climatic variations with a period of tens of thousands to hundreds of thousands of years. Owing to the melting of the continental ice sheets, the sea began to rise from 100 m below its present stand only 12,000 years ago, reaching its present level approximately 4000 years before Christ. The magnetic field of the earth has been reversing once or twice every million years, while the continents have been drifting about so as to change the locations of the continents drastically in a period of one hundred million years.

To us, the landscape appears essentially static, disturbed on only rare occasions by the birth of a volcanic island or by an earthquake that causes a landslide. If we view our planet with a more appropriate time scale, the picture changes. Mountains rise and are eroded. Volcanoes emerge from the ocean, are ringed by coral reefs, and subside again beneath the waves. The continents shift about to produce an ever-changing geography. On a scale of hundreds of millions of years, the solid surface of the earth is as dynamic as the waters of the ocean and the air of the atmosphere are on a scale of days.

It is only because the earth is an active planet that man has been able to evolve. If the earth were inert, the continents would long ago have been wasted away and drowned under the sea. Not only would this have prevented the evolution of life on land, but the life in the sea itself would have found it difficult to survive. Without the erosion of rocks on land, it is difficult to see how the oceans could have retained the minerals necessary for life support. To appreciate the earth as a stage for life, we must transcend our human time-space scale and look at our planet from a new point of view.

Study Questions

1. Assume that you are on a small island. You know that the distance to the mainland is exactly 20 km. If you have a meter stick and a pair of binoculars, how can you measure the radius of the earth?
2. The present world population is 2½ billion. Assume that a man lying down occupies 1 m². If the world population doubles every 40 years, in what year will the people of the earth, lying next to one another, cover the surface of the earth?
3. The distance from the equator to the pole is divided into 90 degrees of latitude, and each degree is subdivided into 60 minutes of arc. The nautical mile is defined as one minute of latitude. How many kilometers is one nautical mile?
4. One knot is a speed of one nautical mile per hour. How many centimeters per second does this represent?

Supplementary Reading

(Starred items require little or no scientific background.)

*Hurley, Patrick M. (1950). *How Old Is the Earth?* Garden City, N. Y.: Doubleday Anchor Books.

3   The History of Life on Earth

The distinctive feature of the earth's surface, including the oceans, is that it is populated by living organisms. But the surface of our planet is not just a stage for life; the oceans and the surface of the land have been modified by life, for life processes interact decisively with the environment. We cannot understand the marine environment if we restrict our study to either the life or the nonlife processes. We cannot consider geology without life, nor can we study life without considering geology. Because of the importance of life processes, let us briefly review what is known about the history of life on earth.

Our solar system, including the planet earth, is about 4.5×10^9 years old. How soon after the earth was formed did life originate?

Organisms have left a record imprinted in sedimentary rocks in the form of fossils. Rocks also contain naturally radioactive elements such as uranium and thorium. These radioactive components decay with time, forming other elements. By studying the radioactivity of rocks, it is possible to infer the time that has elapsed since the rocks formed. By combining information of fossils with absolute-age determinations from adjacent rocks, paleontologists have gradually pieced together the history of life on earth, (Fig. 3-1).

Sedimentary rocks younger than the beginning of Cambrian times (600 million years ago) often contain an abundance of fossil remains, while Precambrian rocks appear to be devoid of fossils. When suitable Precambrian rocks were examined under the electron microscope, however, it was found that life has existed on earth for at least three billion (3×10^9) years. In 1965, Barghoorn and Schopf found bacteria-like microfossils about 0.5×0.25 micron $(10^{-6}$ m) in black, carbon-rich cherts from the Fig Tree formation of South Africa, a rock sequence estimated to be 3.1×10^9 years old (Fig. 3-2). Their contention that these rod-shaped forms, named Eobacterium *isolatum,* were organisms

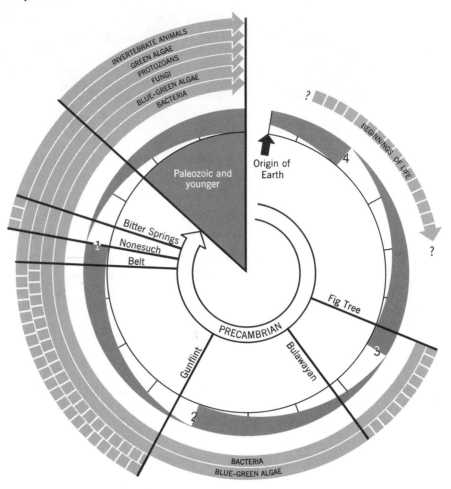

Figure **3-1**   The history of life on earth. (After Schopf, 1967)

and not inorganic artifacts was corroborated when complex organic molecules were found to be associated with the fossils. Future discoveries will undoubtedly bring even older fossils to light.

There are other evidences that life has existed on the earth for at least the last three quarters of its history. The first simple cells capable of photosynthesis, blue-green algae, have been found in the Bulawayan rocks of South Africa, which are 2.7×10^9 years old. In the Great Slave

Lake region of Canada, Hoffman found extensive beds formed by mats of blue-green algae about 2×10^9 years old (Fig. 3-3).

Bacteria and blue-green algae are simple cells that do not contain nuclei. All plants, animals, and most unicellular organisms have cells of complex structure, including a nucleus. The oldest complex unicellular organisms have been found in the 10^9-year-old Belt series of Montana and the Nonesuch shale of Michigan. The slightly younger Bitter Springs rocks of central Australia also contain the remains of green algae. These are the simplest plants that contain a nucleus.

Figure **3-2** Electron micrographs of Eobacterium *isolatum* in Fig Tree chert (3×10^9 years old): (*a*) a well-preserved rod-shaped cell; (*b*) transverse section of the fossil organism showing the preserved cell wall. "Length of line in photo is 1 micron" (Barghoorn and Schopf, 1966). (Photograph courtesy of E. S. Barghoorn)

(a)

polishing
scratch

(b)

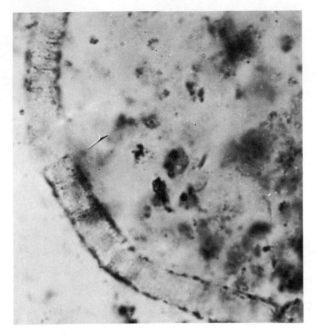

Figure **3-3** Filaments of blue-green algae from the Precambrian Gunflint chert about 2×10^9 years old (Barghoorn and Tyler, 1965). (Photograph courtesy of E. S. Barghoorn)

The next forms of life to appear in the record are impressions of soft-bodied multicellular organisms that have been preserved in the Ediacara sandstone of Australia (Fig. 3-4), which underlies the oldest Cambrian rocks on that continent. The impressions found in this rock resemble jellyfish, worms, and sea pens, as well as other forms that have little likeness to living animals.

The Beginning of the Cambrian

The beginning of the Cambrian is marked by the sudden appearance of various hard-shelled marine invertebrates. Within a period estimated to be about 5×10^6 years, mollusks, sponges, brachiopods, echinoderms, arthropods, and the now-extinct archeocyathids (calcareous organisms resembling sponges) have all been found in the fossil record (Fig. 3-5). Of the phyla of invertebrates, only the bryozoa made their first appearance after early Cambrian times. The introduction of shelled invertebrates is as significant to the paleontologist as the introduction of writing is to the historian.

What caused this sharp discontinuity in the record of life on earth? Why does the fossil record become so rich and diverse within an interval

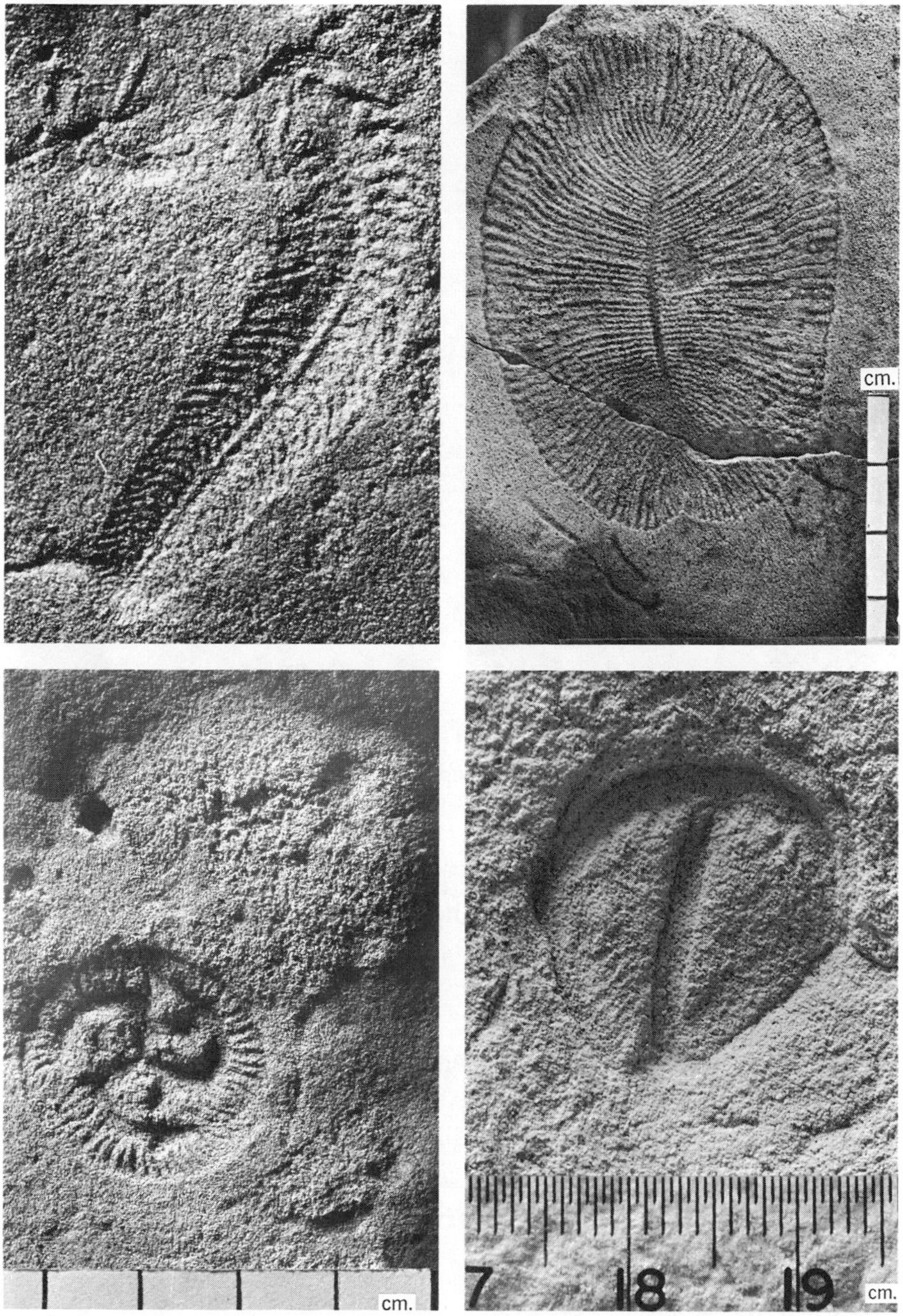

(a)

(b)

(c)

(d)

Figure **3-4** Impression of solf-bodied organisms from the Ediacara of Australia: (*a*)
Spriggina floundersi; (*b*) *Dickensonia costata;* (*c*) *Tribrachidium heraldicum;* and (*d*) *Par-vancorino minchonii.* (Photographs: M. E. Glaessner)

(a) (b)

Figure **3-5** Cambrian fossils: (*a*) a trilobite from northeast Australia. (Photograph: C. G. Gatehouse) (*b*) A Crinoid. (Photograph: Richard Robinson, 1965)

that amounts to only 0.1 percent of geologic time? A possible answer is suggested by a reconstruction of the probable evolutionary history of the marine environment. Only after examining the processes in the present oceans can we profitably discuss this hypothesis (see Chap. 26, pp. 440–442).

The drastic and sudden global change in life style that marks the beginning of the Cambrian period must have been the result of a change in the global environment. As we shall see, this change was, in turn, the result of biological processes. Thus cause and effect do not have a simple linear relationship but form a network of interactions that couple life to the surface of our planet.

Life since the Cambrian Period

Study of the fossil record since the beginning of Cambrian times shows that the history of life on earth is not one of continuous gradual evolution. Changes appear during relatively short periods, with long intervals of relative stability. These "rapid" changes are, of course, rapid only on a geologic time scale.

There have been two episodes when large numbers of diverse species became extinct. Accordingly, geologists have divided geologic time since the beginning of the Cambrian period into three *eras*. The first great extinction, which occurred about 230 million years ago, separates the *Paleozoic* from the *Mesozoic* era; the second about 65 million years ago, separates the Mesozoic from the *Cenozoic*. Although these extinctions have had significant global effects on life, they have not seriously affected the continuity of evolution. There is no evidence that the more complex forms of life on earth ever became extinct, so that evolution had to start over again from simpler forms. Thus, although the environment has been marked by periods that were fatal to many species, conditions on earth have remained compatible with life. Considering the complexity of the environment and the enormity of geologic time, this is a remarkable record.

Lesser changes in the fossil flora and fauna are used to subdivide geologic time into *periods* (Fig. 3-6). As we have noted, the beginning of the Cambrian period is marked by the sudden appearance of various hardshelled marine invertebrates. During early Paleozoic times, life was limited to the sea. Corals and the first vertebrates, fish, appeared about 450 million years ago, in Ordovician times. One of the early fish, the shark, has survived essentially unchanged since Devonian times.

The first evidence of vegetation on land appears in Silurian times, about 410 million years ago. These early plants had no proper roots; the lower part of their stems served as a root. Once the land became vegetated, animals became adapted to venturing out of the sea; thus the first amphibians evolved, about 350 million years ago. By the end of the Paleozoic era, the land harbored trees and reptiles as well.

The end of the Paleozoic era is marked by mass extinctions of marine invertebrates. The plants and animals on land, however, were little affected by the crisis that killed many forms of life in the sea. During the next era, the Mesozoic reptiles were the dominant form of life. Diversification among the reptiles produced the giant dinosaurs (Fig. 3-7).

The last geologic era, the Cenozoic, started about 65 million years ago. During this time the mammals, which originated about 180 million years

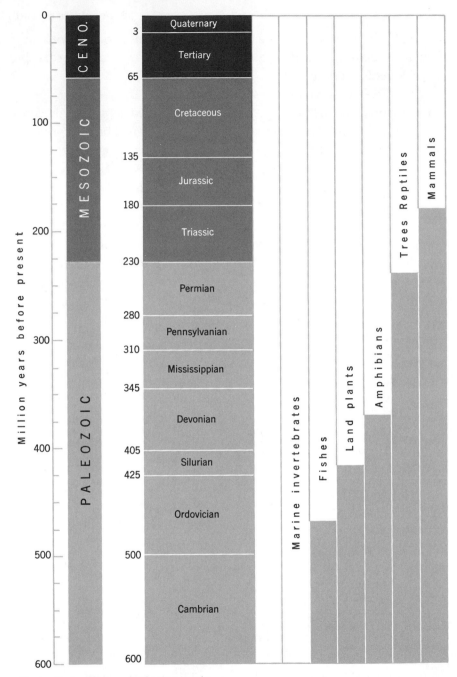

Figure **3-6**   The geologic time scale.

Figure **3-7** Tyrannosaurus and Triceratops, two dinosauria from the Cretaceous. (From a painting by C. R. Knight; courtesy American Museum of Natural History)

ago, became the dominant form of life. Some mammals returned to the sea as porpoises and whales (Fig. 3-8).

In addition to showing the diversification of plants and animals with time, the fossil record also shows that the geography has been dynamic. Marine fossils have been found on high mountains, and fossils of tropical organisms in the far north and on the Antarctic continent. The geographic distribution of fossils through time raises many problems which can be resolved only if we regard the distribution of continents and oceans as changing. But in spite of, or perhaps because of, these

Figure **3-8** The porpoise, a marine mammal. (Photograph: Charles Meyer, from National Audubon Society)

changes in the environment on the surface of our planet, the record of life has been continuous. This brings us to the impact of industrial man on the natural environment.

The Effect of Man on the Environment

All organisms interact with the environment in which they live. Since the industrial revolution, however, man's effect on the environment has been on a hitherto unprecedented scale. As a biologic organism, the average man must daily consume an amount of food having an energy equivalent of 2×10^6 cal and he must drink 1.5×10^3 g of water. In the United States in 1955, the daily per capita power consumption was 2×10^8 cal and the per capita water consumption was 2.6×10^6 g. The contrast between the requirements of biological man and those of industrial man shows dramatically that the impact of man on the environment is unlike that of any other organism in the history of the earth.

If we represent the time since the Cambrian period as one year, the period of man's industrial activity amounts to only five seconds. During these "five seconds" we have driven the fish from many rivers, lakes, and estuaries. Many species of birds and animals have become extinct or are close to extinction. To remain camouflaged in a landscape blackened by industry, some insects have evolved black colorations. In a very short time man has drastically altered the economy of nature.

Are the changes caused by man *unnatural* and basically different from the *natural* processes? By burning fossil fuels we add carbon dioxide and sulphur to the atmosphere, but so do volcanic eruptions and forest fires. We change the regime of rivers by building dams, but rivers can also be dammed by landslides and lava flows. We add poisons to the environment, but the weathering of some rocks also adds toxic substances to the water. We pollute rivers and estuaries, but so do birds. Man is much faster and more thorough than nature in changing the landscape, but then nature has had much more time. For the "five seconds" of man's industrial activity, nature has had "a year" (Fig. 3-9).

When building spaceships, engineers have difficulty designing a life-support system that will last for a round trip to the moon. In contrast, the life-support system of our planet has been functioning for 3×10^9 years. One of the reasons for this is that the earth is an ocean planet. While rivers and lakes are easily polluted, the size of the ocean helps to

Figure **3-9** Man's impact on the environment. The Jersey Meadows. (Photograph: Bruce Davidson, Magnum)

preserve the life in the sea. When we consider the immensity of geologic time, however, the stability of life in the ocean is by no means self-evident. To appreciate that stability, we must study the marine environment. But we cannot limit our inquiry to the ocean, for the sea interacts with the atmosphere and the land. It is only through this interaction that the ocean has been able to support life during the last three billion years.

Summary

Life on earth is almost as old as the earth itself. The first organisms, bacteria and blue-green algae, have existed for about 3×10^9 years. The beginning of the Cambrian, 6×10^8 years ago, is marked by the sudden appearance of most of the shelled invertebrate phyla in the sea. Since then, evolution has led to the emergence of vertebrates and to the population of the land by plants and animals. Since the industrial revolution, man has been altering the environment at an unprecedented rate.

Study Questions

1. Life probably evolved first in the sea. Why is it simpler for life to evolve in an aqueous environment than on the dry land?
2. What would the fossil record be like if life had advanced to the late Precambrian stage, had become extinct, and then had evolved anew?
3. In terms of its impact on the environment, apart from scale, how does a large city differ from an anthill?

Supplementary Reading

(Starred items require little or no scientific background.)

*McAlester, A. Lee (1968). *The History of Life.* Englewood Cliffs, N. J.: Prentice-Hall.

Kummel, Bernhard (1961). *History of the Earth, An Introduction to Historical Geology.* San Francisco: W. H. Freeman.

*Newell, Norman D. (1963). "Crises in the History of Life," *Scientific American,* February.

*Schopf, J. William (1967). "Antiquity and Evolution of Precambrian Life," *McGraw-Hill Yearbook of Science and Technology.* New York: McGraw-Hill.

Woodford, A. O. (1965). *Historical Geology.* San Francisco: W. H. Freeman.

4 The History of Oceanography

The content of oceanography, like that of any other science, depends to a large extent on its history. Let us therefore examine how man has explored the oceans of the world. At first this exploration was largely limited to the surface of the sea and to the mapping of the distribution of land and water. Later man turned to the exploration of the depths of the ocean.

In developing the history of the study of the sea, we shall make no attempt to give a complete account. Rather, we shall try to convey a feeling for how the map of the world was completed, how the geographical exploration of the distribution of land and sea led to an interest in the sea itself, how an oceanographic expedition is conceived and carried out, and how the product of the expedition, its published scientific report, adds to our store of knowledge. Finally, we shall see how puzzling observations led to the development of a dynamic theory, which transformed oceanography from a purely descriptive to an analytical science. Recent discoveries and insights are presented in later chapters, organized by subject matter rather than chronologically.

Prehistory

Man probably evolved in Africa, some two million years ago. By the end of the fifteenth century, when European navigators set out to explore the globe, essentially all the inhabitable areas of the world were inhabited. Obviously the oceans did not prevent the spread of man. Tiny islands in the Pacific, separated by thousands of kilometers from the nearest mainland, were settled by primitive mariners. The ends of the earth, from Greenland in the north to Tierra del Fuego at the southern tip of South America, were populated. The icebound continent of Antarctica was the only major land area in which man had not established residence. Today Antarctica has become a continent of science,

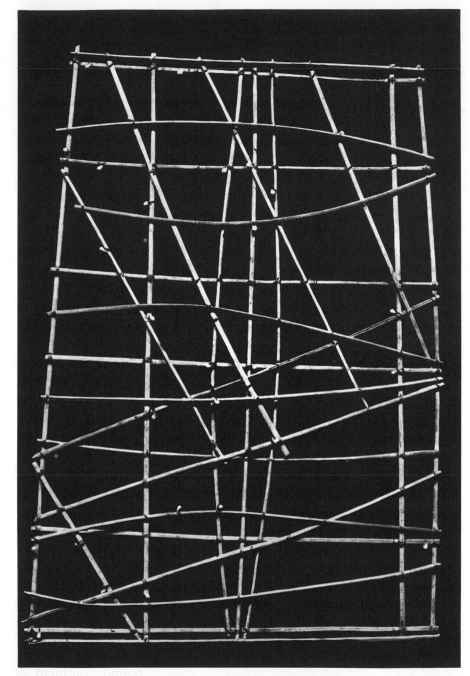

Figure **4-1**   Micronesian sailing chart of the Marshall Islands. A different method for mapping these islands is shown in figure 27-1, p. 450 (British Museum).

with laboratories and observatories established by scientists from many nations. The year-round operation of these facilities is made possible by modern methods of supply.

Although primitive man had explored all the oceans and most of the land areas of the world, his geographic knowledge was localized. The inhabitant of a Pacific atoll knew his immediate environment and had stick charts indicating the locations of neighboring islands (Fig. 4-1). His oral traditions told of the faraway places from which his ancestors had come to settle the island. While there was extensive knowledge of the local geography, there was no geography of the earth as a whole. The gradual evolution of the picture of the world, as seen from Europe, began with the earliest roots of civilization.

The Greeks and Romans

The world of the ancient Greeks consisted of the lands surrounding the Mediterranean Sea. The waters of this ocean enclosed by three continents provided a ready means of transport for goods and soldiers. The master merchant mariners, the Phoenicians, passed through the Pillars of Hercules, the Strait of Gibraltar, into the Atlantic Ocean. They circumnavigated Africa and penetrated north to Great Britain. In spite of

Figure **4-2**   The world according to Herodotus, 450 B.C. (from Challenger office report, Great Britain, 1895).

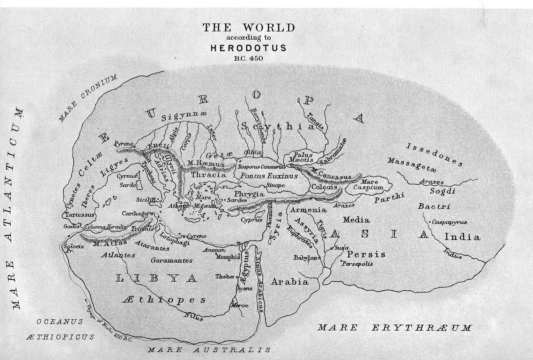

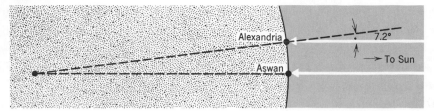

Figure **4-3**   How Eratosthenes determined the size of the earth.

these extensive travels, the Greek philosophers regarded the world as consisting of three concentric circular areas. In the center was the Mediterranean. This was surrounded by continents which, in turn, were surrounded by an external ocean (Fig. 4-2).

Eratosthenes of Alexandria (276–196 B.C.) was the first to determine the size of our planet accurately. He proceeded as follows (Fig. 4-3). He observed the angle the sun made with the vertical at Alexandria at noon on the day of the year when the sun was directly overhead at Aswan, in southern Egypt. He found that this angle was ⅟₅₀ of a circle, or 7.2°. He estimated the distance between the cities to be equivalent to 900 km, giving 45,000 km for the circumference of the earth. Since the actual circumference is 40,000 km, his measurement was only 10 percent in error.

The character of the earth differs markedly from region to region. Some areas are quiescent, and other portions of the globe are often disturbed by earthquakes. The ideas people have about the earth are greatly influenced by the local conditions. Even today, scientists tend to assume that the entire earth is like the areas with which they are familiar. Recently, for example, the presidential address to the Geological Society of America had as its subject "The Quiet Earth." (The earth "replied" with a small earthquake although the meeting was held in an inactive area.) In contrast, the Mediterranean area is very active, with frequent earthquakes and a number of active volcanoes.

Because he lived in an active part of the globe, the Greek geographer Strabo [63 B.C.(?)–A.D. 24] believed that the crust of the earth was in a constant state of flux. He pointed out that earthquakes and volcanic eruptions make the land move vertically and cause the oceans to invade the land. The rivers, by flowing over the land, erode it and transport soil to the sea. Because the eroded material is almost always deposited close to the shore, the oceans do not fill up so rapidly as one might expect. Strabo taught that the landscape is sculpted by the wind as well as by running water. Applying his geographic knowledge to problems in the social sciences, he pointed out that the ocean plays an important role in the development of civilization. Before the introduction of railroads, in the nineteenth century, heavy loads could be carried only by ships. River transport was therefore essential to the ancient kingdoms of the Nile, the Tigris and Euphrates, and the Indus rivers. The Mediterranean Sea was a necessity rather than a barrier for the movement of goods and armies; therefore it was the geographic basis for the Greek and Roman civilizations.

Seneca [54 B.C.(?)–A.D. 30], the Roman rhetorician, was one of the first hydrologists. He noted that because of what is now called the hydrologic cycle, the constant flow of rivers into the sea does not cause the ocean to overflow, nor does the addition of fresh water dilute the saltiness of the sea. Water is evaporated from the ocean, falls as rain, and collects in rivers to return to the sea.

The geographic knowledge of the Romans was summarized by the astronomer Ptolemy about A.D. 150. He introduced the concepts of latitude and longitude and presented a projection of the globe in a map

Figure **4-4**   The world according to Ptolemy, A.D. 150 (from Challenger office report, Great Britain, 1895).

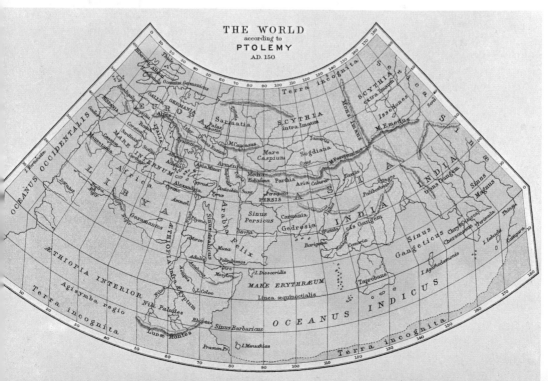

(Fig. 4-4), which showed the Indian Ocean surrounded by land in part unknown. Although it was relatively easy to measure the latitude by noting the angle the pole star makes with the vertical, the ancients had no way to measure longitude directly. Ptolemy therefore estimated distances in the east-west direction from the time required for voyages.

The Middle Ages

During the Middle Ages, geographic knowledge deteriorated severely, as is indicated by the world map of Cosmas, the sixth-century navigator to the Indies (Fig. 4-5). Cosmas insisted that the earth was not a sphere but a quadrilateral, measuring 20,000 × 10,000 km.

While southern Europe was preoccupied with matters of theology, the Norsemen were engaged in journeys of discovery, aided by improved climatic conditions, which reduced the amount of ice in northern waters. Iceland was visited by the Picts and the Celts in 650, and settled by the Celts in 770. In 835 a papal bull referred to Christian settlements in both Iceland and Greenland. The Vikings began to take over these northern lands in about 870. In 982 Eric the Red crossed the Davis Strait from Greenland to Baffin Island, in Canada. Three years later he established a colony in Greenland. His son Leif Ericsson sailed west from Greenland in 995 and spent the winter in Newfoundland.

Because of a deterioration of the climate beginning in about 1200, the

Figure **4-5**   The world map of Cosmas Indicopleustes, sixth century (from Challenger office report, Great Britain, 1895).

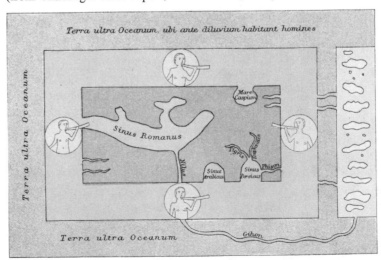

Figure **4-6**   Sigurd Stefansson's map of 1570.

Viking colonies in Greenland became isolated; the Vikings were therefore never able to exploit their discovery of America. Were it not for the readvance of ice in the North Atlantic, the history of America might have been very different. The Vikings' conception of the northern ocean is shown in Sigurd Stefansson's map of 1570 (Fig. 4-6).

While the Vikings were exploring the northern seas, Arab traders were exploring the Indian Ocean and sailed as far as China. A map dated 1030 illustrates the geographic knowledge of the Arabs (Fig. 4-7).

The Arabs brought the lodestone from China and thus introduced the

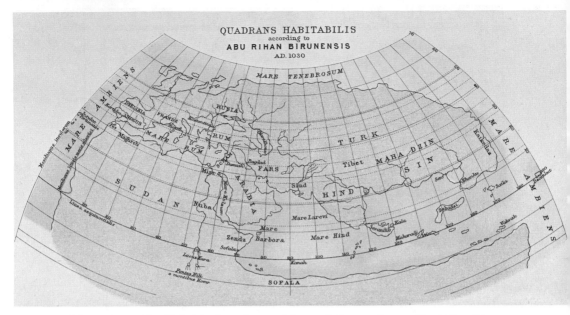

Figure **4-7** The "inhabited quadrant of the earth," according to Abu Rihan Birunensis, A.D. 1030 (from Challenger office report, Great Britain, 1895).

magnetic compass to the West. At first it was viewed with suspicion as being under the influence of some infernal spirit. However, the mariner's need for an instrument with which he could steer a fixed course, regardless of visibility, led to the rapid adoption of the magnetic compass.

The Age of Discovery

The period from 1492 to 1522 is known as the Age of Discovery because geographic knowledge expanded at a very rapid rate during these 30 years. The continents of North and South America were added to the globe, and the earth was circumnavigated. These daring voyages of discovery were brought about by a political event. In 1453 the Sultan Mohammed II captured the capital of eastern Christendom, Constantinople. As a result, the Mediterranean ports were cut off from the riches of the East. At the same time, learned Greeks expelled from Constantinople brought the geographical knowledge of the ancients to Italy, and the introduction of paper permitted the wide distribution of these works. Thus the Turks indirectly revived old knowledge, which made it possible to find new sea routes while creating an economic motivation for exploration.

Meanwhile, the Portuguese and others had been making preparations for their great voyages of discovery. In 1420 Prince Henry the Navigator established a maritime observatory and assembled the best Italian map-makers and pilots to teach navigation to the Portuguese. Until that time the Portuguese had been afraid to sail out of sight of land, and all expeditions to round Africa had turned back at Cape Bojador (27°N). The Cape of Good Hope was finally rounded by Bartholomew Diaz in 1486. In 1498, Vasco da Gama extended the trip around Africa to India.

In 1474, the Florentine astronomer Toscanelli wrote to the King of Portugal suggesting an expedition to explore a route to the spice islands of the East across the Atlantic Ocean. He appended a map which greatly underestimated the distance to the east coast of Asia, placing it at a longitude just off the west coast of America. Later, on inquiry, he sent a copy of this letter to Christopher Columbus.

As every schoolchild knows, in 1492, Columbus set sail westward to reach the Indies. His underestimation of the distance to China caused him to believe that he had reached the Indies when he had in fact discovered what we know as the West Indies; actually he was farther from his goal than when he had left Spain. The Spanish and the Portuguese set out to explore the eastern shores of the Americas and the Indian Ocean. The greatest of the oceans, the Pacific, was not discovered until 1513, when Vasco Nunez de Balboa sighted it from a mountain in Panama.

These early voyages of discovery culminated in the circumnavigation of the globe by Ferdinand Magellan. He left Spain in September 1519 and on the 21st of October of the following year found a passage to the great western ocean at 52°S, now known as the Straits of Magellan. In March 1521 Magellan discovered the Philippines, where he was killed in a fight with the natives. One of his ships, the *Victoria,* under the command of Sebastian del Cano, reached Spain in 1522, thus completing the circuit of the globe. On his trip, Magellan made the first recorded attempt to measure the depth of the open ocean by lowering a weighted line. His ship was able to round the world, but his line was not adequate to measure the depth of the ocean.

Another measuring problem faced by the mariner was how to determine the speed at which his ship was moving through the water. Since there are no fixed reference points at sea, the captain would throw a floating object overboard and time how long it took the object to drift by a measured interval marked off on the deck of the ship. An improved method of measuring speed was introduced by the Dutch near the end of the sixteenth century. This was the so-called Dutchman's log, which has left its traces in nautical jargon.

The Dutchman's log consists of a piece of wood (the log) attached to a reel of string, with knots tied in at equal, fixed intervals. When the log is thrown overboard, an hourglass is inverted. As the sand in the hourglass runs out, the knots that pass overboard are counted. Thus one obtains the speed of the ship in *knots*. The speed is then entered in the *logbook,* with information about the state of the weather and the sea.

A student of the sea whose life was stranger than fiction was Luigi Ferdinando Marsigli (1658–1730). Count Marsigli, a born adventurer, was a fellow of the French and London Royal Societies, a general in the Austrian army, a slave gardener in Turkey, and a pensioner of the Queen of Sweden. Whenever he had the opportunity to do so, he studied the sea. He wrote one of the first textbooks of oceanography, published in Venice in 1711. While at the Bosporus, Marsigli observed the currents flowing between the Black Sea and the Mediterranean. He found that the surface water flows out of the Black Sea, but the deep water flows in the opposite direction. The local fishermen had been making good use of this fact. To travel from the Black Sea to the Mediterranean, a fisherman merely drifted in the surface current. To proceed in the opposite direction, he lowered his net into the bottom current. The large net acted as a sea anchor, dragging the boat toward the Black Sea against the surface current.

When Marsigli studied the deposits brought up from the sea bottom by fishermen, he became interested in the depth of the sea. He investigated the variation of temperature in the Mediterranean and found that it does not change significantly with depth. He measured the density of seawater with a hydrometer and found that the density increases with depth.

The Voyages of James Cook

The early discoverers set out, not to discover the secrets of nature, but rather to find the riches of the world and claim them for their royal sponsors. Their successes were measured in treasure-laden galleons rather than in scientific information recorded in expedition reports.

While the gold of the Aztecs and the Incas has been largely dissipated, the scientific treasures of the newer explorers represent a permanent addition to our store of knowledge.

One of the first to lead a journey intended to produce scientific discoveries was James Cook (1728–1770). The son of an English farm laborer, Cook worked for a shipowner as a boy and rapidly mastered the art of seamanship. In 1755 he joined the royal navy as mate; four years later he became master. His duties involved surveying and mapping the St. Lawrence River and the coasts of Newfoundland. The excellence of the sailing directions he produced and his skill in mathematics and astronomy made him known to the scientific community. Thus when the navy, at the request of the Royal Society, sent a ship to the South Pacific to observe the transit of Venus, Cook was chosen to command the astronomical expedition.

In 1768, Cook set sail on the 370-ton ship *Endeavour,* accompanied by several scientists. On June 3, 1769, having set up their telescopes, they observed the transit of Venus on the Island of Tahiti. Next, Cook set out to search for the great continent that was believed to exist in the south. He discovered the Society Islands; then he charted the islands of New Zealand in the southern hemisphere and explored the east coast of Australia. Returning to England, Cook was put in command of two ships, the 462-ton *Resolution* and the 330-ton *Adventure,* and was ordered to continue the exploration of the South Pacific. Heading south from the Cape of Good Hope, he crossed the Antarctic Circle and sailed east again in search of the southern continent. He failed to sight land until he returned to New Zealand. From New Zealand he headed southeast and penetrated as far as 71°10′S, where floating ice forced him to turn back. Next he explored many islands in the South Pacific, including Easter Island. He was able for the first time to determine accurately the location of these islands.

Cook was the first explorer provided with the proper instruments to determine latitude and longitude accurately. On his second voyage he had four accurate clocks to help in navigation. They had been developed as a result of a naval disaster in 1707, when 2000 men were lost because of faulty navigation. To help avoid such occurrences in the future, Parliament established a prize for a method of determining longitude. The prize offered ranged from 10,000 to 20,000 pounds, depending on the accuracy of the longitude determination. A British carpenter's son, John Harrison, was the first to develop a timepiece that could keep time with sufficient accuracy even on board a moving ship. His chronometer No. 1 was tried at sea in 1735. At the end of the voyage it had an error of only 3° longitude. Since the earth rotates once every 24 hours, the time at noon changes by one hour for every 15° of longitude (360°/24). To be accurate to 1°, a clock must not vary by more than 4 minutes at the end of a voyage.

Harrison continued to improve his chonometer, and in 1761 his model No. 4 was tested. At the end of a two-month trip to Jamaica, the clock was slow by only 9 seconds. When the ship finally returned to England, the total error was less than 2 minutes. In spite of this performance, Harrison only received one-fourth of the prize, and another test was required. This time, at the end of four months, the total error was only 54 seconds.

Cook continued to explore the southern seas in search of the southern continent. He explored south of the tip of South America and finally returned to England after having sailed the equivalent of three times around the equator. In 1000 days at sea, Cook lost only one sailor out of a crew of 118. He was the first to conquer the sailor's disease, scurvy, which results from a lack of vitamin C. By watching the diet of his sailors and giving them citrus juice, Cook showed that long sea voyages were possible without detriment to health. Because they were required to drink lime juice to avoid scurvy, British sailors were thenceforth called Limeys.

In 1776, Cook set out on his third and last voyage. This time his mission was to search for a possible northern passage from the Pacific to the Atlantic through the Arctic Ocean. He sailed eastward around Africa and to New Zealand. Now, however, he went to the North Pacific, stopping in Hawaii and then heading eastward to explore the west coast of North America, from the Oregon coast to the Bering Sea. He got as far as 70°41′N before being stopped by a wall of ice rising 12 feet above the water. After exploring the northern waters off Alaska, he returned to Hawaii. There, on the 14th of February, 1779, he was slain by natives while trying to recover a stolen boat.

Cook's third voyage essentially completed the geographical exploration of the oceans of the world. Only the continent of Antarctica, hidden by a shield of ice, remained to be discovered. (This was accomplished in 1820 by the New England sealing-vessel captain Nathaniel Palmer.) Cook's skill as a navigator set new standards and accurately fixed the location of many islands that had previously been known only vaguely. He showed that long voyages of exploration were possible without endangering the health of the crew. Finally, he demonstrated convincingly that the land was not distributed symmetrically about the equator: the great southern continent did not exist unless it was south of 70°, protected by ice.

From Cook to the Voyage of the *Challenger*

The voyages of geographic exploration undertaken in the nineteenth century focused on the biological and chemical aspects of the sea and on the nature of the land beneath the sea. At the same time, it was realized that a knowledge of ocean currents would help to speed the mariner from port to port. To improve the North Atlantic trade, Benjamin Franklin published a map of the Gulf Stream in 1777, based on his own observations.

The American interest in the practical aspects of oceanography were advanced by the efforts of Matthew Fontaine Maury (1806–1873). Having been injured early in his naval career, Maury was placed in charge of the depot of naval charts and instruments. There he found a collection of logbooks containing a wealth of information about currents and weather at sea. Maury proceeded to analyze these data systematically and from them prepared charts of winds and currents which proved to be extremely useful. In order to obtain even better data, Maury was instrumental in arranging for the first international oceanographic conference. At the Brussels Maritime Conference of 1853, uniform methods of making nautical and meteorological observations at sea were agreed upon. These increased the available data and made them more reliable. In 1855 Maury published *The Physical Geography of the Sea,* a summary of his findings.

The chemistry of seawater was investigated by Johann Forchhammer of Copenhagen, a professor of geology. Over a period of 20 years, Forchhammer analyzed surface samples of seawater brought to him by sailors from all over the globe. When he published his findings in 1865, he demonstrated that while the total salt content of seawater differs from place to place, the relative amounts of the various major salts remain constant.

The question of life in the ocean deep excited the interest of many scientists and led to an error and a controversy that proved to be germinal to the further development of oceanography. The error was the result of an excess of alcohol. In samples brought up from the sea bottom, the biologist Thomas Henry Huxley (1825–1895) discovered a gelatinous ooze containing calcareous particles. This he regarded as the remains of a giant abyssal organism, which he called *bathybius*. This primordial bottom life slime was shown to be an artifact during the *Challenger* expedition of 1872–1876. To preserve the samples from the deep, the scientists had added alcohol. In some cases too much alcohol was added, and as a result, gelatinous calcium sulphate was precipitated from the seawater, forming the slime. This demonstrates how important it is for the scientist to go to sea to obtain and preserve his own specimens.

The controversy arose around the question of whether the deepest part of the ocean contains life or is a zone devoid of living organisms. Does the enormous pressure of overlying seawater kill all organisms, or is the very depth of the sea inhabited? This was a lively topic for arguments at scientific meetings. To find the answer one had to go to sea and find ways to bring up samples from the as yet unmeasured depths of the ocean.

Early masters of the art of deep soundings were two Englishmen named Ross who explored the opposite ends of the earth. The Arctic Ross, Sir John, explored Baffin Bay, in Canada, in 1817–1818. By attaching a "deep sea clamm" of his own design to the end of his sounding line, he was able to measure the depth of the sea and at the same time to bring up samples of the bottom. With this apparatus he dredged up a starfish, as well as mud containing worms, from a depth of 1.8 km. His nephew, Sir James Clark Ross, extended the soundings to greater depth during his Antarctic voyages from 1839 to 1843. He used a 7-km sounding line, and occasionally even this length of line was not adequate to reach the bottom of the ocean. He did extensive dredging and repeatedly brought up animals from below 700 m. He also noted that the same species inhabits the deep waters of the Arctic and the Antarctic. Since these animals were found to be very sensitive to elevated temperatures, he reasoned that the distribution could be explained if the water in the deep ocean had a uniformly low temperature from pole to pole.

A new world of marine life was discovered by a professor of medicine at the University of Berlin, Christian Gottfried Ehrenberg (1795–1876). He traveled widely, exploring the natural history of Egypt, Abyssinia, Arabia, and Russia, all the way to the frontier of China. After returning from these voyages, he examined his collections under the microscope. He discovered that many of the rock samples he had brought back were not inorganic products, as he had thought, but consisted of the remains of countless microscopic animals. In 1836 he showed that many silicious rocks were similarly composed of the remains of diatoms, sponges, and radiolaria. Next he showed that living organisms similar to the ones that make up rocks still inhabit the sea. He reasoned that these rocks are continually forming as a result of the constant rain of dead organisms to the sea bottom. Ehrenberg also showed that the phosphorescence of the sea is due to the presence of microscopic organisms. Thus life in the sea extends from the largest living animal, the whale, to microscopic organisms which are so numerous that their accumulated remains make up thick layers of rock.

Important contributions to the biology of the sea were made by the British naturalist Edward Forbes (1815–1854). Forbes was particularly interested in investigating the vertical distribution of life. He divided the sea into specific depth zones and pointed out that plant life is limited to the zone near the surface. As one proceeds downward, the animals become more and more modified and fewer in number. As one approaches the abyss of the ocean, he thought, only a few sparks of life, if any, remain.

Followers of Forbes went beyond this cautious statement, claiming that a zone of zero life exists in the deep. It seemed absurd to them to suppose that life could exist in the absence of light and air, under the great pressure of the sea. In this belief they ignored the findings of the Rosses and others who had dredged up living organisms from the sea bottom.

The controversy about a lifeless zone in the deep sea helped to stimulate interest in marine biology and thus contributed to the commissioning of the *Challenger* expedition. National rivalry, reminiscent of the current space race, played a role in convincing governments to spend the required funds for scientific research. Spurred by this international competition, in 1871 the Royal Society appointed a committee which recommended that funds be requested immediately from Her Majesty's Government for an expedition with the following objectives:

1. To investigate the physical conditions of the deep sea in the great ocean basins.

2. To determine the chemical composition of seawater at all depths in the ocean.

3. To ascertain the physical and chemical characters of the deposits at the sea floor and the nature of their origin.

4. To examine the distribution of organic life at all depths in the sea as well as on the sea floor.

To carry out these objectives, it was recommended that a sizable ship, a staff of scientists qualified to carry out the desired investigations, and an ample supply of equipment, instruments, and special apparatus be made available. As a result of these recommendations, the first great oceanographic expedition, a model for all subsequent efforts, was organized.

Figure **4-8**   *H. M. S. Challenger* (from Challenger office report, Great Britain, 1895).

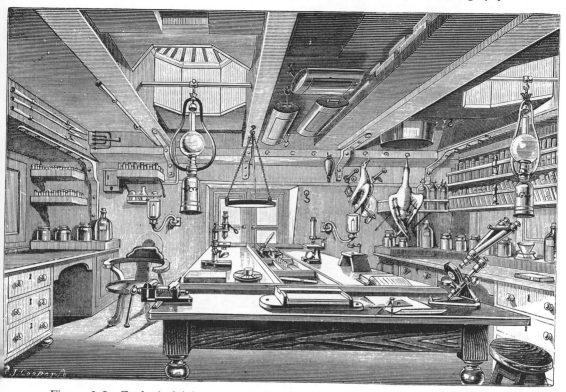

Figure **4-9**   Zoological laboratory on *H. M. S. Challenger* (from Challenger office report, Great Britain, 1895).

The Voyage of *Challenger,* 1872–1876

To carry out the objectives outlined by the Royal Society, the steam corvette H. M. S. *Challenger,* 2306 tons displacement, was refitted. Zoological and chemical laboratories were installed, and the ship was provided with cables for dredging and sounding. To sample the water at depth and measure its temperature *in situ,* special water samplers and thermometers were procured. A staff of six scientists was assembled under the direction of C. Wyville Thompson. The ship, commanded by Captain George Nares, left England in December 1872 and returned in May 1876, after having traversed the Atlantic, Pacific, and Southern oceans.

To get a feeling for the work carried out during the cruise, let us sample the reports:

Figure **4-10** Chemical laboratory on *H. M. S. Challenger* (from Challenger office report, Great Britain, 1895).

STATION 104, August 23, 1873; lat. 2°25′N., long. 20°1′W. Temperature of air at noon, 78.3°F.; mean for the day, 76.6°F. Temperature of water:

Surface	78.0
10 fathoms	78.4
1500 fathoms	36.7
Bottom	36.6

Density at 60°F. at surface, 1.02602; bottom, 1.02601. Depth, 2500 fathoms; deposit, Globigerina Ooze, containing 71.70% $CaCO_3$.

At 8 a.m. shortened and furled sails, and got up steam to sound and trawl. At 9 a.m. brought ship to wind, and sounded in 2500 fathoms. At 1 p.m. lowered cutter to ascertain surface current. At 2 p.m. obtained serial temperatures down to 1500 fathoms. At 5:30 p.m. trawl came up with several specimens.

St. Paul's Rocks distant at noon, 568 miles. Made good 170 miles. Amount of current 20 miles, direction N. 75°W. . . .

Figure **4-11** *H. M. S. Challenger* shortening sail to sound (from Challenger office report, Great Britain, 1895).

Excluding Protozoa, about 20 specimens of invertebrates and fishes were obtained at this station, belonging to about 15 species, of which 10 are new to science, including representatives of 4 new genera; 6 of the new species and 2 new genera were obtained elsewhere. . . .

Yesterday and today Murray put down the surface-net to 100 fathoms. The result was most satisfactory, for the net was full of animals such as are caught on the surface at night, while a similar net on the actual surface yielded next to nothing. The water was perfectly swarming with living animals. It is a great step to have discovered where the surface animals that one catches occasionally at night are to be obtained, and where they live constantly, during the day.

STATION 153, February 14, 1874; lat. 65°42′S., long. 79°49′E. Temperature of air at noon, 33.0°F.; mean for the day, 32.5°F. Temperature of water:

Surface	29.5
100 fathoms	29.0
. . . .	
Bottom	33.0

Density at 60°F. at surface, 1.02413; bottom, 1.02567. Depth, 1675 fathoms; deposit, Blue Mud, containing 3.50% of carbonate of lime, and many rocks and pebbles.

Figure **4-12** *H. M. S. Challenger* passing an iceberg (from Challenger office report, Great Britain, 1895).

At 4 a.m. passed icebergs. At 5 a.m. fifteen icebergs in sight. At 6 a.m. observed open pack ice to S.E. Got up steam to sound and dredge. At 6:30 a.m. shortened and furled sail, and proceeded under steam. At 7 a.m. sounded in 1675 fathoms. Numerous icebergs and pack ice in sight to E. and S.E. At 10 a.m. put dredge over. At 11 a.m. took a series of temperatures at 50, 100, 200, 300, and 500 fathoms. At noon, numerous icebergs and pack ice in

Figure **4-13** Iceberg and pack ice, February 25, 1874 (from Challenger office report, Great Britain, 1895).

sight to E. and S. The carbonic acid was determined in bottom water, and amounted to 82.9 mg per litre. Weather fine; quantity of wash ice about the ship. At 1:15 p.m. commenced heaving in dredge, which came up at 3:15 p.m. with several specimens. At 3:30 p.m. made sail. At 4 p.m. pack ice extending from S. to E. At 6 p.m. 47 icebergs in sight. Pack ice to S.E.; apparently open sea to S.W. Observed several whales during the day, also Cape pigeons, snow-birds, black and white petrels, and penguins. . . .

Excluding Protozoa, over 20 specimens of invertebrates were obtained at this Station, belonging to about 18 species, of which 11 are new to science, including representatives of 5 new genera; 7 of the new species and 3 new genera were not obtained elsewhere. . . .

Whales were seen constantly about the ship. I went away in a boat to shoot birds, and saw a whale blow close by. The spout looks very different from the level of the water in a boat than from the deck of a ship: it appears so much higher, and shoots up into the air like a fountain. In the evening a whale was close alongside, and the expiratory noise in blowing was of a loud, somewhat prolonged, deep bass tone. . . .

During the afternoon of the 16th whales were extremely abundant, both those with the small fin very far back, as well as shoals of a grampus-like Cetaccean with high pointed fins projecting out of the water as they swim, and looking like sharks' fins; on the side, behind the head, they had a white blotch, and a large light transverse patch immediately behind the high dorsal fin, which was placed nearly in the middle of the body.

In the afternoon of the 17th the sea was of a greenish colour, and the water was found to be filled with many little spherical transparent masses, which were identical with those Mr. Murray had observed in the Arctic Ocean. These minute Algae can be seen in the water with the naked eye, when the vessel is held towards the light; they have the surface covered with little dots of a greenish or yellowish tinge, which when examined under high powers were seen to be arranged in groups of four. A few hours later the sea was blue, and these Algae could not be observed in the water. Similar banks of these Algae were passed through on other days when in the neighborhood of the Antarctic ice.

Occasional landfalls offered opportunities to explore islands:

The orange, lemon, and lime, which grow wild all over Tahiti, do not appear to deteriorate at all in quality nor in quantity of the fruit produced in the feral condition, indeed the fruit almost appears finer and better for running wild. The oranges were unanimously pronounced the best ever eaten. The limes lay in cartloads upon the ground, rotting in the woods. It would pay well to make lime juice for export in Tahiti. Some native insect must have adapted itself completely to the blossoms of the orange tribe as a fertilizer, so abundant is the fruit. Vanilla, which is cultivated in the island with success, requires, as everywhere else away from its home, to be fertilized by hand. . . .

The ground just above the shore near Papiete is everywhere burrowed by large land crabs, which are difficult to catch, for they never, in the daytime at least, go far from their holes, but watch a passer-by from near the mouths of their retreats, and bolt in if suspicious of danger, like rabbits. An old marine, named Leary, who acted as a constant assistant to the naturalists whilst collecting on shore, invented a plan by which he caught some of the largest and oldest of the crabs. He tied a bit of meat on the end of a string, fastened to a fishing rod, and by dragging the meat slowly enticed the crabs from their holes, and then made a dash forward, put his foot in the hole, and so caught them. The larger crabs were far more difficult to catch than the younger ones. . . .

On the 3rd October the anchor was weighted and the ship steamed out of Papiete Harbour, the band playing the Tahitian National Air, a quick and

Figure **4-14** Working in the zoological laboratory on *H. M. S. Challenger* (from the Challenger office report, Great Britain, 1895).

lively jig which is characteristic of the place, and sets the Tahitians dancing at once; it is popular with the French also, and as the *Challenger* entered Valparaiso Harbour the band on board a French man-of-war struck up this tune as a greeting to recall the gaiety of the beautiful island left behind.

Challenger returned to England on the 24th of May, 1876, having traversed 68,890 nautical miles, made 492 deep-sea soundings, taken 263 serial water-temperature observations, dredged the bottom 133 times, and trawled the sea 151 times. The completion of the voyage, however, was only a small part of the total effort. The data obtained had to be analyzed, and the collections of sediments and botanical and zoological specimens carefully examined by experts. The final report, issued between 1880 and 1895 under the editorship of John Murray, filled 50 large volumes containing 29,500 pages and more than 3000 illustrations. A total of 76 authors contributed to the report, and numerous other specialists were consulted. No other expedition has made so many important contributions to oceanography.

Oceanography since *Challenger*

Important oceanographic expeditions since *Challenger* include the following:

Ship	Nationality	Period	Oceans Investigated
Gazelle	German	1874–1876	Atlantic, South Indian, South Pacific
Fram	Norwegian	1893–1896	Arctic
Valdivia	German	1898–1899	Atlantic, Indian
Deutschland	German	1911–1912	Atlantic, Antarctic
Discovery	British	1925–1927	Antarctic
Meteor	German	1925–1927	Atlantic south of 20°N
Carnegie	American	1928–1929	Atlantic, Pacific
Dana II	Danish	1928–1930	Atlantic, Indian, Pacific
Discovery II	British	1932–1934	Antarctic
Albatross	Swedish	1947–1948	Atlantic, Indian, Pacific
Galathea	Danish	1950–1952	Atlantic, Indian, Pacific

The expedition of the *Fram* in 1893 was under the direction of the Norwegian polar explorer Fridtjof Nansen. The wooden ship was constructed so that, in Nansen's words, "It should be able to slip like an eel out of the embrace of the ice." Sailing along the northern coast of Russia, the *Fram,* to test the effectiveness of the construction, was frozen into the ice north of the New Siberian Islands. It then proceeded to drift westward with the ice. Because of its unique construction, the ship was not crushed by the ice but rather was lifted up as the ice closed in around it. Thus rafted by the ice, the *Fram* was carried to just north of Spitsbergen, where it once more entered open water and sailed home. Nansen had hoped that the drift would carry the ship across the North Pole; however, when it became apparent that it would miss the pole, Nansen and a companion left the drifting ship to head north, by dogsled. They were forced to turn back at 86° 14', the highest latitude then reached by man, and then made their way south to Franz Josef Land, in the Russian Arctic. They suffered many hardships and were forced to winter on one of the islands. In May of 1896 they set out westward toward Spitsbergen. On the way they met the Jackson-Harmsworth expedition and returned with it, arriving in Norway a week before the *Fram*.

Until the drift of the *Fram*, the Arctic Ocean was believed to be a shallow sea. Nansen, however, had to improvise additional sounding line; thus he showed that the Arctic has oceanic depth. He observed that the polar ice does not drift in the direction of the wind but, rather, at an angle to the right of the wind. This observation led V. Walfrid Ekman, a Scandinavian physicist, to develop a theory of wind-driven ocean currents. Nansen also realized the need for more accurate measurements of salinity and temperature and so developed the Nansen bottle. In these ways he stimulated the study of the sea in Scandinavia. There, through the cooperation of explorers, meteorologists, and physicists, oceanography was transformed from a purely descriptive to an analytical science.

One of the first detailed studies of an ocean was carried out by the German research ship *Meteor,* which made 13 crossings of the Atlantic between 20°N and 60°S from 1925 to 1927. The results of this survey were published in a series of atlases. The *Meteor* was the first expedition to use the echo sounder to measure the depth of the ocean almost continuously along its track. As a result, a more accurate idea of the shape of the ocean floor was obtained than it was possible to form with isolated soundings.

There have been, of course, many more recent contributions to oceanography made by scientists from many countries. Rather than develop the subject chronologically, we shall look, in subsequent chapters, at various aspects of the surface environment of our planet to see how the ocean plays a dominant role in that environment.

How we think about the ocean depends to a large extent on developments in the physical and life sciences. Thus the history of oceanography is part of the more general history of science and technology. As new laws of physics, chemistry, and biology are discovered, we apply these laws to problems of the marine environment. Developments in the basic sciences also lead to new technology, which produces the instruments required to make observations in the sea. Without a steam engine, *Challenger* could not have sounded and dredged the bottom of the ocean. Without developments in electronics, *Meteor* would not have had an echo sounder. Today the use of newly developed instruments, deep submersibles, and satellites is transforming the science of oceanography. It is difficult to predict where these new developments will lead. We can be sure, however, that they will necessitate frequent revision of textbooks such as this one. The sea still holds many secrets to satisfy the curiosity of future generations of oceanographers (Fig. 4-15).

Figure **4-15**   The research vessel *Argo* of Scripps Institution of Oceanography, University of California, San Diego. A 70-m, 2000-ton, converted, former Navy auxiliary rescue and salvage vessel. (Photograph courtesy of Scripps Institution of Oceanography)

Summary

Initially man explored the surface of the planet to map the distribution of land and water. Exploration was largely completed when Magellan completed his circumnavigation of the globe, in 1522. James Cook greatly improved the charting of distant coasts. The *Challenger* expedition gathered the first comprehensive data about the deep ocean. Subsequent expeditions, equipped with newly developed tools, have added considerably to our knowledge.

Study Questions

1. Contrast the voyages of James Cook and the *Challenger* expedition.
2. Compare Ptolemy's map of the world (Fig. 4-4) with a modern map by noting the latitude and longitude of six identifiable features. About 20° of longitude must be subtracted from Ptolemy's scale to obtain the longitude east of Greenwich.
3. The location of Alexandria is 31.20°N and 30°E of the prime meridian at Greenwich. Aswan has coordinates 24.09°N 33°E. The airline distance between the two cities is 840 km. If you assumed that one city was due south of the other, what would you obtain for the circumference of the earth? What is the distance between Alexandria and a point due south of that city on the same parallel of latitude as Aswan?

Supplementary Reading

(Starred items require little or no scientific background.)

* Bailey, H. S., Jr. (1953). "The Voyage of the *Challenger,*" *Scientific American,* May.
* Deacon, G. E. R., ed. (1962). *Seas, Maps, and Men: An Atlas-History of Man's Exploration of the Oceans.* Garden City, N. Y.: Doubleday.
* Guberlet, Muriel L. (1964). *Explorers of the Sea.* New York: Ronald Press.
* Nansen, F. (1898). *Farthest North.* New York: Harper.
Von Arx, William S. (1962). *An Introduction to Physical Oceanography,* Reading, Mass.: Addison-Wesley. Appendix A: Chronological account of events that have influenced thought in the marine sciences.
Wüst, G. (1964). "The Major Deep-Sea Expeditions and Research Vessels 1873–1960—A Contribution to the History of Oceanography," in M. Sears, *Progress in Oceanography,* Vol. 2. New York: Pergamon Press, pp. 1–52.

5 Mapping the Earth

The oceans circle the globe and cover most of its surface. An ever-present problem in depicting the oceans is how to map the earth on paper.

The earth is a sphere; the pages of a book are flat. Therefore it is impossible to portray the features of the earth on a sheet of paper without some distortion. If price were not a consideration, perhaps books about the earth could be published in the form of an onion. Instead of turning pages, one would peel away layer after layer until one got to the core. However, having only two eyes, we are not well equipped to utilize an "onion" library. We would require six eyes on movable stalks to get a simultaneous view of the sphere from all sides and so comprehend the earth as a whole. We would also need a brain able to fuse the six views simultaneously back into a sphere. The study of the earth cannot await the evolution of creatures thus endowed. We must make do with our two eyes and two-sided brain.

Photograph and Projection

The moon, like the earth, is a sphere. In a photograph, we obtain a flat picture of the visible side of the moon. In today's space age, we can also send a camera to the moon and have it take a picture of the earth. Such a picture, transmitted back to earth, is a flat image of almost half of our planet (Fig. 5-1). The central part of the photograph will be a faithful rendition of the earth's surface, but the image will be increasingly distorted as we approach the edge of the visible disk.

If we simulate such a photograph by taking a picture of a globe, as from point *A* in Figure 5-2, we see that equal distances on the globe become shortened as we approach the rim of the picture. The opposite effect would be achieved if we could place a camera *within* the earth. In this case, distances would become enlarged as we approach the periph-

Figure **5-1**   The earth as photographed from the moon. (Photograph: NASA)

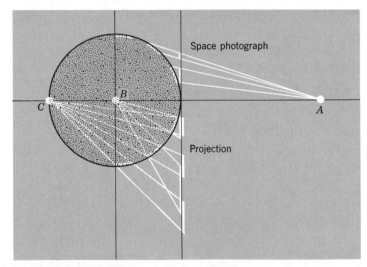

Figure **5-2**   How the surface of a sphere is projected onto a plane surface. In a photograph, the projection is from a point outside the sphere, *A*. To produce a gnomonic projection, the point of projection is at the center of the sphere, *B*. To produce a stereographic projection, the point of projection is on the surface of the sphere, *C*.

Figure **5-3** Latitude and longitude, Washington, D. C.: latitude 38°53′N, longitude 77°0′E.

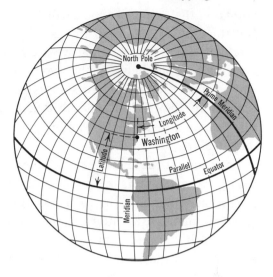

ery. A picture taken from the inside is called a *projection*. The projection obtained if the camera is located at the center of the earth (point *B* in Fig. 5-2) is called a *gnomonic* projection; a projection from the surface (point *C* in Fig. 5-2) is called a *stereographic* projection. To map a hemisphere with a gnomonic projection requires an infinite sheet of paper, whereas the stereographic map will be finite.

Since the time of Ptolemy, about A.D. 150, locations on earth have been specified by giving their latitude and longitude. Globes are provided with a grid of *meridians,* circles running north-south through the poles, and *parallels,* circles running east-west parallel to the equator. The longitude is the angular distance, east or west, from the *prime* meridian, which passes through the Royal Observatory at Greenwich, England. The latitude is the angular distance between the parallel and the equator (Fig. 5-3). The meridians and the equator are called *great circles*. A great circle is formed by the intersection of the surface of the globe with a plane passing through its center. All great circles have the same diameter. Parallels other than the equator, since their planes do not pass through the center of the earth, have smaller diameters.

The Shortest Route

If we wish to travel from one place on earth to another, we need to know the shortest route between them. On a plane, the shortest distance between two points is a straight line. The shortest distance between two points on the surface of the earth is also the straight line joining them. But this line cuts through the earth; to follow this path, we would be forced to dig a tunnel. What we are really looking for, then, is not the shortest distance between the two points but, rather, that path on the surface which represents the shortest distance between the points. If we go from the North to the South Pole, the shortest distance is along the earth's axis, a distance $2R$. A more practical route is along a meridian, a distance πR.

The shortest route between two points on the earth's surface is along the great circle joining the two points. To find the shortest route, we must pass a plane through the earth's center and the two points. The great circle formed by the intersection of this plane and the surface of the sphere is the desired path. On a gnomonic projection, such a plane will appear as a straight line. Therefore, the shortest surface route between two points on the earth is the straight line joining these points on a gnomonic projection. To determine the shortest route between Yokohama and San Francisco, for example, we need a gnomonic chart of the North Pacific. We then join the two ports by a straight line to find the route (Fig. 5-4). Since great circles are straight lines on a gnomonic projection, gnomonic charts are also known as great-circle charts.

San Francisco (38°N) is almost due east of Yokohama (35°N). The shortest route between these cities, however, runs well north of these latitudes, to 48°N. To navigate between the two ports we draw a straight line between the ports on a great-circle chart, as we have shown. But the helmsman must know the direction in which he is to head the ship and the distance over which he is to hold this course. Leaving Yokohama, we

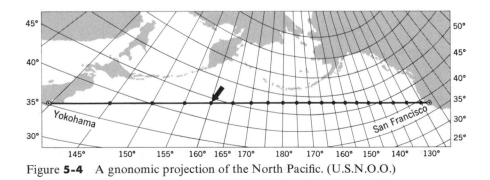

Figure **5-4** A gnomonic projection of the North Pacific. (U.S.N.O.O.)

must head north of due east. Because of the distortion of the latitude-longitude grid on the gnonomic chart, however, it is not possible to determine the exact course.

A Chart for Navigation

How can one construct a chart of the earth from which the heading of a ship can be determined? First, since all meridians point north-south, they will have to be parallel lines intersected at right angles by the east-west-running parallels. Such a chart is obtained by projecting the earth from its axis onto a cylinder having a common axis with the earth (Fig. 5-5). The resulting rectangular grid always has north-south at right angles to east-west.

Figure **5-5** Projecting the earth onto a cylinder.

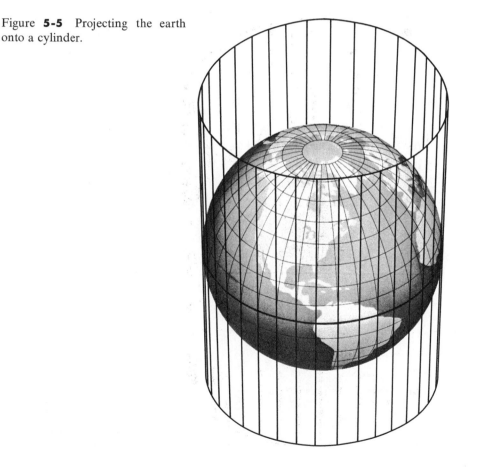

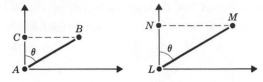

Figure **5-6** Corresponding triangles: *ABC* on the earth and *LMN* on the map projection.

To be useful, however, all directions must be true on the chart. Is it possible to construct such a map? As long as we deal with distances on the earth that are much shorter than its radius, we can consider the earth to be flat. Assume that we are at point *A* on the earth (Fig. 5-6) and wish to go to point *B*. Point *B* is in a direction $\theta°$ east of north. Let *A* and *B* be represented by points *L* and *M* on the chart. In order that the direction from *L* to *M* be the same as the direction from *A* to *B*, it is necessary that the triangles *ABC* and *LMN* be similar. If the map is to show true directions, the ratio of the north-south distance *AC* to the east-west distance *CB* on earth must be the same as the ratio *LN/NM* on the chart.

Now consider two 1° × 1° squares, one located at the equator and one at 60°N latitude. At the equator, a degree of longitude is approximately equal to a degree of latitude. Since the meridian is a great circle, a degree of latitude always has the same length regardless of latitude. A degree of longitude at 60°, however, is only half the length of a degree of latitude. Therefore, the two 1° squares on earth will be as shown on the left of Figure 5-7.

In order to preserve directions, the ratio of lengths on the chart must be the same as on the earth, and since the meridians point north, they must be parallel lines. Therefore a degree of longitude on the map must

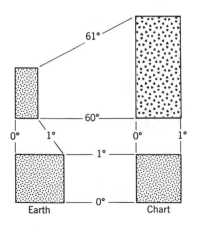

Earth   Chart

Figure **5-7**   Two 1° × 1° squares on the earth and on the Mercator projection.

have the same length regardless of latitude. To maintain the same ratio between E-W and N-S distances, a degree of latitude on the map must be twice as long at 60° as it is at the equator. The two squares, therefore, appear as on the right in Figure 5-7 on the so-called Mercator chart, named after the Flemish mathematician *Gerardus Mercator* (1512-1594), who published a map using this projection in 1568 under the title *Nova et aucta orbis terrae descripto ad usum navigantium accommodata.*

Scale

The scale of a map is the ratio of the distance between two points on the map to the distance between corresponding points on the earth. Thus at a scale of 1 : 1,000,000, 1 cm on the map will represent 10^6 cm or 10 km on the earth. A map showing only a small fraction of the earth can have an essentially uniform scale throughout. If we wish to portray a large portion of the earth, however, the scale must vary over the map. Thus the scale of the navigation chart varies. Specifically, the scale increases (i.e., the map distance divided by the earth distance increases) with increasing latitude.

For navigation, it is necessary to determine distances as well as headings from the chart. This is complicated by the variation in scale with latitude. The navigator can make use of the fact that a degree of latitude on earth always has the same length regardless of the latitude. Therefore, he can use the latitude scale as a measure of length. Actually the earth is flattened slightly at the poles and so is not a perfect sphere. A degree of latitude has a slightly different length at different locations.

Length of 1° Latitude

Latitude	Nautical Miles	Kilometers
0	59.701	110.567
30	59.853	110.848
60	60.159	111.414
90	60.313	111.699

A Mercator projection of the globe is shown in Figure 5-8. As one moves toward the poles, the scale of the map keeps increasing. Since at the pole a degree of longitude has zero length, the scale of the map there becomes infinite; therefore it is not possible to plot the pole on a Mercator projection.

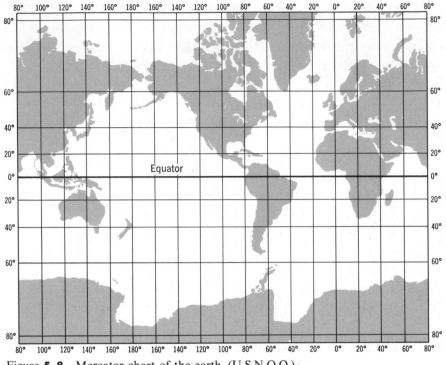

Figure **5-8**   Mercator chart of the earth. (U.S.N.O.O.)

Now let us return to the problem of navigating by the shortest route from Yokohama to San Francisco. In addition to the gnomonic chart (Fig. 5-4), we require a Mercator chart (Fig. 5-9). Using the grid on the gnomonic projection, we now transfer points from the gnomonic chart to the Mercator chart. The straight line of the gnomonic chart becomes

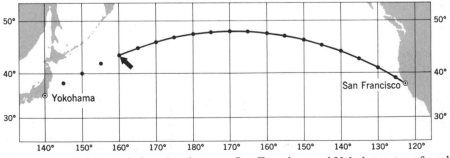

Figure **5-9**   The great circle points between San Francisco and Yokohama transferred from the gnomonic (Fig. 5-4) to the Mercator chart. (U.S.N.O.O.)

a curved line on the Mercator chart. The required heading at any point along the track can be measured with a protractor. To simplify navigation, the curved great-circle track is approximated by a series of straight lines on the Mercator chart.

A Chart of the Globe

We have seen that for navigation we require two projections, the Mercator and the gnonomic. Neither of these permits us to show the entire surface of the globe. There are other types of projections and map grids designed for different purposes, each of which gives a different appearance to the world. If we wish to show only a small fraction of the globe, it is relatively simple to obtain a fair representation. If we wish to encompass the entire globe, however, we must make serious compromises. While the surface of the earth is a continuous surface, our map is bounded at its periphery. Thus points that are close together on earth will appear at opposite sides of the map. Also, as noted, as we proceed from the center of the map to its periphery, the distortion will increase.

One possible solution of the mapping problem is to surround the globe by a plane-faced solid, a polyhedron, and then project from the center of the globe onto the polyhedron. For example, the globe can be enclosed by a cube and projected onto the six square faces. The cube is then unfolded to obtain a flat map. The closer the polyhedron approaches the shape of the sphere, the less the distortion of the final map.

There are five regular polyhedra, solids whose faces are equal regular polygons. These are the tetrahedron, enclosed by four equilateral triangles; the cube, enclosed by six squares; the octahedron, enclosed by eight equilateral triangles; the dodecahedron, enclosed by twelve regular pentagons; and the icosahedron, enclosed by twenty equilateral triangles. The best regular polyhedra to use are the dodecahedron and the icosahedron. It turns out that both will portray the sphere with the same fidelity.

A map of the earth based on the icosahedron was published by Buckminster Fuller (1943) as the "Dymaxion Globe." This was designed to show the continental areas. Here we are more concerned with the oceans of the world. Therefore we reorient the icosahedron to obtain the maps given in Figure 5-10. In order to show the Arctic Ocean better,

62

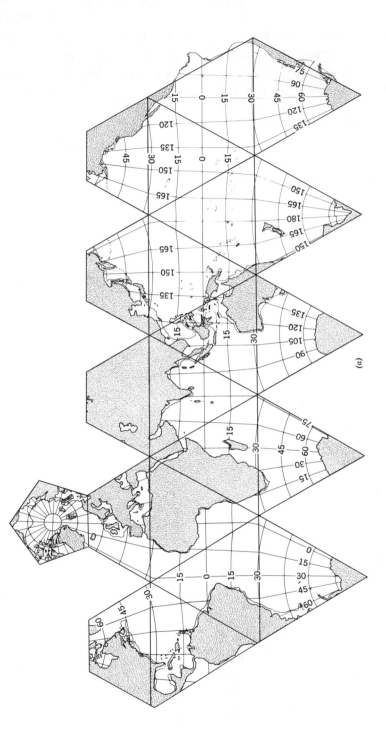

(a)

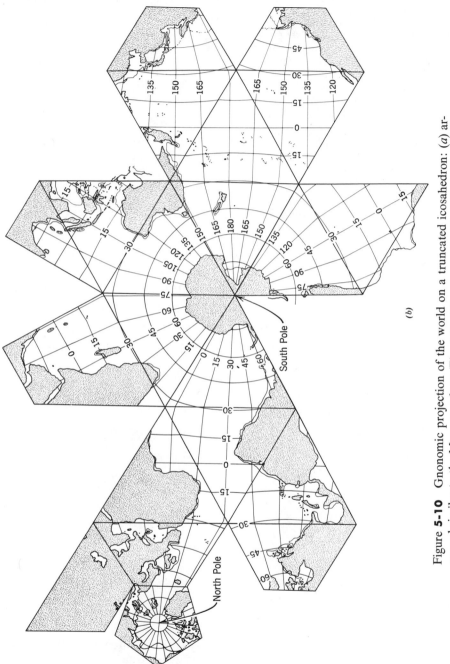

Figure **5-10** Gnomonic projection of the world on a truncated icosahedron: (*a*) arranged similar to the Mercator chart (Fig. 5-8); (*b*) rearranged to show how the oceans merge in the Southern Ocean.

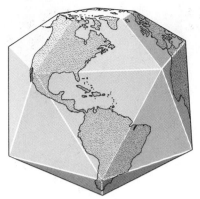

Figure **5-11**   The gnomonic chart of Figure 5-10 assembled into a globe.

the icosahedron in Figure 5-10 is truncated by a regular pentagon at 65°N. The map can be cut out, folded, and assembled into a polyhedral globe (Fig. 5-11). Five equilateral triangles meet at each vertex of the icosahedron. When the polyhedron is unfolded, there is a 60° gap at each corner. The triangles can be arranged in various patterns to show the globe with relatively little distortion. A disadvantage is that the map has a jagged outline. To close the gaps we must introduce additional distortions. Thus one has to choose between faithful ugliness and elegant

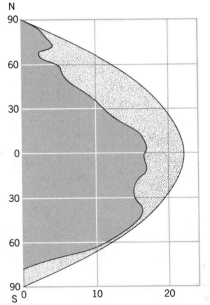

Figure **5-12**   The distribution of land and water with latitude. Units are 10^6 Km^2 per 5° Latitude.

distortion. The only map of constant scale is the sphere, and that brings us back to the notion of an onion-like book with which this chapter began.

The Continents and the Ocean

The distribution of land and water with latitude is shown in Figure 5-12. While the northern hemisphere contains 39 percent land, the southern hemisphere contains only 19 percent. The earth as a whole is 71 percent ocean and 29 percent land. The distribution of land and water is asymmetric about the equator. North of 35°S, the World Ocean is separated by the continents into the Atlantic, the Pacific, and the Indian oceans. Since science has a northern hemisphere bias, we tend to think of these three oceans as separate entities. From the vantage point of the South Pole (Fig. 5-10*b*), however, we see a single ocean circling the Antarctic continent which is separated into three branches in the north.

Study Questions

1. Why does a great circle appear as a straight line on a gnomic projection?
2. The length of a degree of longitude is proportional to the cosine of the latitude. Draw a diagram similar to Figure 5-7 for 1° squares centered at 45° and 80° latitude. What happens as you approach the pole?
3. Would it be possible to make a distortion-free map if the earth had the shape of (*a*) a cone, (*b*) a cylinder, (*c*) an egg?
4. In an atlas, examine a number of different representations of a major portion of the globe. Compare the projections used with Figure 5-10, and discuss the advantages and disadvantages of each.

Supplementary Reading

(Starred items require little or no scientific knowledge.)

Bowditch, *American Practical Navigator,* U. S. Navy Oceanographic Office, Pub. No. 9, Chaps. 2, 3, 8.
* Cortright, E. M., ed. (1969). *Exploring Space with a Camera.* Washington, D. C.: NASA SP-168, U. S. Government Printing Office.
* Greenhood, David (1964). *Mapping.* Chicago: University of Chicago Press, Phoenix Science Series.
Fisher, I. (1943). "A World Map on a Regular Icosahedron by Gnomonic Projection," *Geographical Review,* Vol. 33, pp. 605–619.

Part II
The Earth as a Heat Engine

Our world is one of motion. The earth rotates about its axis and makes a yearly orbit around the sun. The air of the atmosphere and the waters of the ocean are constantly moving. The motions of the earth and those on the earth's surface differ in one important respect. In the vacuum of space the earth moves unimpeded; therefore no energy is needed to maintain the motion. The air and the water, however, are subject to friction. Without a driving force the winds, the waves, and the ocean currents would eventually stop. To maintain the motions of the atmosphere and the ocean a driving force is required.

The energy that powers the winds and the ocean currents, as well as all life processes, is ultimately derived from the sun. Sunlight is intercepted by the atmosphere and the earth's surface, largely absorbed, and then sent back to space as thermal radiation. The earth receives less heat, on the average, at the poles than at the equator. The uneven heating gives rise to movement of the atmosphere and the ocean which carries heat from the equator to the poles. These motions are modified by the rotation of the earth about its axis.

The earth is a heat engine; its "boiler" is the sun, and its heat sink is the cold of outer space. The operating fluid consists of the atmosphere and the ocean. Of particular importance is the interaction between these two fluid realms.

Large cumulus clouds over the Caribbean Sea.

Ships moving through sea ice to carry supplies for the Antarctic research program. (U. S. Navy.)

6 Sunlight

The earth is a satellite of the sun, held in its orbit by the sun's mass. The energy released by nuclear processes within the sun and radiated as light powers the wind of the earth's atmosphere and the currents of the ocean. Until man achieved the first self-sustaining nuclear chain reaction, on December 2, 1942, he was entirely dependent on the sun as a source of energy. The coal and petroleum we burn are nothing more than fossil solar energy. The power of water running downhill to the sea originated as solar heat that evaporated water from the ocean. Only with the advent of nuclear energy has our dependence on the sun been reduced.

The unit of energy is the calorie, one calorie being the heat required to raise the temperature of one gram of water one degree Celsius. The Celsius temperature scale is named after the Swedish astronomer Anders Celsius (1701–1744), who proposed that temperature be designated by calling the freezing point of water 0° and dividing the interval between this freezing point and the boiling point of water into 100°. The Celsius (or centigrade) scale is universally used by scientists, and we shall use it in this text. In contrast, the Fahrenheit scale, devised by the German physicist Gabriel Daniel Fahrenheit (1696–1736) and conventionally used in the United States and Great Britain, has its zero at the coldest temperature that had been observed at the time of its adoption.

The actual zero of temperature occurs at $-273°C$. As substances are cooled, the motions of their atoms are gradually reduced until all motion is frozen at absolute zero. The Kelvin, or absolute, temperature scale has its zero at absolute zero and has the same-size degree as the Celsius scale. A conversion table for the three temperature scales is given in Appendix 2.

Every second the sun radiates 10^{26} calories. If this amount of energy

69

resulted from the burning of coal, a mass of fuel and oxygen equal to the mass of the sun would be used up in only 1500 years. Before the discovery of nuclear physics, therefore, it was difficult to account for the antiquity of the solar system.

If we burn coal in oxygen, each gram of the carbon-oxygen mixture yields 2300 cal. The chemical reaction is:

$$\text{carbon (C)} + \text{oxygen (O}_2) \longrightarrow \text{carbon dioxide (CO}_2) + \text{heat}$$

The mass of the sun is 2×10^{33} g. If the sun were composed of the proper mixture of carbon and oxygen, it could give off a total of $2.3 \times 10^3 \times 2 \times 10^{33} = 4.6 \times 10^{36}$ cal. If this energy were given off at the sun's current rate of 10^{26} cal sec^{-1}, the sun could shine for 4.6×10^{10} sec. Since one year has 3.16×10^7 sec, the sun would burn out in only about 1500 years.

The Solar Energy Received at the Earth

The sun radiates 10^{26} cal sec^{-1} uniformly in all directions. By the time they reach the earth, an average distance of 1.5×10^{13} cm from the sun, the sun's rays are spread out uniformly over an area of $4\pi(1.5 \times 10^{13})^2$ cm^2, or about 3×10^{27} cm^2. (Fig. 6-1.) Thus each square centimeter perpendicular to the rays of the sun at the earth receives 0.033 cal sec^{-1}. The earth, however, is a sphere rather than a flat disk perpendicular to the sun. Since the surface area of a sphere $(4\pi r^2)$ is four times the area of a circular disk of the same radius (πr^2), the average amount of solar energy received at the earth is 0.033/4, or 0.0083 cal cm^{-2} sec^{-1}.

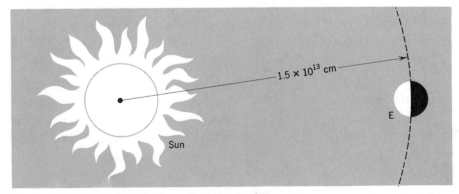

Figure **6-1**   The sun and the earth (not to scale).

Because of the spherical shape of the earth, the average amount of sunlight received decreases from the equator to the poles. To study this variation, first let us assume that the earth's axis is perpendicular to the plane of the earth's orbit and so is always perpendicular to a line from the sun to the earth. In this case, the circle separating the illuminated from the dark face of the earth will pass through the poles. Thus the length of the day will be equal to the length of the night everywhere. While at noon the sun will be directly overhead at the equator, the noon sun will move lower in the sky as we move toward the poles until, at the poles, it remains at the horizon. As the sun moves lower in the sky, the amount of sunlight falling on a square centimeter on the earth will decrease.

To calculate how the amount of sunlight varies with latitude ϕ we must proceed as follows: At latitude ϕ let us paint a 1 cm-wide stripe around the globe (Fig. 6-2). If R is the radius of the earth, the distance at ϕ to the earth's axis is $R \cos \phi$. Thus the total area of our 1 cm stripe is $1 \times 2\pi R \cos \phi$ cm². From the sun, the stripe will look like a painted line with a length of $2R \cos \phi$. Because of the inclination, the stripe will appear to be only $\cos \phi$ cm wide. The total radiation received by the stripe per second will therefore be $0.033 \times \cos \phi \times 2R \cos \phi$ cal. This heat has to warm the total area of the stripe, or $2\pi R \cos \phi$ cm². The daily average solar radiation per cm² of the earth's surface is therefore

$$\frac{0.033 \cos \phi \times 2R \cos \phi \text{ cal sec}^{-1}}{2\pi R \cos \phi \text{ cm}^2} = 0.0105 \cos \phi \text{ cal cm}^{-2} \text{ sec}^{-1}$$

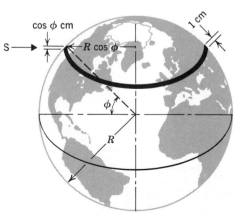

Figure **6-2**   A 1 cm wide band on the earth at latitude ϕ.

Thus the average radiation would vary from 0.0105 at the equator ($\phi = 0$, $\cos \phi = 1$) to 0 at the poles ($\phi = 90$, $\cos \phi = 0$).

Actually, the axis of the earth is inclined 23.5° to the plane of its orbit. The latitude where the sun is directly overhead at noon varies from 23.5°N, the Tropic of Cancer (June 21), to 23.5°S, the Tropic of Capricorn (December 21). Only on March 22 and September 22 is the noon sun directly above the equator. As a result, the northern hemisphere receives more energy between March 22 and September 22 than the southern hemisphere, giving rise to summer in the north and winter in the south. Thus the inclination of the earth's axis produces the seasons.

The amount of sunshine received at the top of the atmosphere varies in a complex manner with latitude and time of year, for it depends on the duration of daylight as well as the elevation of the sun at noon. The effect of the inclination of the earth's axis is to reduce somewhat the latitudinal variations in the average annual heat received. Instead of varying between 0.0105 at the equator and 0.0000 at the poles, the range is from 0.0101 to 0.0042. The main result, then, is a large increase in the average annual energy received at the polar regions (Fig. 6-3).

If the distance between the sun and the earth remained constant, the average annual heat received would be symmetrical *about* the equator.

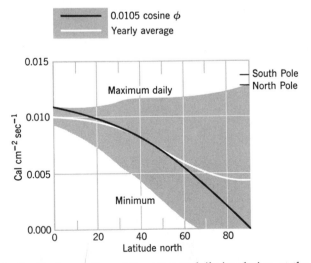

Figure **6-3**   Latitudinal variation of the average daily insolation at the top of the atmosphere for the northern hemisphere.

Actually, the earth's orbit is slightly elliptical, the sun being closest to earth on January 3 and farthest away on July 5. The variation in distance leads to a 7 percent variation in the amount of global heat received between these dates. Since the closest approach is during the northern winter, the annual variation is slightly less in the northern than in the southern hemisphere. Figure 6-3 shows the variation in daily insolation at the top of the earth's atmosphere with latitude for the northern hemisphere. It also shows the maximum insolation at the South Pole. Ironically, the maximum daily amount of sunshine occurs above the South Pole on December 21. This is because the 24 hours of sunshine during the southern summer more than compensate for the lower position of the sun in the polar sky.

Figure 6-3 shows that the difference between summer and winter daily insolation increases markedly as we move from the equator to the poles. Thus we would expect the seasonal differences in climate to be greater as we go to higher latitudes. At the equator, the insolation at the top of the atmosphere varies only 10 percent over the year; at 45°, it varies by a factor of 4. The heat received in midsummer increases slightly from the equator to the poles, but the heat received in midwinter decreases rapidly with latitude. It becomes zero as we enter the zone of total darkness north of the Arctic Circle at 66.5°N.

So far we have discussed only the heat received at the top of the atmosphere. This energy, acting on the atmosphere and the ocean, gives rise to the climate. Before we can investigate the climate at the surface of the earth, we must investigate the nature of the solar radiation.

Thermal Radiation

Every warm body emits radiation. Visible light is one form of thermal radiation. As we heat an object, it begins to glow. We can feel the heat radiated long before the object gives off visible light. As we continue to heat the object, it first glows dark red; gradually the color changes to yellow, and finally to white. This phenomenon is produced by electromagnetic radiation which moves at a speed of 3×10^{10} cm sec^{-1}, the speed of light.

There are two important physical laws that describe the total energy radiated and its distribution in color. The first law, discovered by Josef Stefan (1835–1893) and Ludwig Boltzmann (1844–1906), relates the amount of energy radiated to the temperature of the body. As long as a body is above the absolute zero of temperature (0°K), it will radiate. According to the *Stefan-Boltzmann law,* then, the energy radiated per unit area per unit time, W, is proportional to the fourth power of the absolute temperature, in degrees Kelvin or

$$W = \sigma T^4$$

The Stefan-Boltzmann constant is represented by lowercase *sigma:*

$$\sigma = 1.36 \times 10^{-12} \text{ cal cm}^{-2} \text{ sec}^{-1} \text{ deg}^{-4}$$

The second law concerns the color of the radiation. A hot body does not radiate a single color but, rather, a distribution of colors. The color of the radiation is expressed by stating the wavelength of the light (represented by lowercase *lambda,* λ). The distribution of wavelength radiated by a hot body is shown in Figure 6-4. Note that very little light is radiated at very short and very long wavelengths, and that there is a peak of intensity at an intermediate wavelength λ_{max}. As the temperature of the body increases, the wavelength of maximum emission decreases. This fact is expressed by *Wien's Displacement law* (named after the German physicist Wilhelm Wien). The wavelength of maximum intensity is inversely proportional to the absolute temperature.

$$\lambda_{max} T = 2.9 \text{ mm deg}$$

Since the temperature of the surface of the sun is approximately 6000°K, the maximum intensity of sunlight falls at a wavelength of 5×10^{-4} mm, or 0.5 micron. (Wavelengths are usually expressed in microns; 1 micron $= 10^{-6}$ meter $= 10^{-3}$ mm.) Visible light ranges from

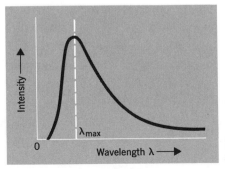

Figure **6-4** The variation in intensity of thermal radiation with wavelength.

violet at 0.4 micron to red at 0.7 micron; the sun's rays, therefore, fall in the middle of the visible region. Invisible radiation shorter than 0.4 micron is called *ultraviolet,* and long radiation beyond 0.7 micron is called *infrared.* Since the average surface temperature of the earth is about 300°K, it will radiate mainly near a wavelength of 10 microns in the infrared.

The earth absorbs radiation from the sun and radiates energy back to space. If the temperature of the earth's surface is to remain constant, it must lose as much energy as it receives. Since the earth is immersed in the vacuum of space, it can lose and gain heat only by radiation. Thus the energy radiated by the earth must equal the solar radiation it absorbs. Knowing the energy the earth receives from the sun, we can calculate the effective radiation temperature of the earth, (*Te*), with the aid of the Stefan-Boltzmann law. Of the total solar energy received, about 35 percent is reflected back to space from the tops of clouds and from the earth's surface. The 65 percent absorbed, on the average, amounts to $0.0054 \text{ cal cm}^{-2} \text{ sec}^{-1}$. Thus

$$0.0054 = 1.36 \times 10^{-12} \, Te^4$$

giving an effective radiation temperature *Te* of 250°K = −23°C. Fortunately, the average surface temperature is considerably warmer, about 15°C.

The simple theory we have just presented works well for the moon, since the moon does not have an atmosphere. Its surface temperature varies from 380°K during the lunar day to 100°K during the lunar night. The earth's climate is more hospitable. The warmer and less variable temperatures on earth are a direct result of the atmosphere and the ocean. Let us see how the fluid envelope moderates our climate and has made life, as we know it, possible.

Summary

The surface of the earth derives its energy from the sun. The variation in the incoming solar radiation produces the seasons and the temperature variations between the equator and the poles. The earth, in turn, gives off long-wavelength infrared radiation to outer space, since the amount of energy received, averaged over the earth's surface, must equal the amount of energy lost.

Study Questions

1. If the earth had no atmosphere, its average temperature would be 250°K. How would this temperature be altered if the solar radiation were to double? What if the solar radiation were reduced by half?
2. If the earth had no atmosphere, and absorbed all the solar energy falling on it, what would be the average annual temperature at the equator, at 30°N, at 60°N, and at the North Pole?
3. The filament of a light bulb has a temperature of 2000°K. What is the wavelength of maximum light emission of the bulb, and into what part of the spectrum does it fall?
4. Using the data on energy emission by the sun and on its temperature, calculate the effective surface area of the sun.

Supplementary Reading

GENERAL REFERENCES ON THE ATMOSPHERE

Donn, William L. (1965). *Meteorology,* 3rd ed. New York: McGraw-Hill.

Pogosyan, K. P. (1965). *The Air Envelope of the Earth,* translated from the Russian by I. Shechtman. Israel Program for Scientific Translations, Cat. No. 1397.

Sellers, William D. (1965). *Physical Climatology.* Chicago: University of Chicago Press.

7 The Climate of the Earth

The amount of sunlight reaching the top of the earth's atmosphere depends only on the latitude and the time of year. The surface temperature, however, is the result of many factors. At a particular place, the temperature varies from day to day, and the weather bureau attempts to predict these changes one or two days in advance. When we average the weather at a particular place and time of year over a long period, we obtain a more regular pattern—the average annual cycle. If we compare such cycles at various places, we find that the annual march of the average temperature is not a function of latitude alone. Rather, the temperature pattern depends to a marked extent on the distribution of land and water on the earth. The ocean has a significant moderating influence on the yearly temperature cycle.

The Distribution of Average Winter and Summer Temperatures

Let us investigate the distribution of the average temperature on the earth during the months of January and July. The temperature distribution is affected by the elevation of the land surface. If we climb a mountain, the temperature of the air drops about 7 degrees for every kilometer increase in height. Thus the actual temperature distribution will reflect to some extent the topography of the land. But our purpose is to investigate the effect of the distribution of land and water on the climate. We must therefore adjust the actual temperature to what it would be at sea level. If we do this we obtain the average adjusted temperature distribution for the months of January and July in degrees Celsius, as shown in Figure 7-1.

Imagine traveling eastward from the Pacific at a latitude of 40°N, the latitude of Denver and Philadelphia. In January, the temperature over the Pacific is about 10°C. As we cross the west coast of the United States, the temperature begins to drop. It continues to decrease until, over the

77

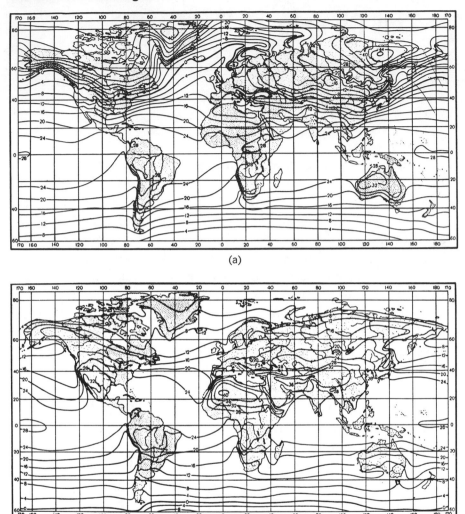

Figure **7-1**   World distribution of the average monthly temperature in degrees Centigrade corrected to sea level: (*a*) January; (*b*) July.

center of the continent, it dips below 0°C. As we cross the east coast, the temperature is still close to 0°C. However, as we cross the Atlantic, it warms up rapidly to slightly over 13°C, dropping again as we approach the coast of Portugal. Over the Mediterranean, the temperature is near 10°C until we reach Turkey. Then, as we head east across Asia, the temperature drops again, reaching a low of − 8°C just before we get to the coast of Korea. As we journey east off the main Japanese island of Honshu, the temperature begins to increase once more, to 13°C. Contrary to expectations, we find considerable variation in the temperature along our path at 40°N. Although the temperatures over the eastern parts of the ocean are above 10°C, the temperatures over the continents fall below the freezing point, with the lowest over eastern Asia.

Let us repeat the journey in July. Over the Central Pacific it is about 8° warmer than in January, or 18°C. As we near the west coast of America, the temperature drops slightly, only to increase rapidly to a maximum of 30°C at Salt Lake City. It drops again to 24°C as we come to the east coast. Over the Atlantic the temperature is constant at slightly above 20°C. Over the Iberian Peninsula, the temperature peaks above 28°C. It drops again slightly over the Mediterranean. As we pass over Turkey, it warms up once more, reaching a peak of 34°C over the Sinkiang Province of China. When we reach the Sea of Japan, the temperature has dropped to 24°C; once we come to the open Pacific, the temperature is back to 19°C.

Our two trips around the world demonstrate that the temperature does not depend on the latitude alone. Further, they show that it is much more comfortable to be an oceanographer than an explorer of the land (provided, of course, that one does not get seasick). While the temperature on the open ocean varied only from 10 to 18°C, over the central United States it went from a low of 0°C in winter to a high of 30°C in summer. Over Asia the variation was even greater, from a low of −8°C in winter to a high of 34°C in summer. The coastal areas, particularly the western shores of the continents, enjoyed a much milder climate. Imagine what our trips would have been like if there were no oceans! We could expect even larger variations than we found over the Asian continent. On the other hand, if we make the same trips at 40°S, where there is very little land, the seasonal variation in temperature is much less, ranging only from 8 to 20°C.

In Figure 6-3 (p. 72), we saw that the annual variation in the heat received from the sun increases as we go from the equator to the poles. Looking at the January-July differences in average temperatures (Fig. 7-1), we no longer see a simple variation with latitude. Rather, we find a range of variations at each latitude. For example, in our two trips at 40°N, the temperature over the open Pacific ranged from 10°C in

January to 18°C in July. Over Asia, we found a range from −4°C in winter to 34°C in summer. Thus the range is about 8° over the Central Pacific and about 48° over central Asia. Off the west coasts of the continents, the range is slightly less than over the mid-Pacific, only about 6°.

Let us now look at the range of the annual temperature variations at different latitudes. We find that the air over the ocean along the equator varies by only about 1°, while in Siberia we obtain an annual temperature range in excess of 60°. Figure 7-2 is a summary of the annual ranges as a function of latitude.

In graphing a particular relationship, such as the dependence of the annual temperature range on latitude, one can use different kinds of

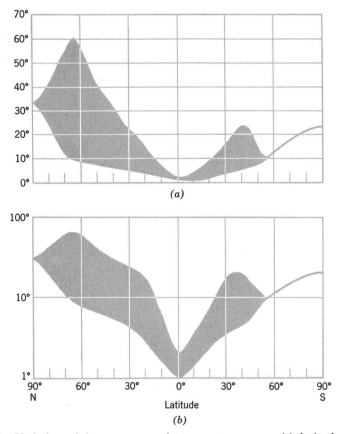

Figure **7-2** Variation of the mean annual temperature range with latitude: (a) linear temperature scale; (b) logarithmic temperature scale.

scales. Most conventional graphs use linear scales, in which the distance from the origin, the zero-zero point on the graph, is proportional to the magnitude graphed, as in Figure 7-2a. Because of the considerable range in the annual temperature variation (from 1 to 60°), the linear graph shows the large values in detail, but it does not give good resolution for small values of the annual range. What we are interested in primarily is the relative value of the annual temperature range. On a linear scale, the difference between 50 and 51 receives the same emphasis as the difference between 1 and 2 although the former represents a 2 percent change while the latter change is 100 percent.

In order to give more emphasis to relative size, one can use a logarithmic scale, on which equal intervals represent increases by a common factor. Thus, on a logarithmic scale, the distance between 1 and 10 is equal to the distance between 10 and 100. In Figure 7-2b, the annual range in temperature is plotted on such a scale. Logarithmic scales are particularly useful if we wish to plot parameters that have a large range in size. Variables such as populations or gross national product tend to change by a fixed percentage per year. If the trend is a straight line when we plot the population on a logarithmic scale against a linear time scale, it means that the percentage increase is constant.

Now let us examine Figure 7-2. Since the mean annual temperature range at a given latitude varies with longitude, we have to indicate a range of values. The maximum range will occur over the continents, while much smaller ranges will be encountered over the ocean. The maximum range, which we find at 65°N, is about six times the minimum range, over the ocean. In the southern hemisphere, which has considerably less land (see Fig. 5-12, p. 64), the maximum range occurs at about 40°S. Here the ratio between maximum and minimum is a factor of 4.5. Note how much more clearly the logarithmic plot displays the contrast between the maximum and minimum ranges. It shows that the ocean moderates the annual temperature cycle. Near the poles, the earth's surface is covered by ice, floating sea ice in the Arctic and land covered by ice sheets in the Antarctic.

The average yearly temperature as a function of latitude is shown in Figure 7-3. According to the Stefan-Boltzmann law, we should expect a global average temperature of 250°K. We see that the temperature gets this low only over the North Pole and the Antarctic. The climate

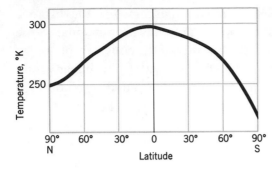

Figure **7-3** Mean average annual temperature as a function of latitude.

on earth, then, differs in two respects from a simple picture based on the incident solar energy. It is warmer, and the annual temperature variations are much greater over the continents than over the oceans. To understand the climate we must look at what happens to the solar energy in more detail.

The Effect of the Atmosphere on Sunlight

On the moon the disposition of solar energy is very simple. Of the solar energy received, the moon reflects about 7 percent. Moonlight is nothing more than this reflected sunlight. If the moon were perfectly white so that it reflected all the sunlight, it would appear 14 times as bright as it actually does. The 93 percent of the incident solar energy absorbed by the moon is reradiated as invisible, long-wave, infrared radiation (Fig. 7-4).

Because of the earth's atmosphere, the disposition of the solar radiation on earth is much more complex. The atmosphere consists of a mixture of gases, clouds (i.e., small water droplets), and dust. Although the atmospheric gases are nearly transparent to visible light, some of them absorb strongly in the infrared and ultraviolet parts of the spectrum.

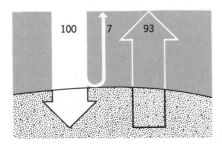

Figure **7-4**   Heat balance of the moon.

Therefore, our eyes are not adequate to explore the light transmission of the atmosphere. We must use instruments to look at the transparency as a function of wavelength.

The atmosphere consists mostly (78%) of *nitrogen*, N_2, plus a lesser amount (21%) of *oxygen*, O_2, and a bit (1%) of *argon*, A. These gases are almost 100 percent transparent to radiation from the short-wavelength ultraviolet to the long-wavelength infrared. The oxygen, however, interacts with the short-wavelength ultraviolet light. When an oxygen molecule, O_2, absorbs ultraviolet light of sufficient energy, it breaks up into two oxygen atoms:

$$O_2 + UV \text{ light} \longrightarrow O + O$$

The atomic oxygen produced in this way is very reactive and attaches itself to an oxygen molecule to form a molecule of ozone:

$$\text{atomic oxygen (O)} + \text{molecular oxygen } (O_2) \longrightarrow \text{ozone } (O_3)$$

The ozone molecule strongly absorbs ultraviolet light. As a result, an ozone layer forms about 20 km above the earth's surface. This layer absorbs the ultraviolet light from the sun, thus shielding the surface of the earth from these rays. Without this ultraviolet filter, life as we know it would not be possible on the earth's surface, for these rays are inimical to life. The ultraviolet rays absorbed by the ozone in the upper atmosphere amount to about 3 percent of the incoming solar energy.

If our atmosphere contained only the major gases, nitrogen, oxygen, and argon, it would be largely transparent to infrared radiation. As a result, the backradiation from the surface of the earth would be able to escape unhindered through the atmosphere. However, the air contains small amounts of carbon dioxide (0.03%) and water vapor, as well as the three major gases. Both the carbon dioxide and the water vapor in the atmosphere strongly absorb infrared radiation. In addition, condensation of water vapor produces clouds which reflect and diffuse the incoming sunlight.

Figure 7-5 shows the absorption of radiation by CO_2 and water vapor as a function of wavelength. Absorption does not occur over the entire spectrum but is concentrated in a series of wavelength bands. When we combine the absorption by ozone, carbon dioxide, and water vapor, we obtain the transparency of the atmosphere. The atmosphere is trans-

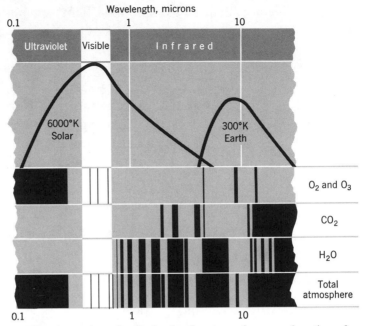

Figure **7-5**   The absorption of radiation by the atmosphere as a function of wavelength.

parent from 0.3 to 0.7 micron. In the infrared spectrum, there remain a few regions below 13 microns, the infrared windows, where the atmosphere is transparent. These windows are important for infrared photography. While the atmosphere is transparent for most of the incoming solar radiation, it is almost opaque for thermal radiation emitted at a temperature of 300° K. To study the effect of the atmospheric absorption on our climate, we must now consider the in- and outflow of radiant energy from the earth.

The Radiation Balance of the Earth

On the average, the earth receives 8.3×10^{-3} cal sec^{-1} cm^{-2}. Setting this value equal to 100 percent, let us see how, on a global average, this energy is disposed of. The ultraviolet, comprising 3 percent of the 100 units of incoming sunlight, is absorbed mostly by the ozone of the upper atmosphere (Fig. 7-6). About 40 of the remaining 97 units interact with clouds. Of these, 24 are reflected back to space, two are absorbed by the clouds, and 14 are scattered, reaching the earth's surface as diffused radiation.

The water vapor, dust, and haze in the atmosphere interact with 32 units of the incoming radiation. Thirteen of these units are absorbed, seven are reflected back to space, and 12 reach the earth's surface as diffused sunlight. Out of the original 100 units, then, the surface receives 25 units as direct sunlight and 26 units as scattered, diffused light. Of this total, four units are reflected from the earth's surface back to space. Thus the total amount of sunlight reflected back to space equals 35 percent of the incident sunlight. Of the 65 percent absorbed by the earth, 3 percent is absorbed in the upper atmosphere, 15 percent in the lower

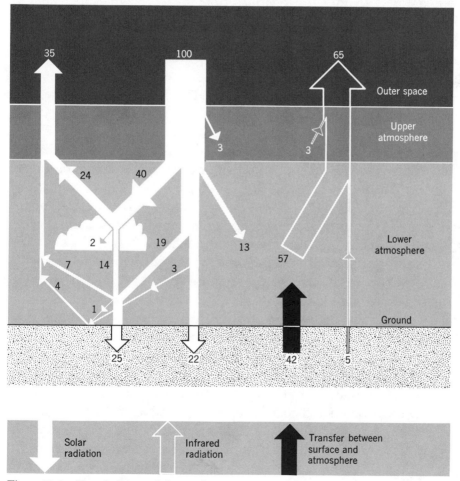

Figure **7-6**   Heat balance of the earth.

atmosphere, and 47 percent by the surface of the earth, both land and sea.

We have noted that in order to maintain a constant temperature, the earth must radiate back to space the same amount of energy it absorbs. Not only must the earth as a whole be in balance as regards to energy, but each layer—the upper atmosphere, the lower atmosphere, and the surface—must itself be in balance. Altogether, the earth must radiate 65 units back to space as thermal radiation. Three of these units come from the ultraviolet light absorbed by the ozone of the upper atmosphere. If the atmosphere were transparent to infrared radiation, we would expect 15 units to be derived from the lower atmosphere and 47 units to be radiated to space from the ground. Actually, because the atmosphere is almost opaque to infrared radiation, only five units are radiated from the surface to space in the wavelength regions comprising the atmospheric window. The remaining 57 units are radiated from the upper regions of the lower atmosphere. This leaves 42 units of energy to be transferred from the surface to the atmosphere.

The transfer of heat from the earth's surface to the atmosphere takes place by three processes. Some energy is transferred as heat radiation. Some of the transfer takes place by heating the air that comes in contact with the ground. However, the largest transfer (24 units) takes place by the evaporation of water. The water vapor, as it rises in the atmosphere, condenses to form clouds and precipitation (rain and snow), and thus returns the heat of vaporization to the atmosphere.

The absorption of infrared radiation by the atmosphere and its heating by the condensation of water vapor impede the loss of heat from the earth's surface. On the moon, on the other hand, there is no impediment to the escape of radiant energy; therefore its average surface temperature is lower than the earth's and its surface cools much more rapidly during the lunar night than the cooling of the earth at night, particularly on a cloudy night. On earth, water plays an important role as a heat shield because of its absorption of the infrared radiation and the mechanism of evaporation and condensation. Over the dry desert regions this effect is reduced, and thus we find a greater daily and annual range in temperature in these areas. The moist ocean areas, on the other hand, have the smallest variation in temperature. In addition, because it has a large heat capacity compared to land, the ocean stores great quantities of heat and thus reduces the annual variation of temperature still further.

The Latitudinal Variation of the Radiation Balance

So far we have looked only at the global-average energy budget of the earth. We have seen that the amount of incoming sunlight decreases from the equator to the poles; now we must investigate how the ingoing and outgoing radiation varies with latitude.

On the moon, lacking an ocean and an atmosphere, there is no way to transport heat horizontally. Thus the annual average radiation given off by each area must equal the annual average solar radiation received. On the earth, however, the situation is more complex, for the air and the ocean currents can transport heat from place to place. In general, the currents of the ocean and the atmosphere carry heat from low to high latitudes. Therefore, an excess of heat must be absorbed near the equator and reradiated to space near the poles. The *global* average absorption, however, must still equal the backradiation.

First let us look at the latitudinal variation of the incoming net radiation—that is, the sunlight incident at the top of the atmosphere minus the amount of sunlight reflected back to space by clouds, dust, haze, and by the surface. The yearly average incident radiation as a function of latitude is shown in Figure 6-3 (p. 72). From this we have to subtract the reflected radiation. Although the worldwide average reflectivity is 35 percent, the reflectivity of individual regions varies from more than 60 percent in the Arctic, where the sky is usually obscured by clouds and the ice cover of the ocean is highly reflective, to less than 10 percent in the relatively cloud-free tropics, where the open ocean absorbs most of the incident light.

Figure 7-7 shows an estimate of the variation with latitude of the net incoming solar radiation (I) for the northern hemisphere. Since there is much more area on earth at latitudes near the equator than near the poles, the latitude scale has been drawn so that equal distances on the horizontal axis represent equal areas on the earth. An estimate of the variation of the average annual outgoing long-wave infrared radiation (O) as a function of latitude is also indicated in Figure 7-7. If there were no heat transport across latitude circles, the two lines would be identical, showing that each individual latitude belt was in thermal equilibrium. In fact, we see that there is an excess of incoming radiation near the equator and a deficit in high latitudes. Only at 38°N is the net incident solar radiation equal to the outgoing long-wave radiation.

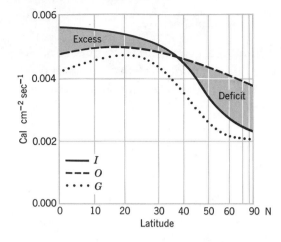

Figure **7-7** Latitudinal variation of the mean annual incoming radiation (I), outgoing radiation (O), and radiation absorbed by the ground (G) for the northern hemisphere.

To maintain the energy balance on earth, therefore, there must be a poleward transport of heat. Slightly less than one-third of this heat is transported by ocean currents; the rest is transported by the atmosphere. The differential heating of the earth causes it to act like a heat engine. In a steam engine, for example, the transfer of heat from a high temperature in the boiler to a lower temperature in the condenser produces motion. On earth, the differential heating between the equator and the poles produces a transport of heat toward the poles, resulting in the motion of the oceans and the atmosphere. Heat transport takes place

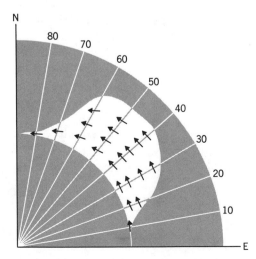

Figure **7-8** The poleward flux of heat for the northern hemisphere.

by the poleward flow of warm air and ocean currents and by the flow toward the equator of cool air masses and currents.

Note that the areas of excess and deficit shown in Figure 7-7 are equal, so that the northern hemisphere is roughly in balance. Averaged over the year, there is very little energy transport across the equator. The poleward transport of heat reaches a maximum at about 40° latitude and becomes zero at the poles (Fig. 7-8).

The latitudinal variation of the annual solar radiation absorbed by the ground (G) is also shown in Figure 7-7. In order for the surface temperature to remain constant, this heat must be transferred upward into the atmosphere, where it is reradiated as heat radiation to outer space. A fraction of this heat is also transferred poleward along the earth's surface by ocean currents. The greatest vertical transfer, which takes place near 20°N, is effected mainly by the evaporation of water vapor from the surface of the ocean. About 140 cm of water from the ocean are evaporated at this latitude per year.

Summary

The climate of the earth not only depends on latitude; the annual temperature range is much greater over the continents than over the ocean. The incoming solar radiation interacts with the atmosphere, with clouds, and with the earth's surface. Energy is transported from the equator toward the pole by the winds of the atmosphere and by ocean currents, which are driven by the differential heating of the earth's surface. Water plays an important role in the energy balance of the earth.

Study Questions

1. Using Figure 7-1, contrast the summer and winter temperatures for the east and west coasts of North America.
2. Discuss the relationship between Figure 7-2 and Figure 5-12 (p. 64).
3. How may the industrial burning of fossil fuels, resulting in an increase in the CO_2 content of the atmosphere, alter the climate on earth?
4. Using the data in Figure 7-6, make a separate heat balance for the upper atmosphere, lower atmosphere, and the ground, and show that the heat absorbed by each is equal to the heat given off.
5. How would the heat balance of the earth be affected if the earth were dry?

8 Water

Over 70 percent of our globe is covered by seawater. In addition, water forms an important part of our atmosphere as vapor, water droplets, and ice crystals. Water is also the major constituent of plants and animals. The total inventory of water on the surface of our planet is as follows:

As seawater in the oceans	1.4×10^{24} g
As ice on the continents	1.5×10^{22} g
As underground water	5.0×10^{20} g
In lakes and rivers	10^{20} g
In the atmosphere	1.7×10^{19} g

This substance (H_2O), a combination of the lightest element, hydrogen, with oxygen, has remarkable properties, which vary with pressure and temperature. We are interested in these properties from a pressure close to zero, at the top of the atmosphere, to 1000 times normal atmospheric pressure, at the bottom of the deepest parts of the ocean. The temperature range with which we are concerned extends from about $-50°C$ at the center of the Antarctic to about $40°C$ over the hottest parts of the continents in summer. Before we examine the properties of seawater, our primary interest here, let us first study pure water.

The Phases of Pure Water

Water exists in three states on the surface of the earth—as solid ice, as liquid water, and as gaseous water vapor. The pressure and the temperature determine the state in which water will exist. *Pressure* is the force acting on a unit area. In centimeter-gram-second (CGS) units, the unit of force is the dyne (1 dyne = 1 g cm sec^{-2}). The pressure in CGS units is measured in dynes per square centimeter. The average pressure exerted by our atmosphere at the surface of the earth is 1,013,200 dynes cm^{-2}. This is close to 10^6 dynes cm^{-2}, which is equal to the commonly used pressure unit, the *bar*. The standard pressure of the atmosphere at the earth's surface, then, is 1.013 bars. This is equal to the pressure exerted by a column of mercury 76 cm high or a column of water approximately 10 m high. In the sea, therefore, the pressure increases 1 atm, or approximately 1 bar, for every 10 m of depth.

Temperature is the property of a body that determines how it will transfer heat to another body. Heat always flows from a body of higher temperature to one of lower temperature. On the Celsius scale, the freezing point of water at 1 atm is taken as zero, and the boiling point is 100°C.

Let us put some water in a cylinder with a tightly fitting piston so that we can independently alter the temperature and the pressure and find the ranges in temperature and pressure over which water exists in the liquid, solid, or gaseous state. When we plot the boundaries of the three realms on a temperature-pressure diagram, we obtain the *phase diagram* for water (Fig. 8-1). The field of the gas is separated from that of the liquid by the boiling curve; the liquid is separated from solid ice by the freezing-melting curve. Note that the freezing point of water drops as the pressure is increased. At low pressures, water will pass directly from the solid to the gaseous state without melting. This process is called *sublimation*. Ice, therefore, is separated from water vapor by the sublimation curve. The three phases coexist at the so-called triple point (0.01°C, 0.006 bar).

The pressure in Figure 8-1 is plotted on a logarithmic scale, while temperature is drawn on a linear scale. The boiling curve is approximately a straight line on this diagram. The pressure of the vapor in

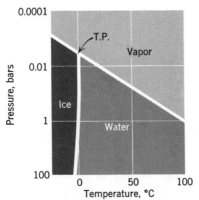

Figure **8-1**   The three phases of water; T.P., triple point.

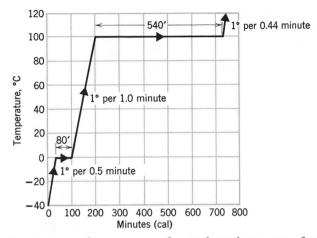

Figure **8-2**   Heating curve for one gram of water heated at a rate of one calorie per minute.

equilibrium with the liquid increases approximately by a factor of 2 (i.e., it doubles) for every 10°C increase in temperature.

To transform water from one phase to another requires energy. This can best be seen by slowly adding heat to a gram of ice, at a constant rate of 1 cal min^{-1} (Fig. 8-2). Starting at a temperature of $-40°$C, it takes 20 min for the ice to reach the melting point, 0°C. While the ice is melting, the temperature remains at 0°C until all the ice is melted, 80 min after the melting point was first reached. Then the water is warmed up once more, this time at a rate of 1° min^{-1}. When 200 min have passed since the start of the experiment, the water will have reached the boiling point, 100°C. Once again, the temperature will not rise until all the liquid has been converted into steam by boiling. To convert 1 g of water to steam takes a total of 540 min. Then the temperature of the steam rises once more, this time at a rate of 1° per 0.44 min.

We can summarize the results of this experiment by noting the heat capacity of the various phases (the amount of heat required to raise the temperature of 1 g by 1°) and the heats of phase transformation (the heat required to transform 1 g from one phase to another at constant temperature).

Heat Capacities

Ice $= 0.5$ cal g^{-1} deg^{-1}

Water $= 1.0$ cal g^{-1} deg^{-1}

Water vapor $= 0.44$ cal g^{-1} deg^{-1}

Heat of Phase Transformations

Fusion of ice $= 80$ cal g^{-1}

Vaporization of water $= 540$ cal g^{-1}

Had we performed the heating experiment with almost any other substance, the temperature would have increased much more rapidly. For example, the only other drinkable liquid, ethyl alcohol, has only half the heat capacity of water (0.5 cal g^{-1} deg^{-1}). Its heat of fusion is only 50 cal g^{-1} and its heat of vaporization is only 200 cal g^{-1}. Our bodies would heat up more rapidly if we were full of alcohol, and a planet whose ocean is purely alcoholic would have a more variable temperature. Thus water is an ideal temperature-stabilizing medium on the surface of our planet.

The unique thermal properties of water are not its only remarkable feature. Let us see how the volume of a gram of H$_2$O changes as we change its temperature and pressure.

The Specific Volume of Water

The volume per gram of a substance is known as its *specific volume.* The manner in which the specific volume of water varies with temperature and pressure determines the vertical motion of water in lakes and in the sea. Just as the less dense (larger specific volume) oil will float on top of water, so water of larger specific volume will float on top of denser water.

Specific volume, units: cm^3 g^{-1}

Density $= 1/$specific volume; units: g cm^{-3}

To determine the variation in specific volume, let us take 1 g of ice at a pressure of 1 bar and a temperature of $-40°$C and measure the change in volume as we warm the ice. Since the volume changes will involve only a small fraction of the total volume, the experiment will have to be done with extreme precision. Our 1 g of ice at $-40°$C will occupy a volume of 1.085 cm^3. As the ice is heated, its volume will expand 0.014 percent per degree, so that when the temperature reaches 0°C the ice will occupy 1.091 cm^3. As we melt the ice, we find that the volume decreases to 1.0001 cm^3. Most solids expand when they melt. Thus another remarkable property of water is that ice is less dense than liquid water. As a result, ice floats on top of water. When water freezes inside the cracks in a rock, it expands and may break the rock apart.

The fact that ice floats on water protects life in lakes and in the sea. The layer of ice that forms insulates the underlying water from the cool air above and so reduces the rate of further freezing. If ice were denser than water, the ice formed at the surface would sink to the bottom of the lake, permitting more ice to freeze at the surface. Before long, the lake would be frozen solid, and the fish in the lake would be killed.

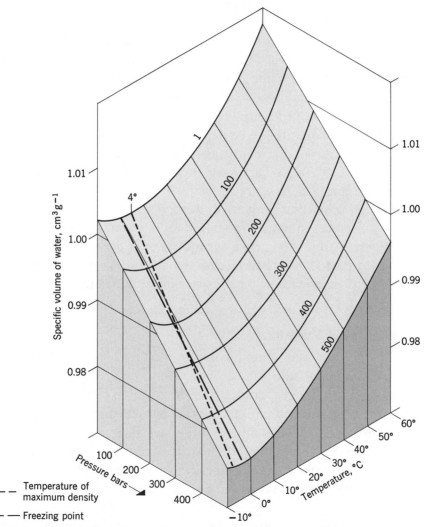

Figure **8-3**   The specific volume of pure water as a function of temperature and pressure.

As we begin to heat the water, its volume shrinks very slightly until we reach 4°C, at which time the volume is almost exactly 1.0000 cm³. Now, as we heat the water above 4°C, the volume expands, at first very slowly, and gradually at a more rapid rate per degree of warming. The change in specific volume of water with temperature is shown in Figure 8-3. The fact that fresh water at 1 atm pressure has a minimum specific volume, hence a maximum density, is important in lakes. In winter, as the water at the lake surface cools, the cooler water becomes more dense than the underlying water and so sinks to the bottom of the lake. Once the lake reaches 4°C, however, further cooling will not lead to sinking. The water at the bottom of a lake in winter will remain at 4°C, although there may be cooler water, and even ice, at the surface of the lake.

If we increase the pressure of the water, its volume shrinks. To reduce the volume by about 1 percent requires a pressure of approximately 200 bars, corresponding to a lake depth of about 2 km. The application of pressure has two other consequences: it lowers the freezing point of water about 1° per 200 bars, and it lowers the temperature of maximum density at an even faster rate.

Another important property of water is that it is an excellent solvent. Most substances dissolve to some extent in water, and many substances are highly soluble. The prime example of salts dissolved in water is seawater. In Part IV we shall discuss the chemistry of seawater in detail. Now let us examine how the presence of salt alters the physical properties of water.

The Salinity of Seawater

Seawater is a solution of a complex mixture of salts in water. Fortunately, the major salts in seawater always exist in approximately the same proportions. Evaporation and the addition of rain and river water alter the total salt content of seawater but not the relative amounts of the various salts. We can describe seawater, therefore, by giving its total salt content without having to analyze each sample in detail. That the ratios of the major chemicals in seawater are constant regardless of the source of the water was one of the important findings of the *Challenger* expedition.

It might seem to be relatively easy to measure the total salt content of seawater. We need merely to take a given weight of seawater—say,

1 kg—boil it to drive off all the water, and weigh the residue. (The residue in 1 kg of seawater will weigh approximately 35 g.) Although a crude measure can be obtained by this method, oceanographers need to determine the salt content of seawater to an accuracy of better than 0.01 g kg^{-1}. When we attempt to perform the drying experiment with this degree of accuracy, we encounter many difficulties. Some of the water combines chemically with the salts. In order to drive off this water and decompose the hydrated salts, we have to heat them well above 100°C. However, at this temperature some of the sea salts decompose, and we lose some of the more volatile components in addition to the water. Therefore, to perform the experiment accurately, we must carefully define the procedure we use so that repeated determinations will give the same result.

The fact that the major constituents of seawater exist in constant ratios provides us with a solution to the dilemma of determining salinity. We need to measure the concentration of only one of these constituents in order to determine the total salt content. The major constituent that can be measured most easily is the chloride ion Cl$^-$. The weight fraction of chloride in seawater, called the *chlorinity,* is usually expressed in grams of Cl$^-$ per kilogram of seawater (‰). Because of the constant proportions, the following relationship exists between chlorinity and total salt content:

$$\text{salinity} = 1.80655 \times \text{chlorinity}$$

To enable oceanographers all over the world to determine the chlorinity of seawater with equal accuracy, an international seawater office has been established at Copenhagen, where large batches of seawater are analyzed. Samples of this water, labeled with the result of the analysis, are carefully sealed in glass ampules and distributed to laboratories throughout the world. The chlorinity of other water samples can then be compared with that of the standard water sample. In this way we make sure that salinities determined by different laboratories around the world are all on the same international scale. A label from a standard seawater ampule is shown in Figure 8-4.

Although the chloride content of seawater can be determined with great accuracy, the measurement requires a great deal of care and is time-consuming. It is difficult enough on shore in a stationary labora-

Figure **8-4**   Label from a standard sea water ampule.

tory. To maintain this accuracy on a moving ship calls for a major effort. Nevertheless, huge numbers of such analyses have been carried out. Recently, however, the job of the oceanographer has been simplified and the accuracy of the salinity determination improved. Instead of measuring the chloride content of seawater, we now measure its electrical conductivity, using a modern conductance-measuring instrument, the *salinometer* (Fig. 8-5). The measuring cell of the salinometer is filled with the unknown sea water and a reading is taken. The cell is then filled with standard Copenhagen water and another reading is taken. From a knowledge of the salinity of the standard water, we can determine the salinity of the unknown sample to better than 0.01‰. The measurement is so accurate that we can determine the increase in salinity

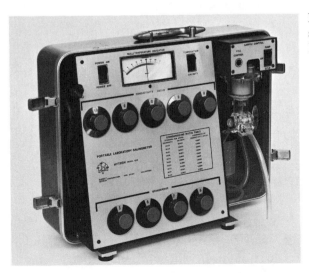

Figure **8-5**   Laboratory salinometer (Bisset and Behrman).

due to evaporation by merely pouring the water from one bottle into another.

The Specific Volume of Sea Water

The specific volume of pure water, as we have seen, depends only on the temperature and pressure; the volume per gram of seawater, however, depends on the salinity as well. Since the density of salts is greater than the density of water, the density of seawater exceeds that of pure water and increases as the salt content is increased. Adding 1‰ salt increases the density of water by roughly 0.8‰. Thus seawater at 0°C and 35‰ salinity has a density of 1.028 g cm^{-3}. To simplify the notation for the density of seawater, oceanographers specify it by writing 1000 times the difference of the density from 1. The density so specified at a pressure of 1 bar is called σ_T (sigma-T). Thus a density of 1.028 g cm^{-3} is written $\sigma_T = 28$. To determine the motion of seawater, oceanographers must determine the density to 0.01 sigma-T units.

The manner in which the density of seawater at 1 bar depends on the temperature and salinity is shown is Figure 8-6. For pure water, the density is maximal at 4°C. The effect of salt is to depress the temperature of maximum density. The freezing point is lowered at the same time, but not so rapidly. As a result, for salinities greater than 24‰ the temperature of maximum density falls below the freezing point. Thus seawater always expands as the temperature is raised. However, at low temperatures the rate of expansion is very small, and it increases markedly as the temperature is increased.

Raising the pressure increases the density of seawater. In addition, as in the case of pure water, increasing the pressure lowers the temperature of maximum density. As a result, at a given temperature, water under high pressure expands more rapidly for a given temperature increase. The change in the density of seawater with salinity, temperature, and pressure has been carefully measured in the laboratory, and detailed tables have been prepared on the basis of these measurements. To determine the density distribution at sea, we measure the distribution of temperature and salinity as a function of depth and then determine the density distributions from tables. To see how this is done at one location in the ocean, let us board an oceanographic research ship while it is making measurements.

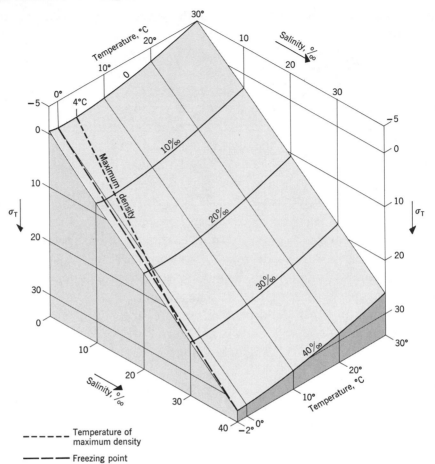

Figure **8-6** The density of sea water as a function of temperature and salinity at a pressure of one atmosphere: $[\sigma_T = (\text{density} - 1) \times 1000$. e.g. $\sigma_T = 15$, density $= 1.015$ g cm^{-3}].

An Oceanographic Station

At Greenwich, England, at $0°$ longitude, it is 2016 hours (8:16 P.M.), on March 3, 1965. At that time, we are in the Northeast Pacific Ocean at $32°00'$N, $140°53'$W. Since the sun circles the earth once every 24 hours, it appears to move from east to west at a rate of $15°$ per hour. Our local time is therefore 9 hours and 24 minutes earlier than Greenwich time; locally, then, it is 10:52 A.M. Our watches, which are still on Pacific standard time, indicate 12:16 P.M.

Our research vessel, the *Yaquina*, of the Department of Oceanography at Oregon State University, has just arrived at the 29th station of this cruise. The air temperature is 62°F (16.7°C), and the relative humidity is about 70 percent. There is a strong wind blowing from the northwest at 35 knots. The captain has just stopped the main engine. However, since the wind will tend to move the ship, he has turned on a special engine which drives a propeller, called the bow thruster, mounted near the bow of the ship. He lowers this propeller out of the well into which it had been retracted while the ship was cruising. The bow thruster can be pivoted 360° to keep the ship in place and point it in a direction which will minimize the ship's roll and pitch. In spite of this maneuverability, however, the ship is rolling in fairly heavy seas.

Our purpose in stopping is to measure the temperature and salinity of the seawater to a depth of 1000 m. Using a steel wire, wound on the spool of the hydrographic winch, we shall lower water samplers and thermometers into the sea. To protect our instruments from scraping against the side of the ship, the wire is run over a pulley which is extended about three feet from the side of the ship by a hydraulic frame. In order to reach the wire, we also extend a small platform, called the "hero platform," out to the side.

A heavy weight is attached to the end of the hydrographic wire and lowered into the sea to make the wire hang straight (Fig. 8-7a). The wire is now ready to receive the samplers. To determine the distribution of salinity with depth, we have to take a water sample at several succes-

Figure **8-7**   Making a Nansen bottle cast: (*a*) general view showing hydrographic winch and "hero platform" with technician attaching Nansen bottle to hydrographic wire; (*b*) close-up of technician attaching bottle to wire. (Photographs: Oregon State University, Department of Oceanography)

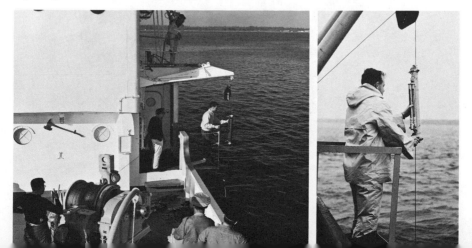

sive depths and bring each sample back on board without contaminating it with other water. To accomplish this, we use a number of Nansen bottles. This device, originally devised by the Danish explorer Fridtjof Nansen, is a cylinder fitted at each end with a valve. The bottle is attached to the wire in an open position, with a clamp at the bottom and a catch at the top (Fig. 8-7*b*). When a cylindrical weight, called the "messenger," is dropped along the wire and hits the catch, the top of the Nansen bottle is released and it flips over, closing the valves and sealing off the water sample (Fig. 8-8). At the same time, the bottle releases another messenger to trip the bottle beneath it.

Although we can measure the salinity at a given depth after the water sample is brought on board, the temperature of the water in the bottle will have changed on its way to the surface. Therefore we must measure the temperature in place. A reversing thermometer (Fig. 8-9) is used for

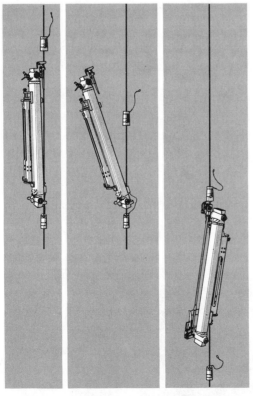

Tripped   After tripping

Figure **8-8** Nansen bottle, showing its mode of operation.

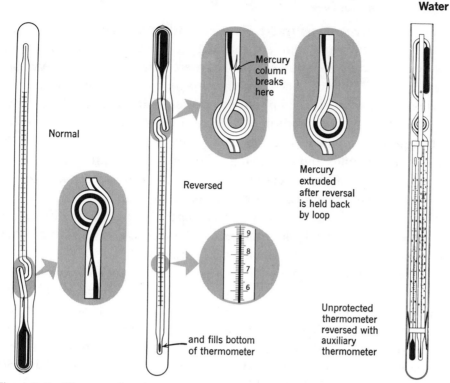

Normal

Reversed

Mercury column breaks here

Mercury extruded after reversal is held back by loop

9
8
7
6

and fills bottom of thermometer

Unprotected thermometer reversed with auxiliary thermometer

Figure **8-9**   The reversing thermometer. The thermometer is shown first in its normal position as it is mounted on the hydrographic wire. When the Nansen bottle is tripped at depth, the thermometer is reversed, the extruded mercury column breaks, fills the bottom of the thermometer, and the temperature is read in this position. Pictured on the right is an unprotected reversing thermometer with its auxiliary thermometer. (After U.S.N.O.O.)

this purpose. As long as the Nansen bottle is in the open position, the reversing thermometer acts like a precision thermometer with its mercury bulb at the bottom. Above the mercury bulb there is a constriction followed by a loop. As the Nansen bottle is reversed, the mercury extruded into the stem breaks off at the point of constriction and slides to the bottom of the thermometer. Any mercury later extruded by the bulb is held back by the loop in the tubing. The temperature at the time of reversal can be read by noting how much mercury had been extruded at the time of reversal. A second auxiliary thermometer is provided to correct the reading for the expansion of the glass and the mercury at the time the reading is taken.

Two kinds of reversing thermometers are attached to the Nansen bottle in a thermometer frame. The first kind, which is used to determine the temperature at depth, is protected from the high pressure of the water by a heavy glass tube. A pool of mercury between the heavy glass wall and the bulb of the thermometer acts as a conductor of heat to bring the mercury in the bulb to thermal equilibrium with the external seawater. A second, unprotected thermometer is exposed to the full pressure of the water. The pressure compresses both the glass of the thermometer bulb and the mercury. Since glass is more compressible than mercury, the net effect of the pressure is to squeeze an excess of mercury out of the bulb at a rate corresponding to an apparent tempera-ture increase of 0.01°C per meter of depth. By noting the difference between the protected and unprotected thermometers, we can calculate the depth at which the thermometer was reversed.

Thus the Nansen bottle with its attached thermometers permits us to obtain a sample of the water at depth, to measure the depth at which the sample was taken, and to obtain the temperature at the time of sampling. Since we have 20 Nansen bottles available to sample the interval from the surface to a depth of 1 km, we decide to place the bottles on the wire corresponding to the following depths: 0, 50, 100, 130, 160, 190, 220, 250, 300, 350, 400, 450, 500, 550, 600, 650, 700, 800, 900, and 1000 m.

The man on the hero platform begins to load the wire, starting with the bottom of the cast, the bottle which will obtain a sample at 1000 m. The operator of the winch zeros a register which records the meters of wire that have been unreeled. When the first bottle is securely clamped on the wire, he lowers it 100 m into the sea so that the next bottle, the one that will sample at 900 m, can be placed on the wire. This bottle is fastened to the wire and a messenger is attached to its clamp to release the next (1000-m) bottle when the 900-m bottle is tripped. The wire is lowered another 100 m, and the 800-m bottle is placed on the wire with its messenger to trip the bottle below it.

While the bottles are being attached to the wire, the ship is rolling, and the hero platform is alternately inundated by water and suspended high above the sea. The operator has to be sure that he attaches the Nansen bottles properly. Should he accidentally drop a messenger, it will prematurely trip all the bottles that are already on the line. If this hap-pens, the wire has to be reeled in, the bottles detached, and the cast started over again. While the wire is being loaded, the captain keeps an anxious eye on it. If the ship should drift, the wire will hang at an angle, and the bottles will not be at the proper depth. Even worse, if the ship should drift over the wire, the bottom might scrape off a bottle or part the wire. Without the Nansen bottles it would be impossible for us to learn about the sea beneath the ship and the scientific objectives of the cruise could not be attained. Using the bow thruster, the captain keeps the wire as straight as possible. The winch operator also has to watch the roll of the ship when he lowers the bottles below the keel, for, because of the heavy seas, the wire touches the side of the ship on every roll.

Our man on the hero platform continues to load the wire with bottles at predetermined intervals until the last bottle, the one that will be nearest the surface, is on the wire. He waits about ten minutes to permit the thermometers to come to equilibrium with the temperature of the sea and then places a messenger on the wire to set off and reverse the first bottle. This and each successive bottle on the string releases a messenger to trip the bottle below it.

When the last messenger has had time to trip the bottle at 1000 m, the string of bottles is ready to be retrieved. The top bottle is brought to the height of the operator on the platform. He unbolts it from the wire and hands it to a technician, who brings it into the laboratory. As the wire is raised, the winch operator watches it carefully to make sure that the next bottle is stopped when it reaches the height of the operator on the platform. Should the winch not be stopped on time, the clamped bottle would hit the pulley, and the force of the winch would tear the steel wire. As a result, the string of wire with its attached bottles and thermometers would sink to the bottom.

When the last bottle has been taken aboard, the weight is lifted out of the water, and the hero platform is pulled in. The captain then heads for the next station, 120 nautical miles due north. Since the ship travels at about 10 knots, we have about 12 hours before the ship will stop again. Now the real work begins. The thermometers have to be read and the salinities of the water samples determined. Meanwhile, the captain has carefully determined our position and has recorded the state of the weather and the sea.

The bottles of seawater are standing in a special rack in the laboratory. The water is transferred from the Nansen bottles to special glass bottles used for the salinity determination. In transferring the water we must be careful to avoid evaporation. One end of a rubber hose is attached to an opening in the Nansen bottle; the other is inserted in the bottom of the sample bottle. A stopcock is opened and water is admitted to the glass bottle. The bottle is first rinsed twice with the water from the Nansen bottle, then filled to the top and sealed. We note the number of the sample bottle and of the Nansen bottle from which it was filled. In this manner water samples are drawn from all the Nansen bottles and carried into the adjacent chemistry laboratory.

In the chemistry laboratory, the salinometer has been turned on and allowed to warm up. Using an ampule of standard seawater from Copenhagen, we standardize the salinometer. Then one of the sample water bottles is placed next to the salinometer, and water from this bottle is pumped into the cell of the salinometer. The water from the first and second filling is used merely to rinse out the cell. On the third and fourth filling readings are taken. From the average of these readings and the calibration tables of the salinometer, the salinity of the water in the bottle is obtained. For example, sample bottle Q22, which was filled from the Nansen bottle reversed at 1000 m, gives us a salinity of 34.395‰. A second sample from the same Nansen bottle gives a salinity of 34.400‰. In this manner all the water samples are analyzed for salinity. Finally, the calibration of the salinometer is checked against another sample of standard seawater.

Meanwhile, technicians, using magnifiers, are taking a reading of each protected, unprotected, and auxiliary thermometer on each of the Nansen bottles. First one technician reads off the data while his colleague records them. Then the technicians exchange roles and the thermometers are read a second time, to prevent errors. Once the reading is completed the Nansen bottles are drained and reversed to ready them for the next station.

Using special calibration curves for each thermometer, we correct the readings of the protected thermometers to obtain the true temperature at the time of reversal. Then the temperaure readings of the unprotected thermometers are corrected. From the difference and the calibration factors for the thermometers, we calculate the true depth at which the thermometer was reversed. Although we know how the bottles were placed on the wire, we must confirm the depth from the readings of the unprotected thermometers. Because of the ship's drift and possible ocean currents, the wire may not be perfectly vertical. Furthermore, it is always possible that a messenger might have been released before the bottles were at their proper depth. On the other hand, a messenger might have become stuck on the wire and been released only as the wire was being hauled in. Since we cannot observe the bottles while they are in the sea, we can be sure that they have been tripped at their proper depth only by measuring the depth of reversal with the unprotected thermometer. When we make all the calculations, we are relieved to find that everything checks and that all bottles were released within 1 m of the desired positions.

When we return to our base, all the observed data are fed into an electronic computer which calculates the salinity and temperature at a number of fixed levels and then computes the density in sigma-T units for each depth. The computer then prints out all the data in a standard format (Fig. 8-10). The heading gives information about the station. The columns at the left are the observations—depth, temperature, and salinity—while the columns at the right give information at the interpolated standard depth. These data are then sent to the National Oceanographic Data Center in Washington, D. C., where they are stored together with data obtained in a similar manner on other cruises by ships from all over the world. Thus we have added another bit of information about the physical condition of the ocean. Any future oceanographer

SICS29 32 0 N 140 53.0 W DATE 03 MAR 65 2016 GCT WIRE 03 DRY 62.0 WET 56.1
WIND DIRECTION 33 VEL 35 KTS BAR 12 SWELL DIRECTION 35 H 8 T 5 CLOUD 0 AMT 3 WEATHER 01

DEPTH m	TEMP. °C	SAL. °/oo			DEPTH m	TEMP. °C	SAL. °/oo			SIGMA-T
0	17.56	34.498	0	0	0	17.56	34.50	0	0	25.02
50	17.56	34.492	0	0	10	18.11	34.52	0	0	24.90
100	16.87	34.460	0	0	20	17.97	34.51	0	0	24.93
130	15.24	34.250	0	0	30	17.84	34.50	0	0	24.96
160	13.39	34.108	0	0	50	17.56	34.50	0	0	25.02
190	12.09	34.135	0	0	75	17.42	34.48	0	0	25.04
220	11.28	34.163	0	0	100	16.88	34.46	0	0	25.16
250	10.72	34.168	0	0	150	13.99	34.14	0	0	25.55
300	9.67	34.113	0	0	200	11.78	34.15	0	0	25.99
350	8.77	34.061	0	0	250	10.72	34.17	0	0	26.21
400	7.74	34.020	0	0	300	9.67	34.12	0	0	26.34
450	6.52	33.990	0	0	400	7.74	34.02	0	0	26.57
499	5.94	33.989	0	0	500	5.93	33.99	0	0	26.79
549	5.57	34.045	0	0	600	5.03	34.07	0	0	26.96
599	5.04	34.070	0	0	700	4.42	34.16	0	0	27.10
649	4.56	34.101	0	0	800	4.10	34.26	0	0	27.22
699	4.42	34.155	0	0	1000	3.57	34.40	0	0	27.38
799	4.10	34.261	0	0						
899	3.80	34.330	0	0	DEPTH m	TEMP. °C	SAL. °/oo			SIGMA-T
999	3.57	34.400	0	0						

DEPTH m TEMP. °C SAL. °/oo

OBSERVED

INTERPOLATED

OCEANOGRAPHIC STATION DATA

Figure **8-10** Computer printout of hydrographic station data.

who requests data about this part of the Pacific from the Oceanographic Data Center will receive a printout including the information gathered on our cruise.

Other Properties of Seawater

By boarding an oceanographic research vessel, we have seen how the distribution of temperature and salinity in the sea is measured. The presence of the salt has the effect of increasing the density of the seawater. In addition, the presence of salt affects the other properties of the water. We have already noted that seawater has a lower freezing point than fresh water (Fig. 8-6). At a salinity of 35‰, the freezing point is depressed to $-1.9°C$. At the same time, the boiling point of seawater is raised somewhat, to $100.55°C$. The presence of salt slightly reduces the tendency of water molecules to leave the liquid as water vapor. As a result, we have to heat the water an additional $0.55°C$ before the vapor pressure of water vapor is equal to 1 atm.

As seawater freezes, most of the salt is rejected by the ice and remains in the liquid. Sea ice therefore has a much lower salinity than the seawater from which it froze. By repeatedly freezing seawater, it is possible to purify it to produce fresh water. Each time, the salt in the ice is reduced by about 70 percent. As seawater in a container freezes, the salinity of the remaining water continually increases, so that the freezing point continues to decrease. If we freeze fresh water, all of it turns to ice at $0°C$. However, if we start with seawater having a salinity of 35‰, it will not begin to freeze until the temperature reaches $-1.9°C$. As the ice rejects salt, the remaining water becomes saltier. By the time half the water has been frozen, the salinity will be near 67‰ and the freezing point will be lowered to $-3.8°C$.

The presence of salt also slightly lowers the heat capacity of water. Whereas pure water has a heat capacity of 1 cal g^{-1} deg^{-1}, seawater of 35‰ salinity has a heat capacity of 0.96. The presence of 3.5 percent salt lowers the heat capacity of water by 4 percent. Therefore, adding salt makes it easier to heat water.

Summary

Water has a large heat capacity; that is, a large amount of heat is required to turn water into water vapor and to melt ice. Ice is less dense than water. Fresh water at a pressure of 1 bar attains maximal density at a temperature of 4°C. The addition of sea salt increases the density of water and lowers the freezing point and the temperature of maximum density. The salinity of seawater is determined by measuring either its chloride content or its electrical conductivity. To determine the density distribution in the sea, we must measure the temperature in place with reversing thermometers and take water samples in order to determine the salinity. From these data we can then calculate the density distribution from the known properties of seawater.

Study Questions

1. Using the data observed at station 29 (Fig. 8-10), plot the salinity and the temperature as functions of depth.
2. Using the computed data, plot the value of sigma-T against depth. (Ignore the values at 10, 20 and 30 m.) The temperature and salinity are maximal at the surface. The high temperature tends to reduce the density while the high salinity tends to increase it. Which is the dominant factor?
3. With depth, the salinity goes through two minima. At the depths of these minima, does the density also have minima? If not, why not?

9 The Motion of Fluids

The differential heating of the earth by the sun results in the motion of its fluid envelope, the oceans and the atmosphere. Since the moon's surface is devoid of fluids, it is relatively static, although its surface undergoes large monthly temperature changes. Only the occasional impact of a meteorite or a volcanic event disturbs the lunar landscape. The ceaseless motion of the fluids on the earth's surface, on the other hand, makes our landscape more dynamic. The action of air and water breaks down the hardest rock to produce sand and soil. The small particles thus formed are then blown by the wind and carried by moving water. At the same time, the motion of fluids redistributes the radiant energy received from the sun to make the climate more equitable. Air and water modify our environment because of their motion. We must therefore review the physical laws that govern the motions of fluids.

Newton's Laws

The basic laws that govern the motion of all matter, be it solid, liquid, or gas, were formulated by Sir Isaac Newton near the end of the seventeenth century. Newton's first law concerns the motion of matter that is not subject to any net external forces. Such an object, if at rest, will remain at rest; if it is in motion, it will continue to move in a straight line at a uniform speed:

$$\text{velocity} = \text{a constant}$$

If, on the other hand, a force acts on the object, then its motion will change according to Newton's second law. This states that a force will cause an object to accelerate (change its velocity) so that the magnitude of its acceleration is equal to the external force divided by the mass of the object:

$$\text{force} = \text{mass} \times \text{acceleration}$$

Since acceleration is the rate at which the velocity changes, the acceleration is a time rate of change of velocity, and its units are therefore velocity/time, or cm sec^{-2}. If an object initially at rest accelerates at a rate of 1 cm sec^{-2}, after 2 sec its velocity will be 2 cm sec^{-1}. Since $F = ma$, the units of force are g cm sec^{-2}, and 1 g cm sec^{-2} is called 1 dyne.

Vectors and Scalars

Quantities that have only magnitude are called *scalars;* those that have direction as well as magnitude are called *vectors.* Scalars—for example, mass, temperature, and energy—can be specified completely by a number—a mass of 5 g, a temperature of 20°C, an increase in energy of 7 cal. However, to state that a ship is traveling at a velocity of 10 knots describes its motion only partially. The direction in which the ship is moving must also be specified. Thus velocity is a vector quantity. Similarly, acceleration and force are vector quantities, for one accelerates in a particular direction and a force acts in a specific direction.

Newton's second law contains three quantities, two of them vectors and one a scalar. The force, a vector, is equal to the mass, a scalar, multiplied by the acceleration, another vector. The expression $F = ma$ implies that the two sides of the equation are equal in both magnitude and direction. Thus the direction of the acceleration is in the direction of the force, and its magnitude is equal to the magnitude of the force divided by the mass.

Let us now consider an object moving at a constant speed v in a circle of radius R. In this case, the rate (cm sec^{-1}) of the object's motion does not change with time; however, the direction of the motion changes, for the object will have described a complete circle of 360° after traveling a distance of $2\pi R$. Since there is a constant change in direction, the object is constantly being accelerated (Fig. 9-1). The acceleration, which has the effect of slowly rotating the velocity vector, acts at right angles to the velocity, and its magnitude is equal to the square of the velocity divided by the radius of the circle

$$a = v^2/R$$

Thus to keep an object of mass m moving in a circle requires a force toward the center equal to mv^2/R. This is the force that counteracts the so-called centrifugal force. The centrifugal force, however, is not really

Figure **9-1** An object moving in a circle at constant speed.

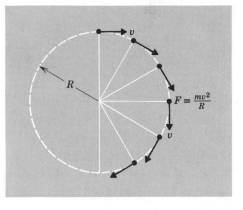

a force but is merely the result of the natural tendency of the object to continue to move in a straight line in its original direction.

Forces in Fluids—Pressure

Solids retain their shape even in the presence of forces that attempt to change that shape. For example, if we pull on both ends of a steel bar, it will resist being stretched until the tension exceeds the strength of the steel and the bar breaks. While the bar is under tension, the forces in it are different in different directions. We can readily see this by cutting the bar apart. If we cut at right angles to the direction of the tension, the bar will part. If, on the other hand, we cut in the direction of the tension, nothing happens (Fig. 9-2). While the bar was under tension in one direction, there were no forces acting at right angles to the direction of pull.

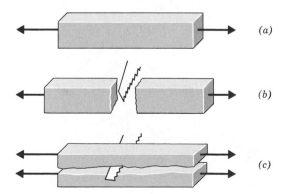

Figure **9-2** (*a*) A steel bar under tension; (*b*) a cut perpendicular to the tension will cause the bar to part; (*c*) a cut parallel to the tension does not interfere with the bar's ability to withstand the tension.

In contrast to a solid, a fluid cannot resist being deformed. If put under tension, it will flow. It is unable to maintain a shape but takes on the shape of its container. Therefore, the forces within a fluid must be the same in all directions; otherwise the fluid would yield until the forces were equalized.

Now consider the force acting on a unit area 1 cm². Such a force is a pressure. In the steel bar, the pressure varies with the orientation of the area. Perpendicular to the external pull, the pressure tends to pull the bar apart, as in Figure 9-2b. If the area is rotated parallel to the tension, the force is zero. The negative pressure or tension in the steel, therefore, depends on the direction. In a fluid, however, the pressure at any one place is independent of direction. The slightest tendency for the pressure to become anisotropic (not the same in all directions) will cause the fluid to deform.

Thus the pressure in a fluid is a scalar quantity. To describe the forces acting in a liquid, we need to know only the distribution of the scalar pressure. To describe the state of stress in a solid, however, we must specify the direction as well as the magnitude of the forces acting per unit area.

The Motion of a Fluid in Response to Pressure

How is the motion of a fluid governed by the distribution of pressure in that fluid? Within a fluid—for example, the ocean—let us imagine a coordinate system having three axes at right angles to each other, with the positive z axis directed downward (Fig. 9-3). The value of z, there-

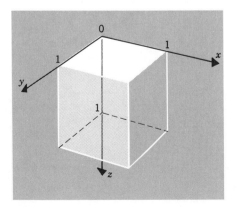

Figure **9-3**   The unit cube in the coordinate system.

fore, will increase with depth. At every point (x, y, z) the pressure will have a particular value.

We wish to learn what the distribution of the pressure must be in order for the fluid to remain at rest. Let us consider a centimeter cube of the fluid lying between $x = 0$ and $x = 1$, $y = 0$ and $y = 1$, and $z = 0$ and $z = 1$. The forces on the cube will be the pressures acting perpendicular to the six faces of the cube. In addition, the earth's gravitational attraction will exert a downward force in the positive z direction. This force is equal to the mass of the cube multiplied by the acceleration of gravity (a_g),

$$a_g = 980 \text{ cm sec}^{-2}$$

The mass of the centimeter cube will be equal to the density of the fluid, ρ, in g cm^{-3}. Thus if the fluid is seawater, with sigma-T $= 27$, the cube will have a mass of 1.027 g and the force of gravity in the positive z direction will be

$$1.027 \times 980 = 1006 \text{ dynes}$$

The other vertical forces acting on the cube will be a downward force acting on the upper face equal to the pressure, P, at $z = 0$ and an upward force acting on the lower face equal to the pressure at $z = 1$ cm (Fig. 9-4). The total downward force is therefore

$$\text{total force} = a_g \times \rho + P(\text{at } z = 0) - P(\text{at } z = 1 \text{ cm})$$

In order that the cube not be accelerated, the total force must be zero;

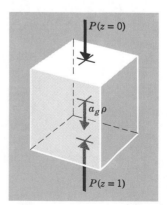

Figure **9-4**   The vertical forces acting on the unit cube.

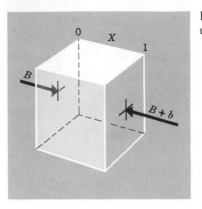

Figure **9-5**   The horizontal forces acting on the unit cube.

therefore the pressure must increase with depth at the following rate:

$$\text{pressure increase with depth} = a_g\rho \text{ cm}^{-1}$$

For seawater this requires a pressure increase of about 1000 dynes cm^{-3}, which amounts to an increase of approximately 1 bar per 10 m increase in depth.

The horizontal forces acting on the cube are the pressures on the four vertical faces. If the pressures on opposite faces are equal, there will be no net force acting on the cube of fluid. If, however, the pressure at a given depth changes in the x or y direction, it will result in a horizontal force, since the pressures on opposite faces will no longer cancel each other out. For example, if the pressure is B at $x = 0$ and $B + b$ at $x = 1$ cm, there will be a net force of b in the negative x direction (Fig. 9-5). Since the cube contains a total mass equal to the density of the fluid, the acceleration will be

$$\text{acceleration} = b/\rho \text{ cm sec}^{-2}$$

in the negative x direction.

The horizontal motion of fluids is the result of horizontal pressure variations. The absolute value of the pressure is unimportant, for if it did not vary horizontally, the equal pressures on the opposite sides of the cube would cancel out. If the pressure does vary in a horizontal direction, the higher-pressure face experiences a greater force. Thus the fluid tends to move from a region of high pressure to one of low pressure.

The Weather Map

Weather stations throughout the world continuously measure various aspects of the weather, such as temperature, wind, precipitation, and atmospheric pressure. These data are then transmitted to the national weather bureaus, where they are combined to give a synoptic view of the weather which is then published in the form of a weather map. Such a map for the United States at 7 P.M. E.S.T. on July 13, 1967, is shown in Figure 9-6. The observed distribution of atmospheric pressure as corrected to sea level is shown by plotting lines of constant pressure. In addition, the direction and speed of the wind and the surface temperatures are shown for each weather station.

The pressure distribution on this particular day had a high of 1025 millibars (mb) over North Dakota. This high-pressure region was surrounded by areas where the pressure dipped below 1016 mb. A region of very low pressure (1004 mb) was centered just east of Hudson Bay.

From our discussion of the motions of fluids in response to horizontal variations in pressure, we would expect the winds to blow from the high-pressure regions to the regions of low pressure, since this is the direction of the forces acting on the air. Thus we would expect the wind to blow across the lines of constant pressure from the high to the low. When we look at the observed wind directions, however, we find that this is not the case. Instead of blowing across the lines of constant pressure, the winds seem to blow *along* these lines. Thus the wind is roughly perpendicular to the direction of the force produced by the pressure variations. The winds in the United States tend to blow clockwise around regions of high pressure and counterclockwise around regions of low pressure.

The data recorded on the weather map seem to contradict the theory of fluid motion we have just developed. The theory is not incorrect, but incomplete, for we have thus far neglected the effect of the earth's rotation. If the earth were not rotating about its axis, the winds would indeed blow from regions of high to regions of low pressure. Let us investigate the effect of the rotation of the earth on the motion of fluids.

The Effect of the Rotation of the Earth

Consider water at rest in a circular dishpan. The level of the water in the pan will be horizontal. We rotate the dishpan counterclockwise to

Weather Reports Throughout the Nation

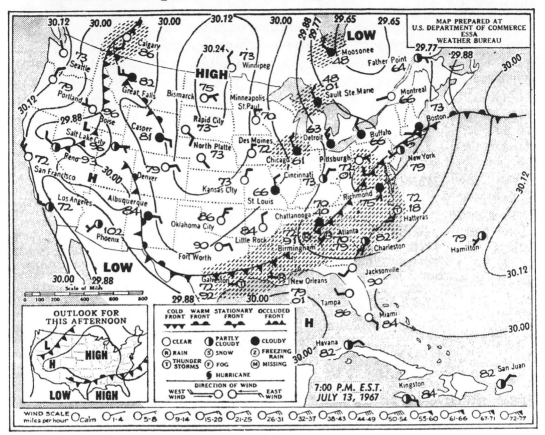

Figure beside Station Circle indicates current temperature (Fahrenheit); a decimal number beneath temperature indicates precipitation in inches during the six hours prior to time shown on map.

Cold front: a boundary line between cold air and a mass of warmer air, under which the colder air pushes like a wedge, usually advancing southward and eastward.

Warm front: a boundary between warm air and a retreating wedge of colder air over which the warm air is forced as it advances, usually northward and eastward.

Occluded front: a line along which warm air has been lifted by the action of the opposing wedges of cold air. This lifting of the warm air often causes precipitation along the front.

Shading on the map indicates areas of precipitation during six hours before time shown.

Isobars (solid black lines) are lines of equal barometric pressure and form pressure patterns that control air flow. Labels are in inches.

Winds are counter-clockwise toward the center of low-pressure systems, and clockwise and outward from high-pressure areas.

Pressure systems usually move eastward, averaging 500 miles a day in summer, 700 in winter

Figure **9-6**   Weather map for the continental United States for 1900 h EST, July 13, 1967. A pressure of 30.00 inches corresponds to 1.016 bar and the contour interval corresponds to 0.004 bar. Thus 30.29″ = 1.025 bar and 29.85″ = 1.011 bar.

118

represent the rotation of the earth at the North Pole. Because of the centrifugal force, the water will be pushed away from the center of the pan toward the rim. The surface of the water will take on a concave shape so that the slope of the water counteracts the centrifugal force. Just as the level of water is depressed in the center of the dishpan and raised at the periphery, so the shape of the earth is flattened at the poles and widened at the equator. The surface of the Arctic Ocean is 21 km closer to the center of the earth than the ocean surface at the equator.

If an object floats on the water in a position which is stationary relative to the rotating dishpan, the forces are in balance; that is, the poleward slope of the water surface exactly balances the centrifugal force on the object. What happens if we attempt to move the floating object relative to the rotating dishpan? First, let us move the object counterclockwise to the east (Fig. 9-7a). Since this is equivalent to a more rapid rotation about the pole, the centrifugal force will increase as a result. And since this increase is not balanced by the shape of the water surface, the object will tend to move outward toward the equator. The object will therefore be deflected to the right. Next, let us move the object in a clockwise direction, to the west (Fig. 9-7b). Now the centrifugal force will be reduced so that the object will move toward the pole on the sloping water surface. Note that the deflection is again to the right.

Now let us move the object toward the wall of the pan, to the south (Fig. 9-7c). As it moves outward, the angular velocity of the object about the pole will be slower than that of the water, so that the object will lag behind the rotation and be deflected in a clockwise sense. The object is again deflected to the right. Finally, let us move the object from

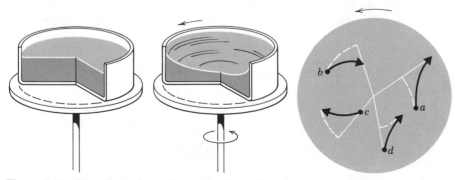

Figure **9-7**   The effect of rotation on horizontal motion.

the periphery toward the pole, to the north (Fig. 9-7d). The angular velocity of the object about the pole will be faster than that of the water so that the object will be deflected in a counterclockwise sense. Again, the deflection of the object is to the right of its direction of motion. Regardless of how we move the object, it is always deflected to the right of its sense of motion.

If we had wanted a model of conditions at the South Pole, we would have had to rotate the dishpan in a clockwise sense, and all the deflections would have been to the left. On the equator, on the other hand, the effect on horizontal motions disappears, for here the centrifugal force is entirely vertical. The effect of the earth's rotation is to deflect moving objects to the right in the northern hemisphere and to the left in the southern hemisphere. This effect is known as the *Coriolis acceleration,* after the French physicist Gaspard Gustave de Coriolis, who published the basic paper on the motion of fluids on the rotating earth in 1835. The Coriolis acceleration always acts at right angles to the velocity and has a magnitude of 1.5×10^{-4} times the velocity v in cm sec^{-1} at the poles. This magnitude decreases as the sine of the latitude, ϕ, to zero at the equator (sine $0° = 0$):

$$\text{Coriolis acceleration} = 1.5 \times 10^{-4}\, v \text{ sine } \phi \text{ cm sec}^{-2}$$

The Coriolis acceleration is difficult to understand without the aid of vector algebra. For the benefit of those who are familiar with vector notation, the vector form of the Coriolis acceleration is given below.

Let Ω be the rotation vector of the earth, with the vector pointing in the direction of rotation of a right-hand screw—that is, northward, along the earth's axis. Since the earth turns once in 24 hours, the magnitude of Ω is $2\pi/24 \times 60 \times 60$, or 0.729×10^{-4} rad sec^{-1}. If **V** is the velocity, then

$$\text{Coriolis acceleration} = -2\Omega \times \mathbf{V}$$

where $\times$ represents the vector product between the rotation Ω and the velocity **V**.

If u and v are the east and north components of the velocity **V**, the corresponding components of the Coriolis acceleration C_u and C_v are

$$C_u = \quad 1.46 \times 10^{-4} \sin \phi\, v$$
$$C_v = -1.46 \times 10^{-4} \sin \phi\, u$$

where ϕ is the latitude.

The Weather Map Reconsidered

Now we are ready to return to the weather map. Let us look in detail at the winds between Bismarck, North Dakota, and Detroit, Michigan. Imagine a tube of air, 1 cm² in cross section, extending between these cities. The pressure on the end of the tube is 1025 mb at Bismarck and 1011 mb at Detroit. The pressure difference between the ends thus pushes the tube eastward, toward Detroit, with a force of 14 mb, or 14,000 dynes, for the 1 cm² cross section (Fig. 9-8). Since the distance between the two cities is 1400 km and the specific volume of air is 860 cm³ g⁻¹, the tube contains about 1.6 × 10⁵ g of air. The force per unit mass on the air column, therefore, is (14,000)/(160,000), or 0.088 cm sec⁻² to the east.

Both cities are at a latitude of about 45°N, and the sine of the latitude is 0.7. The Coriolis acceleration is therefore about 1.0×10^{-4} times the velocity in cm sec⁻¹. We note from the weather map that the movement of air between Bismarck and Detroit is from the north. Since the Coriolis acceleration in the northern hemisphere acts to the right of the velocity, it would tend to push the air to the west. This is in direct opposition to the pressure force, which tends to move the air to the east.

Now let us calculate how fast the wind would have to blow from the north for the Coriolis acceleration to be just equal and opposite to the force per unit mass due to the pressure difference:

$$0.088 = 1.0 \times 10^{-4} \, v$$

Therefore $$v = 880 \text{ cm sec}^{-1}, \text{ or 20 mph.}$$

The wind velocities between Bismarck and Detroit, as indicated on the weather map, are between 9 and 14 mph. Because of the retarding effect

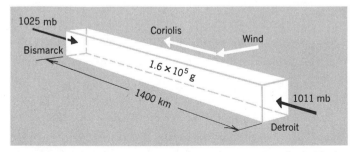

Figure 9-8   The forces on a 1 cm² tube of air between Bismarck and Detroit at 1900 h EST, July 13, 1967.

of friction near the ground this is somewhat slower than would be required for the Coriolis acceleration to balance the force of the pressure. If the pressure force is to be balanced by the Coriolis acceleration, the wind has to blow along the lines of constant pressure. Since the deflection is to the right in the northern hemisphere, the high pressures must be to the right of the wind. If the two forces are exactly balanced, there will be no acceleration, and the wind will therefore blow with a steady velocity.

The approximate balance between the Coriolis and pressure-difference accelerations is not limited to the atmosphere; the same forces dominate the movement of ocean currents. In the ocean, however, the velocities are lower and the density of the medium is greater. Let us see what sort of pressure difference is required to counteract the Coriolis acceleration of a current such as the Gulf Stream. Typically, this stream is about 50 km wide; its speed ranges up to about 200 cm sec^{-1} (4 knots), with an average of about 100 cm sec^{-1}. At this average speed, the Coriolis acceleration will be about 10^{-2} cm sec^{-2}. Consider again a 1 cm^2 cross-section tube across the stream. This tube will contain about 5×10^6 g of seawater. To withstand an acceleration of 10^{-2} cm sec^{-2} an opposing force will be required of 5×10^4 dynes cm^{-2} or a pressure difference of 50 mb—the pressure exerted by a column of water 50 cm high. To compensate for the Coriolis effect, the water would have to stand 50 cm higher on the right side of the stream than on the left. Thus the sea surface would slope about 1 cm km^{-1}.

The Magnitude of the Accelerations

The magnitude of the Coriolis acceleration is very small. Typically, it is about 0.1 cm sec^{-2} for the winds and about 0.005 cm sec^{-2} for ocean currents. In contrast, the acceleration of gravity is about 10^3 cm sec^{-2}. For experiments on a laboratory scale, the Coriolis effect is negligible. For example, if we drive a car at 90 mph (4000 cm sec^{-1}) the Coriolis acceleration will tend to push it to the right with an acceleration of 0.4 cm sec^{-2}. Meanwhile, gravity will be acting downward with an acceleration of 980 cm sec^{-2}. As a result, the car will lean to the right by 4 parts in 10,000, or 1.4 minutes of arc. We do not have to slope our freeways to compensate for the Coriolis effect. In laboratory experiments with fluids, the Coriolis effect is masked completely by frictional forces.

To do model experiments on fluids in rotating systems, we have to increase the speed of rotation greatly over the one rotation per day of the earth. Why, then, do these very weak forces control the motions of the ocean and the atmosphere?

The other force we encounter in fluid experiments is the force of friction. After we stir water in a glass, the force of friction gradually slows down the motion, and the water comes to rest. The friction in a fluid depends on the scale of the motion. Water in a small glass comes to rest much more rapidly than water in a bathtub. This is because fluid friction depends on how fast the velocity changes in relation to distance. At the wall of the glass, the fluid is at rest; in the center of the glass, the fluid is still moving. The velocity thus changes significantly over a distance of 1 cm. In the bathtub, the change in velocity with distance is much less, since the same velocity differences are spread over distances of tens of centimeters. As a result, the motion dissipates more slowly.

When we are dealing with the ocean and the atmosphere, we are concerned with vertical distances measured in kilometers and horizontal distances in thousands of kilometers. As a result, the frictional effects are reduced by a factor of at least 10^4 (1 km to 10 cm). Thus friction, which dominates the laboratory experiments, becomes insignificant in observations on a larger scale, permitting the remaining weak accelerations to control the outcome. Small pressure differences and the Coriolis effect become dominant, and friction, while always present, becomes relatively unimportant. Because of the change in scale, different forces control the behavior of fluids on a man-size and a global scale. Our intuition, which is based largely on experiences with a man-size scale, is therefore a poor guide to processes in the ocean and the atmosphere.

Potential Energy

The motions of the ocean and the atmosphere are controlled by pressure differences and the effect of the rotation of the earth. The pressure differences, in turn, result from variations in the density of air and seawater. The effect of such density variations can best be studied by considering a kind of energy called *potential energy*.

Rivers always run downhill. In a hydroelectric station, we can generate electric power by making the downward-running water turn a turbine. Thus the water must contain more energy when it is in the reservoir

above the dam than when it comes out of the turbine at a lower elevation. The greater the difference in level, the more energy we can extract per gram of water (Fig. 9-9).

The increase in energy with height results from the gravitational attraction of the earth. Energy has to be invested to lift water up a mountain; when we let the water flow down again, we can recover a portion of that energy. If we lift a mass of A g a vertical distance h, we have to expend an amount of energy equal to Aa_gh, where a_g is the acceleration of gravity (980 cm sec^{-2}). Thus the potential energy of the object has been increased by an amount Aa_gh. If we do not hold the object up, it will fall back down again, reducing its potential energy. The natural tendency is for the potential energy to become a minimum. Water flows downhill because doing so reduces its potential energy.

If we lower 1 cm^3 of water a distance of 1 cm, its potential energy is reduced by 980 g cm^2 sec^{-2}, or 980 ergs. Actually this is only approximately true, for we operate in the atmosphere rather than in a vacuum. What we have really done is to interchange 1 cm^3 of water with 1 cm^3 of air 1 cm apart. We have thus increased the potential energy of the air and decreased that of the water. Since the 1 cm^3 of air weighs only 1/860 as much as 1 cm^3 of water, the net change is a decrease in potential energy. In the case of air and water, we can ignore the air. Now, however, let us consider what happens if the interchanged fluids differ only slightly in density.

What we wish to consider are columns of seawater or air that have small vertical differences in density. As an illustration, let us consider water and oil, the oil being slightly less dense than the water. If we pour some water and oil in a narrow tube, stoppered at the bottom, we can place either the water or the oil on top (Fig. 9-10). If we repeat the experiment in a glass, we have no trouble pouring the oil on top of the water. If we attempt to pour the water on top of the oil, however, the water will sink to the bottom and the oil will float on top. The natural situation is for the denser fluid to be at the bottom.

We must now consider the potential energy. Let us take a tube 1 cm^2 in cross-sectional area. In it we place 10 g of water of density 1 g cm^{-3}

◀ Figure **9-9**   Bridal Veil Fall, Yosemite National Park, California. (Photograph: Ansel Adams)

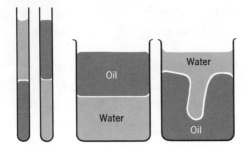

Figure **9-10**   Oil and water in a narrow tube and in a beaker.

and 10 g of oil of density 0.833 g cm^{-3}. Thus we have 10 cm^3 of water and 12 cm^3 of oil in our tube. Regardless of how we fill our tube, it will be filled to a height of 22 cm. Let us calculate the potential energy of the liquids in our tube. The masses will be distributed over a distance of 22 cm. In calculating the potential energy, however, we can consider that each fluid is concentrated at its midpoint. The two calculations are shown in Figure 9-11. With the water on the bottom, the total potential energy is 210 a$_g$, while with the oil on the bottom it is 230 a$_g$. Thus the potential energy is less with the denser fluid on the bottom, and so this represents the stable state of the system.

When the oil was at the bottom in the small tube, the system was unstable. Because of the narrowness of the tube, this unstable state was preserved. If we had used a slightly wider tube, and if we were careful, we might still have been able to pour the water on top of the oil. Any jarring, however, would make the water flow to the bottom. The unstable situation is similar to standing a coin on its edge. Its potential energy is greater in this position than if it is lying flat, since a slight disturbance will make it fall over to a position of least potential energy. In nature, of course, there are always disturbances and the fluids are not confined, so that the densest materials will be at the bottom and the density will decrease upward.

The simple relationship between density and potential energy holds only for fluids that are incompressible. In our calculation we ignored the change in density with pressure as we interchanged the oil and the water. For oil and water this was a good approximation. If we consider columns of air, however, we must take the compressibility and the internal energy into account. The change in internal energy results from

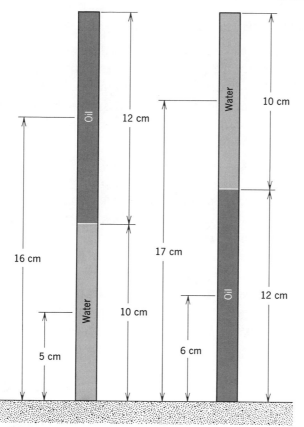

Figure **9-11**   The total potential energy m a_g h.
Oil on top of water: $10 \times a_g \times 5 + 10 \times a_g \times 16 = 210\ a_g$ ergs.
Water on top of oil: $10 \times a_g \times 6 + 10 \times a_g \times 17 = 230\ a_g$ ergs.

the work done in compressing and decompressing the interchanged fluids. As we interchange unit masses of air between two levels, we must calculate the change in potential and internal energy of the air column produced by the interchange. If the total energy increases, the system was stable; a decrease in total energy indicates that the system had been unstable and would tend to overturn. The same situation holds in the case of seawater. If the interchange of unit masses of water results in an increase in the sum of the potential and internal energies, the water was originally stably stratified. A decrease in the total energy implies an unstable stratification.

Horizontal Density Variations

Locally, the surface of the earth is usually stably stratified, with the density increasing as we proceed downward from air to seawater to rocks. Laterally, however, the earth is not in a state of minimum energy, for the dense rocks on the tops of mountains are above the level of the sea. An equilibrium earth would offer a monotonous seascape, and the depth of the ocean would be the same everywhere. The heat generated within the earth by radioactive decay counteracts the tendency toward lower potential energy by causing mountains to rise and portions of the sea floor to sink. The sun-powered water cycle, in turn, tends to erode the relief by moving matter from the highlands to the bottom of the sea. Thus the earth is "alive" and there is life on earth.

Let us look at the effect of the horizontal differences in density by considering a column of oil and one of water in two vertical tanks interconnected by pipes at various levels (Fig. 9-12). The right-hand tank is filled with water to a depth of 10 m. In order for the pressure at the bottom of the two tanks to be equal, we must fill the left-hand tank with

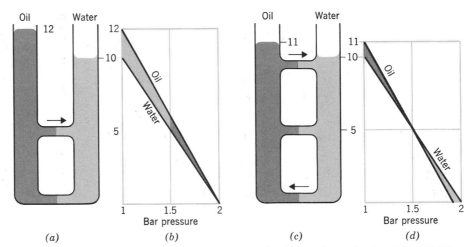

Figure **9-12**   Pressure differences between a tank of oil and a tank of water: with 12 m of oil in the left tank (*a*), the pressures are equal at the bottom (*b*). As a result, if the tanks are connected above the bottom, flow will take place from the oil tank to the water tank. With 11 m of oil in the left tank (*c*), the pressures are equal 5 m above the bottom (*d*). Flow is from the oil to the water near the top, while below 5 m the flow is from the water to the oil.

12 m of oil. Interconnecting the tanks at the bottom will not lead to a transfer of fluid, since the pressure at the bottom of both tanks is 2 bars (1 bar of fluid pressure in addition to the 1-bar pressure of the atmosphere). The pressure in the two tanks as a function of the distance above the bottom varies as shown in Figure 9-12b.

As we move upward, the pressure difference between the two tanks gradually increases. At a height of 10 m, the difference is 2/12, or 0.166, bar. If we interconnect the tanks above the bottom, there will be a flow of oil into the water. Because of the different rates of increase of pressure with depth, the two tanks can be in equilibrium only at one depth—the bottom.

If we had put only 11 m of oil in the left-hand tank (Fig. 9-12c), the pressures in the two tanks would be equal at 5 m above the bottom (Fig. 9-12d). Above this level, the pressure in the oil is greater than that in the water; below 5 m, the pressure in the water is greater. If the tanks are interconnected anywhere except at 5 m, flow will ensue until at equilibrium both tanks would have 5 m of water in the bottom with 5.5 m of oil above.

Horizontal Density Variations and the Rotation of the Earth

Now consider the combined effect of horizontal density variations and the rotation of the earth. Let the counterclockwise-rotating dishpan represent the northern hemisphere of a water-covered earth. The water will be cold near the poles and warm near the equator; thus the density will be higher in the polar region. Let us represent the polar sea by water and the less dense tropical sea by oil. The water, being denser, will also fill the bottom of the ocean. Our model then consists of a bottom layer of water overlaid by a cylinder of water near the pole and surrounded by a layer of oil (Fig. 9-13).

The only place where horizontal pressure differences exist is at the latitude that separates the oil from the water. Because of its lower density, the surface of the oil will stand above the water. The pressure near the surface in the oil will therefore be greater than that in the water, and so the oil will have a tendency to flow poleward and override the water. The effect of the rotation will be to deflect this motion to the right, causing the oil near the boundary to flow to the east.

Replace the two-fluid model by a single fluid, seawater. The sharp dis-

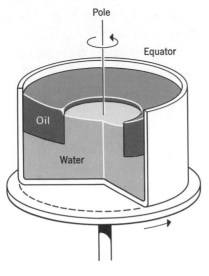

Pole

Equator

Oil

Water

Figure **9-13** Oil and water in a rotating dishpan.

continuity in density now becomes blurred. Instead of an abrupt rise in the sea surface at the fluid boundary, there will be a gradual increase in elevation as the density decreases. In the region of density change there will be a horizontal change in pressure. The tendency of the water to flow to the north will be deflected to the right by the rotation of the earth, resulting in an eastward current. Note that the current produced as a result of the joint action of the pressure variation and the Coriolis effect flows in the region of the surface density change. Thus it is neither a warm nor a cold current but, rather, a current formed by and flowing along the region of rapid temperature change.

The distribution of ocean and continents on the real earth makes the actual currents more complex than in our simple "ocean only" model. In the North Atlantic and the North Pacific, however, we can identify the predicted eastward-flowing currents between the cool and warmer water as the North Atlantic and the North Pacific currents (Fig. 9-14). These are the eastward extensions of the Gulf Stream and the Kuroshio Current. Without rotation, the warm tropical waters would override the cooler polar waters and so reduce the temperature variation from the equator to the poles. By deflecting the water motion, the Coriolis effect helps to maintain a greater temperature difference. It acts like a barrier preventing the poleward drift of the warm water and so allows it to

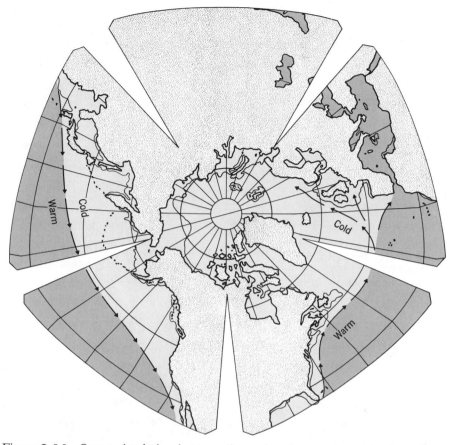

Figure **9-14** Ocean circulation between the cool and warm waters in the northern hemisphere. The strong currents at the western margins of the oceans are the Kuroshio in the Pacific and the Gulf Stream in the Atlantic.

become even warmer. This "bottling up" of the warm water is not wholly efficient. Because of fluid friction and mixing, we do not have an exact balance between the Coriolis and the pressure-difference accelerations. As a result, there is interchange between the polar and equatorial waters. However, this interchange is partially inhibited and so results in a larger temperature difference than would exist without it.

The rotation of the earth thus plays a very significant role in shaping our environment. The ocean acts as if it were heated in low latitudes and cooled in high latitudes. Consider a rectangular tube filled with water that is heated at one end to the same extent as it is cooled at the other

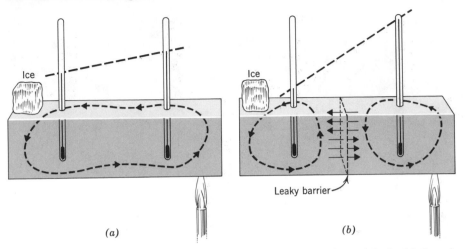

Figure **9-15**   Water convection between a heat source and a heat sink. In (*a*) there is no impediment to the motion of water between the warm and cold ends. The Coriolis effect acts like a barrier to the heat exchange, hence increases the temperature difference (*b*).

end. In other words, the same amount of heat is added to one end as is extracted from the other. Because of the density differences produced, the water in the tube will begin to convect, carrying the heat from the equator to the pole. A temperature difference will be set up which depends on the ease of convection. The faster the water moves between the ends, the less difference in temperature is required to carry the heat (Fig. 9-15*a*).

The Coriolis acceleration now acts as if we had placed an imperfect barrier between the pole and the equator (Fig. 9-15*b*). As a result, the poleward motion of the water is somewhat impeded; if the water is to carry the same amount of heat, the temperature difference must be increased. There will be rapid circulation on each side of the barrier but a relatively slow exchange of water, and therefore heat, across the barrier. This Coriolis thermal-barrier effect takes place in both the ocean and the atmosphere. In the ocean, the situation is complicated by the geography of land and water. In the atmosphere, the effect of geography is less, since only the lower portion of the atmosphere is affected by the presence of mountain barriers. We must now investigate the properties of air and the atmospheric circulation that results from the differential heating of air by the radiation from the sun.

Summary

The fluids of the ocean and the atmosphere move in response to horizontal variations of pressure. Because of the rotation of the earth, the horizontal pressure variation is balanced by the Coriolis force. Thus the winds do not blow from regions of high to regions of low pressure but flow along the lines of equal pressure, with the high pressure on the right in the northern hemisphere. The pressure variations result from changes in density. The sea level of the less dense tropical water is higher than that of the denser subarctic water. In the northern hemisphere, the slope of the sea surface produces an eastward-flowing current in the region of density change. Thus the Coriolis effect reduces the horizontal mixing between waters of different densities.

Study Questions

1. Obtain a copy of a weather map from a newspaper and discuss the distribution of pressure and the directions of the wind as shown on the map.
2. How would the ocean currents behave at the boundary between the warm and cold water in the southern hemisphere? (Check your deduction against Fig. 12-5.)
3. If the earth did not rotate about its axis, how would the motions of the ocean and the atmosphere be affected?
4. How would the latitudinal variation of the average yearly temperature be altered if the earth did not rotate, and why?

10 Air and the Circulation of the Atmosphere

The motions of the fluid envelope of the earth result, as we have seen, from horizontal variations in pressure and the Coriolis acceleration. In the atmosphere, as in the ocean, horizontal pressure differences arise from variations in density. Since seawater is a liquid, its density depends primarily on the temperature and the salinity and varies only slightly with pressure. Air, however, is highly compressible, as indeed are all gases. The specific volume of air is directly proportional to the absolute temperature and inversely proportional to the pressure.

The Specific Volume of Air

In a gas the molecules move about freely. The higher the temperature, the faster is the movement of the individual molecules. The pressure the gas exerts on the walls of its container results from the collisions of the individual gas molecules with the walls. The more molecules that collide with the walls in a given time, the greater the pressure they exert. The higher the temperature, hence the greater the velocity of the gas molecules, the greater will be the impact of the molecules with the walls, hence the greater will be the pressure the gas exerts. In order for the molecules to exert no pressure, they must be standing still. This happens at absolute zero, corresponding to a temperature of $-273°C$. At any temperature above absolute zero, the molecules are in motion and exert a pressure. To describe the behavior of a gas most easily, we use the absolute temperature scale, the Kelvin scale. On this scale, absolute zero = $0°K$, the freezing point of water is $273°K$, and the boiling point of water is $373°K$. This is the same temperature scale we used to describe the thermal radiation of a body. At absolute zero the molecules are motionless and are unable to emit thermal radiation.

If we consider a given mass of gas, say 1 g, we would expect the pressure to increase with the temperature. The larger the volume that the gram of gas occupies, the farther apart the molecules will be, hence the fewer collisions they will make with the walls in unit time. Thus the larger the volume per gram, the lower we would expect the pressure to be. The actual relationship between the specific volume, the temperature, and the pressure is very simple:

$$\text{specific volume} = \frac{\text{constant} \times \text{absolute temperature}}{\text{pressure}}$$

Thus the pressure is proportional to the absolute temperature and inversely proportional to the volume. For air we obtain:

$$\text{specific volume of air} = \frac{2.89 \times T(^\circ\text{K}) \text{ cm}^3 \text{ g}^{-1}}{P(\text{bar})}$$

At 1 bar and a temperature of 300°K (27°C), a gram of air therefore occupies a volume of 870 cm³, or roughly 1000 times the volume of a gram of water.

We can derive the equation for the specific volume of air from the general gas law, which states: For any ideal gas, the product of the pressure and the volume of the gas is equal to the product of the number of moles of the gas (N), the gas constant (R), and the absolute temperature. (The number of moles is obtained by dividing the mass of the gas in grams by its molecular weight.)

$$PV = NRT$$

The gas constant $R = 83$ bars cm³ deg⁻¹ mole⁻¹.

This law holds for single gases as well as for mixtures such as air. In the case of a mixture we must determine the average molecular weight.

Air	Composition (%) (1)	Molecular Weight (2)	(1) × (2) (3)
N₂(nitrogen)	78	28	21.8
O₂(oxygen)	21	32	6.6
A(argon)	1	40	0.4
Average molecular weight of air			28.8

Thus 1 g of air contains 1/28.8, or 0.0348, mole. The specific volume of dry air is therefore:

$$\text{specific volume} = \frac{83 \times T(^\circ K)}{28.8 \; P(\text{bar})} \; \text{cm}^3 \, \text{g}^{-1}$$

The Specific Heat of Air

When we measured the specific heat of water, we merely placed a gram of water in an open container and noted how fast the temperature of the water increased as we added heat. However, if air is to be confined in a container, the container must be sealed. As we heat the air, its pressure will increase. We can perform the heating experiment in two ways. We can seal the air in a container so that the volume of the air remains constant while its pressure increases (Fig. 10-1a) or we can contain the air in a cylinder fitted with a piston that exerts a constant pressure while we allow the volume to increase (Fig. 10-1b).

In the first experiment we obtain the heat capacity at constant volume (C_v). We find that it takes 0.171 cal to raise 1 g of air 1°C. In the second experiment, as the piston moves out to maintain a constant pressure, it does work against the atmosphere. Therefore we must add an amount of heat to compensate for the work done by the air against the atmosphere.

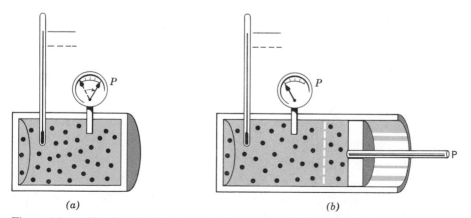

(a)   (b)

Figure **10-1**   Heating air: (a) at constant volume and (b) at constant pressure.

This amount of work, equal to the change in $P \times V$, is

$$PV \text{ (at } T + 1) - PV \text{ (at } T) = 2.89 \, (T + 1) - 2.89T$$
$$= 2.89 \text{ bar cm}^3 \text{ deg}^{-1} \text{ g}^{-1}$$

The pressure-times-volume units of energy can be converted to calories by the relationship

$$41.9 \text{ bar cm}^3 = 1 \text{ cal}$$

Therefore, the additional heat required amounts to 2.89/41.9, or 0.069 cal deg^{-1} g^{-1}, giving:

Specific heat of air at constant pressure $(C_p) = 0.171 + 0.069$
$$= 0.240 \text{ cal g}^{-1} \text{ deg}^{-1}$$

The Change of Pressure with Height

In the preceding chapter we saw that the pressure in a fluid increases per centimeter at a rate equal to the product of the acceleration of gravity and the density. Since water is only slightly compressible, its density does not increase significantly with pressure; therefore the pressure in a water column increases at a roughly uniform rate of 1 bar for every 10 m of depth. The density of air, on the other hand, is directly proportional to the pressure and inversely proportional to the absolute temperature:

$$\text{density of air} = \frac{P}{2.89T} \text{ g cm}^{-3}$$

Therefore the change in pressure per centimeter is given by

$$\text{change in pressure per centimeter} = \frac{a_g P}{2.89T} \text{ dyne cm}^{-3}$$

Thus, as we move downward in the atmosphere, the pressure increases at a rate that is proportional to the pressure and inversely proportional to the absolute temperature.

Near the ground, the pressure drops rapidly with elevation. The reduction in pressure at high elevations, where the pressure is low, is much more gradual. The total atmosphere contains about 1 kg of air per square centimeter. At a pressure of 1 bar, 1 g of air occupies about 10^3 cm^3. Thus a column of air 10 m high, near the ground, will contain

Figure **10-2**   The change of atmospheric pressure with elevation for a constant temperature of 300°K.

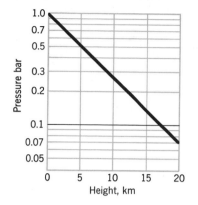

0.1 percent of the total atmosphere. The pressure will drop by 10 percent during the first kilometer to 0.9 bar. The pressure in an isothermal atmosphere continues to drop at a rate of 10 percent per kilometer, leading to a linear trend on the log pressure-height plot (Fig. 10-2). The pressure continues to drop with increased height but never becomes zero; thus the atmosphere gradually blends into the vacuum of outer space. At a height of 17 km, the pressure is 0.1 bar; at 34 km, only 1 percent of the surface pressure remains.

The Change in Temperature Due to the Vertical Motion of Dry Air

So far we have investigated the change in pressure with elevation, assuming that the temperature remains constant. We know, however, that the temperature decreases as we climb a mountain. In Chapter 7, we had to make allowances for this fact when we considered the temperature distribution on the earth. To understand why the temperature must decrease with elevation, consider what happens if we lift 1 g of air from the earth's surface to an elevation of about 1 km, where the pressure has been reduced to 0.9 bar. As a result of the decrease in pressure, the air will expand by about 10 percent. As it expands, the air does work against the atmosphere. Per gram, this work is equal to the pressure multiplied by the change in volume, or about 86 bar cm³. This amounts to an expenditure of energy of about 2 cal g⁻¹ of air. If this heat is not supplied from outside, the air must cool off as it rises and expands, to compensate for the work done against the atmosphere. Since the heat capacity at constant pressure is 0.24 cal g⁻¹ deg⁻¹, the air will cool by

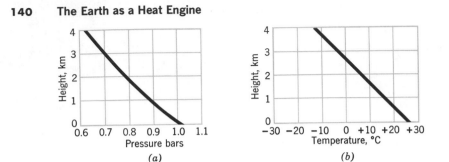

Figure **10-3** Change of pressure (*a*) and temperature (*b*) for dry air with elevation.

about 8.5°. The temperature is therefore reduced from 300 to 291.5°K. The variation in pressure and temperature with height for dry air having a temperature of 300°K at the surface is shown in Figures 10-3*a* and *b*.

As we have noted, water is much less compressible than air, and its specific heat is higher. Nevertheless, as water sinks in a deep basin, we would expect its temperature to increase because of the work done in compressing it. In most places in the ocean, this effect is so small that it is masked by other processes. However, in some isolated deep basins, such as the Philippine Trench, the increase of temperature with depth due to compression is clearly evident (Fig. 10-4).

The Moisture Content of Air

So far we have considered the properties of dry air only; however, air contains water vapor. Water vapor considerably alters the transparency of the atmosphere for infrared radiation, and the presence of clouds, droplets of condensed water, changes the heat balance of the earth.

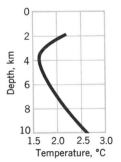

Figure **10-4** Variation of seawater temperature with depth in the Philippine Trench.

When we studied the three phases of water (Fig. 8-1, p. 92), we saw that at a given temperature a certain pressure of water vapor is in equilibrium with liquid water or ice. At 100°C the pressure of the water vapor is 1 atm, and for each 10° decrease in temperature, the pressure of the water vapor is halved. In molecular terms, we can think of this equilibrium between water and water vapor as follows. The molecules in the liquid water are in constant motion. As they reach the surface of the water, a few will be traveling fast enough and in the right direction to leave the water for the space above. The higher the temperature of the water, the faster the molecules are moving about, hence the more molecules will be able to escape across a unit area in unit time.

At the same time, molecules of water vapor in the space above the liquid will collide with the water surface, stick, and be converted to the liquid state. The number of collisions will increase as the pressure of the vapor increases. Thus there is a constant exchange between the liquid and the vapor. For the two phases to be in equilibrium, the rate at which water molecules escape from the liquid must equal the rate at which water molecules are captured from the vapor. As the temperature of the water increases, the rate of escape goes up rapidly; therefore, the pressure of the water vapor must increase to bring the rate of return of water molecules back in balance.

When we considered the phases of water (pp. 91–93), our cylinder was filled with only liquid water and water vapor. Now let us consider what will happen if we also have air in the cylinder, so that the total pressure of air and water vapor is equal to 1 bar. Now the water is bombarded by both molecules of air and water vapor. The molecules of the air—nitrogen, oxygen, and argon—that hit the water surface are mostly turned back. To a very slight extent, however, the gases of the air dissolve in the water. While this does not significantly alter the physical properties of the water, the dissolved oxygen gas in the water is very important for life in the sea and in lakes.

The molecules of water vapor that hit the water will stick, just as if no air were present. Also, the rate at which the water molecules leave the liquid depends only on the temperature of the water and is not altered by the extra pressure of the air above it. To maintain equilibrium between the liquid and the vapor will therefore take the same number of water-vapor molecules per unit volume as if the air

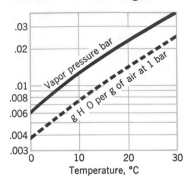

Figure **10-5** Variation of the moisture content of saturated air with temperature. The moisture content can either be expressed as the vapor pressure of the water vapor or as the mass fraction of water vapor in the air at a pressure of 1 bar.

were not present. We can specify the required concentration of water-vapor molecules by giving the *partial pressure* the water vapor exerts. The total pressure, then, is the sum of the partial pressure of water vapor and the partial pressure of the air. The partial pressure of the water vapor in equilibrium with the liquid will be virtually the same as the total pressure of water vapor in the absence of air. The higher the temperature, the greater the partial pressure of water vapor, and it will approximately double for each 10° temperature increase.

The partial vapor pressure of water in air in equilibrium with water is shown in Figure 10-5 as a function of the temperature. Air in equilibrium with liquid water is called saturated air and has a relative humidity of 100 percent. We can also specify the moisture content of the air by giving the mixing ratio in grams of water per gram of air. This is indicated at a pressure of 1 bar in Figure 10-5. Note that this ratio increases from 0.0037 at 0°C to 0.026 at 30°C, a factor of 7 or roughly 2^3. If air rises without gain or loss of water vapor by condensation or evaporation, the mixing ratio will remain constant. If, on the other hand, we have air in equilibrium with water at a constant temperature, the vapor pressure of water vapor will remain constant.

For air and water in equilibrium at 20°C and at a pressure of 1 bar, the partial pressure of water vapor is 0.024 bar, and the mixing ratio is 14 g water per kilogram of air. If we now reduce the pressure to $\frac{1}{2}$ bar while maintaining the air at 20°C and in equilibrium with water, there will be half as much air for the same partial pressure of water, and so the mixing ratio will double to 28‰. If, on the other hand, the air is decompressed without being in contact with water, the mixing ratio will remain the same, but since the water vapor has also been expanded, its vapor pressure will be only 0.012 bar.

The relative humidity is the water-vapor content expressed as a percentage of the saturated content at the same temperature. Thus air of 50 percent relative humidity at 20°C would have a partial pressure of 0.012 bar. At 1 bar this will be equivalent to a mixing ratio of 7‰; at $\frac{1}{2}$ bar the mixing ratio would be 14‰.

The Vertical Motion of Moist Air

The water vapor contained in the air of the atmosphere has a marked effect on the behavior of rising and falling air masses. Near the ocean, the air will be roughly saturated with respect to water vapor. As this air rises, it expands and thus cools off. The cooled air, however, still contains the same mass of water vapor per unit mass of air; as a result, the air will become supersaturated in water vapor. The supersaturation leads to a phase transformation from water vapor to droplets of liquid water. At the same time, the phase transformation from vapor to liquid will liberate the heat of vaporization, thus warming the air. This warming due to condensation will counteract the cooling that results from the expansion of the upward-moving air, leading to a smaller drop in temperature with elevation than would result from the elevation of dry air.

To illustrate the effect of water vapor on the vertical temperature gradient in air, let us consider a quantitative example. Let us start with air at a pressure of 1 bar and 27°C and raise this air to a level where the pressure is 0.8 bar, corresponding roughly to a height of 2 km. If the air is dry, the drop in pressure will result in a temperature drop to 8°C. If the air is saturated, it contains 23‰ of water vapor at 27°C. As the air expands and cools, its ability to hold moisture is reduced. At 0.8 bar, the expanded air will be saturated at a temperature of 19°C with a moisture content of 17‰. Thus the condensation of 6 g of water vapor per kilogram of air has reduced the cooling of the air by 11°.

If the air at 1 bar had been cooler, its ability to hold moisture would have been less, and therefore the amount of moisture condensed would also be smaller. As a result, the temperature drop on expanding would be larger.

What would have happened if the 27°C air at 1 bar had been only partially saturated with moisture? Assume that the original air had a relative humidity of 78 percent; the moisture content would then be 18‰. As the air is decompressed and cools, the relative humidity increases until it reaches 100 percent, at a pressure of 0.953 bar and a temperature of 22°C. At this point condensation takes place. At a pressure of 0.8 bar, the air has a temperature of 16°C with a final water-vapor content of 14.5‰.

These results are summarized below:

	Initial relative humidity		
	0	78%	100%
Temperature at 0.8 bar	8°C	16°C	19°C

As air rises without the addition of external heat the presence of water vapor has two effects. It reduces the rate of temperature decrease, and it leads to the condensation of water, hence to the formation of clouds. Water vapor also changes the specific volume of air. The average molecular weight of air is 28.8, while the molecular weight of water is only 18. Thus moist air at the same temperature and pressure will have a slightly lower density, since the lighter water molecules are replacing some of the molecules of air. At the same temperature and pressure, therefore, moist air rises relative to dry air. Should moist and dry air rise together, the moist air will cool off more slowly, and thus its density will decrease relative to the dry air.

The properties of the air are therefore ideally suited for it to interact with the sea. As dry air sinks, it comes in contact with the sea surface. There, water molecules leave the sea and enter the air to saturate it with moisture. The saturated air then rises, allowing more unsaturated air to sink and come in contact with the sea surface, where the energy from the sun evaporates the seawater to saturate the air again.

Horizontal Motions of the Air

Vertical motions of the air are important for the exchange of water between the ocean and the atmosphere. Most of the energy transfer between the ocean and the atmosphere is in the form of heat of vaporization; thus the vertical motions of the air have a large effect on the energy budget of the sea. The air motions with which we are familiar, the winds, are primarily horizontal. We have seen that the winds depend on the distribution of pressure and the rotation of the earth. If we have a weather map showing the distribution of atmospheric pressure at sea level, we can infer the speeds and directions of the winds. What, however, determines the circulation of the atmosphere? Although the immediate cause is the distribution of pressure, the pressure distribution itself must ultimately result from the solar radiation received and the backradiation of infrared to outer space. The winds are a response to the differential heating and cooling of the atmosphere.

Hadley's Theory

A simple theory explaining the winds was originally proposed by George Hadley in 1735. He reasoned that since the earth receives more heat near the equator than near the poles, air heated in the tropics must rise, drift toward the poles, and, being cooled there, sink and return to the equator along the surface (Fig. 10-6). If the earth were not rotating, this would produce winds blowing from the poles to the equator along the earth's surface. Because of the rotation of the earth, the air moving toward the equator is deflected to blow from east to west. Thus Hadley pointed out the effect of the earth's rotation exactly 100 years before Coriolis published his theory.

Hadley's paper was written in order to explain the trade winds. These steady winds, with a speed between 500 and 750 cm sec^{-1}, occupy belts between latitudes 25° and 5° on either side of the equator. North of the equator they blow from the northeast; in the southern hemisphere their direction is generally from the southeast. Although Hadley's theory did

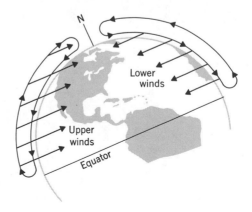

Figure **10-6**   The circulation of the atmosphere according to Hadley.

explain the trade winds, it went too far, for it would require the winds at all latitudes to blow in the same direction as the trades. Before we investigate how Hadley's theory must be modified, let us first look at the average distribution of the surface winds on earth.

The General Circulation

While the trade winds blow with steady speed and direction, the winds in temperate latitudes tend to blow in all directions as regions of high and low pressure form, drift along the surface, and dissipate. We saw an example of this in the weather map (Fig. 9-6, p. 118). If we average the wind vectors over a period of time and along each parallel of latitude, we obtain a climatic average, the *general circulation.* Averaging causes the north-south components of the wind largely to disappear, leaving us with E-W and W-E winds. The average wind system is associated with a system of average pressures, as shown in Figure 10-7.

Along the equator, the atmospheric pressure tends to be low and the winds weak. This is the region of the *doldrums,* where sailing vessels can make but very little headway. About 5° to either side of the equator we enter the region of the steady trade winds, which extend to a latitude of about 25°. These reliable winds were a boon to early explorers and traders.

Moving poleward from the region of the trade winds we enter a region of low wind-velocity and high pressure, the *horse latitudes.* In this region,

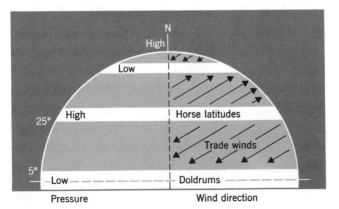

Figure **10-7** Generalized distribution of pressure and winds for the northern hemisphere.

from 30° to 35° latitude, the winds are unpredictable. It may be calm
and then the wind may start blowing from almost any direction. Poleward of the horse latitudes, from 35° to 55°, we come to a region
of variable winds whose average velocity is from the west. Finally we
come to a subpolar region of low pressure. As the pressure increases
from about 65° to the pole, the winds, on the average, blow from the
east.

The Three-Cell Circulation

We see that the general circulation of the atmosphere does not correspond to Hadley's simple model. Between the easterly trades and the
polar winds is a broad region where the winds blow predominantly from
the west. The error in Hadley's model resulted from his assumption that
the heating and cooling of the atmosphere are restricted to the earth's
surface. Actually, as we saw in Figure 7-6 (p. 85), the upper regions of
the atmosphere are cooled by emitting infrared radiation to outer space.
This cooling amounts to between 1 and 2°C per day. As a result, by the
time the poleward-flowing upper air has reached a latitude of about 30°,
it begins to sink toward the surface. As it sinks, it is warmed by compression. Once it reaches the surface, it spreads out to the north and
south. Thus it forms the high-pressure region of the horse latitudes.

The air that rises at the equator does not circulate all the way to the
pole but forms a closed circulation cell between the equator and 30°, as
shown in Figure 10-8. The air that subsides at 30° and heads poleward
will produce winds from the west because of the rotation of the earth.

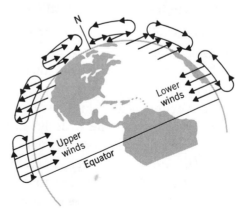

Figure **10-8**   The three-cell circulation
model.

Near the poles there is rapid cooling of the upper air, causing it to sink. As the air flows toward the equator along the ground, it is deflected to blow to the east. The subpolar low-pressure region, leading to an upward flow, is formed where the cold easterly polar winds meet the warm westerlies.

Thus we see that Hadley's single-cell circulation model is too simple. Instead of a single cell, we end up with three circulation cells.

The Dishpan Experiment

Insight into the general circulation of the atmosphere was provided by laboratory experiments carried out by D. Fultz (1961) at the University of Chicago in the early 1950's. In these experiments, the atmosphere of one hemisphere was modeled by water in a rotating dishpan. The pan was heated at its periphery and cooled in the center to model the differential heating between the equator and the poles. When the pan is stationary, water rises at the rim (the equator) and flows on top toward the center (the pole), where it sinks to return to the periphery along the bottom of the pan. If the pan is now rotated slowly, the north-south currents are deflected to give the circulation postulated by Hadley.

As the speed of rotation is increased, the single Hadley cell breaks up into a series of vortexes that move about resembling the high- and low-pressure regions of the weather map. The faster the rotation, the greater the number of circulation cells that form in the model. The rotation thus breaks up the simple circulation and produces warm and cool whirls that move into mid-latitudes from the equatorial and polar regions.

Eddies and the General Circulation

Because of the breakup of the simple circulation into eddies, the mid-latitude region of winds from the west is a battleground between high- and low-pressure areas. Do these vortexes remove energy from the westerly circulation, or do they feed energy into the general circulation? When we pump water through pipes we try to avoid turbulence. If eddies form, they disturb the orderly motion of the fluid and so dissipate energy, forcing us to pump harder in order to move the fluid at the same rate. The aircraft designer is careful to avoid producing turbulence in the wake of the moving plane, for this would seriously decrease its speed.

In the atmosphere, however, the situation is different. Cool air masses moving into the mid-latitudes will tend to sink while warm air masses will tend to rise. As a result, potential energy is transformed into the kinetic energy of the wind. Victor P. Starr, of the Massachusetts Institute of Technology, showed that the high- and low-pressure regions that move into the mid-latitudes do not dissipate the energy of the westerly winds. Rather, the eddies feed energy into the westerly winds. Thus the disorderly eddies of warm and cold air in the mid-latitudes, whose motions the meteorologists can predict for only short periods, give rise to the general circulation of air from west to east.

The Effect of Vertical Motion on Evaporation and Precipitation

The vertical component of the general circulation of the atmosphere, as depicted in Figure 10-8, should have a strong effect on evaporation and precipitation at the earth's surface. As we have seen, when air is moving upward, the cooling due to expansion will lead to supersaturation of the water vapor in the air, hence to precipitation. Downward-moving air, on the other hand, as it is heated by compression, will become undersaturated in water vapor. As this air flows over the ocean, we can expect considerable evaporation of seawater in order to reequilibrate the air with water vapor.

Estimates of the latitudinal variation of evaporation and precipitation for the northern hemisphere are shown in Figure 10-9. The centimeters

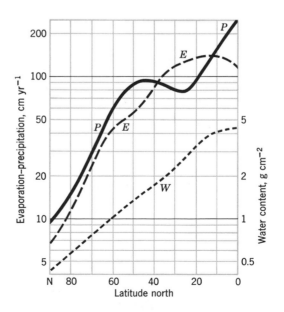

Figure **10-9** The latitudinal variation of the average rate of evaporation (E), precipitation (P), and the average water content (W) for the northern hemisphere.

of water evaporated or precipitated per year are shown on a logarithmic scale against latitude. The amount of water that can be precipitated or evaporated depends on the capacity of the air for water vapor. There is a marked increase in both precipitation and evaporation from the pole to the equator, because as the temperature increases, the air is able to hold more water (Fig. 10-5). The total water content of a vertical column of the atmosphere, also shown in Figure 10-9, increases from about $0.4\,\mathrm{g\,cm^{-2}}$ at the North Pole to over $4\,\mathrm{g\,cm^{-2}}$ at the equator.

To get a better picture of the vertical motion, we should look at the ratios of the evaporation and precipitation rates to the water content (Fig. 10-10). These ratios will give us the number of times per year the water column is replaced by evaporation or emptied by precipitation. The ratio of precipitation to water content has two peaks, at 55° and at the equator, where the turnover of water vapor by precipitation is about once every six days. These peaks correspond to the regions of upward motion in our three-cell circulation model (Fig. 10-8).

The ratio of the evaporation rate to the water content has a broad peak near 35°N, corresponding to the downward motion of air in the horse latitudes. We would expect a second peak at the North Pole; however, this expectation is not borne out by measurements. With this exception, the data on evaporation and precipitation confirm the vertical component of the three-cell circulation model depicted in Figure 10-8.

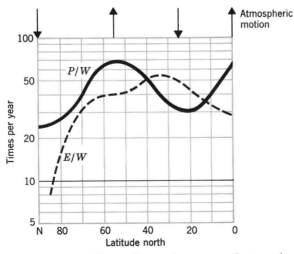

Figure **10-10**   The exchange frequency of atmospheric water vapor by evaporation (E/W) and precipitation (P/W) for the northern hemisphere.

Pressure Field and Prevailing Winds on the Earth's Surface

If the earth were entirely covered by water, the three-cell circulation model would correspond rather closely to the actual circulation. However, the average pressure and the wind system are strongly affected by the distribution of land and water. In Chapter 7 we saw that the geography of the earth has a strong effect on the average summer and winter temperatures. Now let us examine the average distribution of pressure and wind in January and July.

First let us look at conditions in January (Fig. 10-11a). In the northern hemisphere, the winter hemisphere, there is a dominant high-pressure area over Asia and a minor high over North America, while the pressures over the oceans tend to be lower. In the southern hemisphere, the pressure is lower over the continents than over the oceans. Because of the absence of large landmasses, the pressure distribution is more even in the southern hemisphere. In July (Fig. 10-11b) the situation is reversed, with the northern oceans having higher pressures than the continents.

The annual variation of pressure over Asia produces an interesting phenomenon over India. In winter the winds blow from India toward the southwest, over the Indian Ocean. The outflow of dry continental air leads to significant evaporation over the land and the adjacent ocean. In July the situation is reversed; the winds blow from the Indian Ocean toward the northeast onto the continent. The moist oceanic air hitting the updrafts over the warm continent leads to excessive rainfall, the monsoon rains.

Summary

Air is highly compressible. As it rises, the expanding air does work against the atmosphere and cools off. At the same time, its ability to hold water vapor decreases, resulting in the condensation of water as clouds and raindrops. The condensation of water vapor, by liberating the heat of vaporization, reduces the vertical temperature gradient in the atmosphere. The general circulation of the atmosphere of each hemisphere approximates three vertical circulation cells that, because of the rotation of the earth, give rise to the average wind system. Regions where the

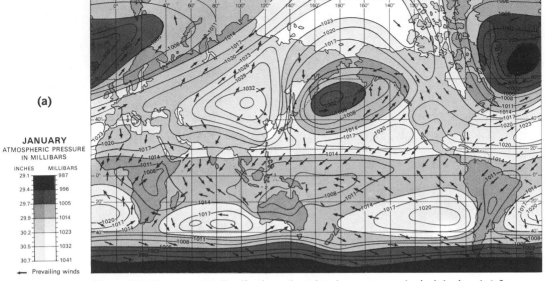

(a)

JANUARY
ATMOSPHERIC PRESSURE
IN MILLIBARS

INCHES	MILLIBARS
29.1	987
29.4	996
29.7	1005
29.9	1014
30.2	1023
30.5	1032
30.7	1041

← Prevailing winds

Figure **10-11**   Average distribution of sea level pressure and wind during (*a*) January and (*b*) July. (From Strahler, 1969, *Physical Geography*.)

Figure **10-12**   The cloud cover over the earth as revealed by the ESSA III satellite on January 6, 1967. See also pg 2. (ESSA). (Photograph: E.S.S.A.)

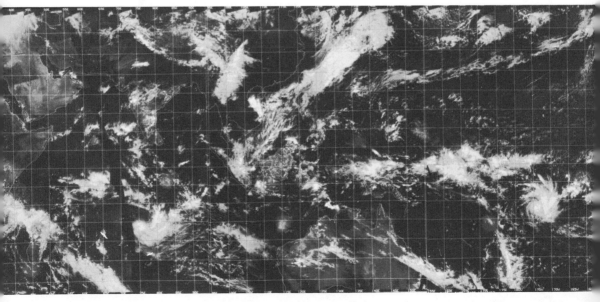

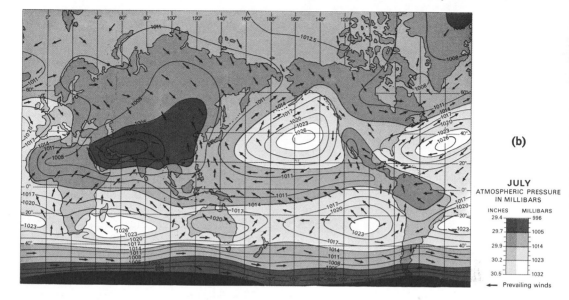

(b)

JULY
ATMOSPHERIC PRESSURE
IN MILLIBARS

INCHES	MILLIBARS
29.4	996
29.7	1005
29.9	1014
30.2	1023
30.5	1032

⟵ Prevailing winds

upward motion of air dominates have an excess of precipitation over evaporation, while sinking air masses lead to an excess of evaporation. The actual circulation is complicated by the distribution of land and water. The greater annual temperature variations over the continents give rise to the annual change in the monsoons over the northern Indian Ocean.

Study Questions

1. Why does the boiling point of water drop with elevation? Using the data in Figure 10-2 and the fact that the vapor pressure of water is approximately halved for every 10° drop in temperature, what is the boiling point of water at 5 km?
2. Air at 0°C holds 3.7 g of H_2O per kilogram at 1 bar (Fig. 10-5). If saturated air at 0°C is heated to (*a*) 10°C and (*b*) 20°C, how much water must be added per kilogram at each temperature to keep the air saturated with water vapor?
3. Why do clouds usually have a well-defined, level lower boundary?
4. Draw the three-cell circulation (Fig. 10-8) for the southern hemisphere.
5. Using Figure 10-11, plot the pressure for the central Atlantic Ocean as a function of latitude, and indicate the direction of the wind for January and July. Compare your results with the simple model depicted in Figure 10-7.

Supplementary Reading

(Starred item requires little or no scientific background.)

* Starr, Victor P. (1956). "The General Circulation of the Atmosphere," *Scientific American,* December.

11 Air-Sea Interaction

The surface of the sea is where the action is. This interface between the ocean and the atmosphere is not a rigid boundary between the liquid and gaseous envelopes of the earth. Rather, it is a transfer station for the exchange of matter and energy. It is the link that couples the ocean with the atmosphere. Just as we cannot understand the atmosphere without considering its interaction with the ocean, so we cannot comprehend the ocean without considering its interaction with the atmosphere through the sea surface.

Radiant Energy

When we considered the heat balance of the earth (Fig. 7-6, p. 85), we saw that 47 percent of the incident solar radiation was absorbed by the ground. Of this, 22 percent was direct sunlight and 25 percent was diffused sunlight, scattered by air, dust, and clouds. What happens as this radiation enters the sea? We generally think of water as transparent to visible light, for a glass of pure water appears to be perfectly transparent. Seawater, particularly if it is obtained near the coast, may look murky; however, if we filter it, the seawater will be just as clear as pure water. The suspended matter that caused the cloudiness will be our main concern when we consider the geology and biology of the sea.

If instead of merely contemplating a glass of water we make scientific measurements of the percentage of light of a given wavelength absorbed by 10 m of pure water, we obtain the results shown in Figure 11-1. Note that there is 100 percent absorption at both short (ultraviolet) and long (infrared) wavelengths. The absorption is minimal near 0.47 micron, in the blue part of the spectrum, but even the amount of blue light is cut in half by only 47 m of water.

Ten meters of seawater taken from the open ocean absorb between 33 and 80 percent of the blue light, depending on the clarity of the water.

155

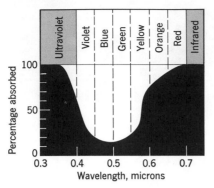

Figure **11-1** Light absorption by 10 m of pure water as a function of wavelength.

Average ocean water absorbs about 55 percent of the blue light in 10 m, while 10 m of coastal water will absorb almost all of it (between 94 and 99+ percent).

The penetration of radiant energy into the sea is shown in Figure 11-2. To point up the attentuation of light, we have indicated the percentage transmission on a logarithmic scale against a logarithmic depth scale. Even in the clearest ocean water, only 1 percent of the light energy penetrates below 100 m. Sunlight is adequate for vision only in the upper 100 m; more than 97 percent of the oceans volume is in perpetual dark-

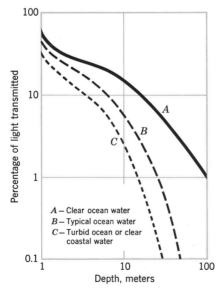

Figure **11-2** Variation of light transmission with depth for various waters: *A*, clear ocean water; *B*, typical ocean water; *C*, turbid ocean or clear coastal water.

ness. Seawater, then, is far from transparent. As we descend into the depths, the sunlight disappears rapidly, and the color of the light changes from white to blue. Many of the creatures inhabiting the sea floor of this region display brilliant colors. However, these are invisible under natural illumination; they become apparent only when we carry artificial lights into the depths.

The light absorption of seawater greatly limits the visibility from submersibles. One might think that, given a suitable craft, man could explore the deeps of the ocean just as astronauts are able to explore the surface of the moon. However, even from a submarine equipped with a searchlight as bright as sunlight, we would be able to see only a short distance from the craft. In clear ocean water, we can obtain sharp photographs of objects a few meters from the camera, but because of light absorption and scattering by particles in the water, picture quality deteriorates rapidly at distances as short as 10 m. In many inshore areas, the water's lack of clarity makes optical observations impractical.

The sunlight that enters the ocean is a form of energy. As the light is absorbed by the seawater, it is converted to heat and warms the water. As we shall see in Part V, a small fraction of the incident light is utilized by plants in photosynthesis to convert inorganic substances to organic matter. When we consider the energy balance of the sea, we can safely ignore this tiny part of the total incident light. When we study the biology of the sea, however, we must take it into account, for this small amount of light provides the energy for all the life in the ocean.

The average amount of solar energy received at the top of the atmosphere is 8.3×10^{-3} cal cm^{-2} sec^{-1}. The earth's surface absorbs 47 percent of this, or 4×10^{-3} cal cm^{-2} sec^{-1} (1.2×10^5 cal cm^2 yr^{-1}). If this amount of energy were used to heat the upper 100 m of the ocean surface uniformly, it would lead to an annual increase in temperature of 12°C. On a global average, the same amount of heat that is absorbed by the sea must be returned to the atmosphere and to outer space; otherwise the temperature of the ocean would increase steadily. In Figure 7-6 (p. 85), we saw that 5 percent of the total incident radiation is radiated back to space in the infrared part of the spectrum by the earth's surface. Since seawater is essentially opaque in the infrared, the backradiation is emitted by the sea surface. By measuring the intensity of this radiation from an earth-orbiting satellite through one of the atmospheric infrared windows (Fig. 7-5), it is possible to determine the temperature of the sea surface—provided that there are no clouds between the detector and the surface of the sea, and that a correction is made for water-vapor radiation from the atmosphere itself.

Only five of the 47 units of heat energy absorbed by the sea are radiated back to space. The other 42 units are transferred to the atmosphere by radiation, by heating the air directly, and as latent heat of water vapor. The atmosphere then radiates this energy back to outer space.

Transfer of Energy between the Ocean and the Air

The air and the surface water of the ocean are in contact at the sea surface. If the air is warmer than the water, heat will be transferred from the atmosphere to the water. If the water is warmer than the air, the transfer will take place in the opposite direction. The tendency is always to equalize the temperatures. There are only a few places in the ocean where we have a detailed record of the seasonal variations in temperature. One location is ocean station P, a weather ship operated by the Canadian government at 50°N and 145°W, in the northeast Pacific Ocean. The monthly mean air and sea temperatures for 1957 are shown in Figure 11-3.

The annual range of the monthly mean temperatures at ocean station

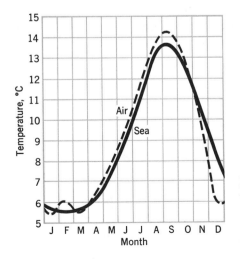

Figure **11-3** Monthly mean air and sea temperature at Ocean Station *P* during 1957. Dashed line indicates air temperature; solid line, sea temperature. (After Dodimead, Favorite, and Hirano, 1962)

P is about 8°C. From April to October, the average air temperature is slightly higher than that of the water, so that heat is transferred from the air to the water. From November to March, the average air temperature is slightly cooler than that of the water, so that the water will tend to warm the air.

The transfer of heat from the ocean to the air takes place by conduction of heat through the sea surface. If the water is warmer than the air, the air in contact with the sea surface is warmed until the temperatures are equal. The greater the temperature difference, the more rapid will be the transfer of heat. To heat 1 g of air 1°C at constant pressure, 0.24 cal have to be transferred. In order for the air to be in equilibrium with the seawater, it is not sufficient that the two temperatures be equal, In addition to a transfer of *sensible heat* to warm the air, there is a transfer of water vapor. Only if the air has the same temperature as the water *and is saturated with water vapor* are the two in equilibrium. As the air is heated by contact with the warmer water, it will need more water vapor in order to remain saturated.

To transform liquid water to water vapor requires energy, the heat of vaporization. This *latent heat* is transferred to the air, to be released in the atmosphere when the moisture recondenses and falls back to the ocean as rain or snow. Thus the heat transferred from the water to the atmosphere consists of two parts, sensible heat and latent heat. Since the heat capacity of air is roughly independent of temperature, the sensible heat required to equalize a given temperature difference for a unit mass of air is independent of the temperature. The moisture content of the saturated air, however, increases steeply with temperature (see Fig. 10-5, p. 142). As a result, the latent-heat requirement increases as the air becomes warmed. The heat required to warm 1 g of air 1°C *and* to keep it saturated with water vapor is shown in Figure 11-4 as a function of temperature.

To heat 1 g of saturated air from 0 to 1°C requires an expenditure of 0.24 cal of sensible and 0.12 cal of latent heat. To do so between 29 and 30° takes 0.24 cal of sensible heat and 1.0 cal of latent heat. Thus, to warm air in contact with the sea requires much more heat per degree in the tropics than in the polar regions.

Although the air in contact with the ocean surface will always be near saturation, this is not true of the air over land. Because the moisture to

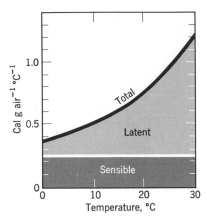

Figure **11-4** Heat required to warm 1 g of saturated air 1°C (sensible heat) and maintain it saturated with water vapor (latent heat).

saturate the air over land must be drawn from the soil or from vegetation, the air at the surface, particularly in desert regions, will have a low relative humidity. Therefore little latent heat is added to the air. Thus it can be warmed much more easily, and the summer temperatures over land are higher than those over the ocean (see Fig. 7-1, p. 78).

The average annual ratio of the latent to the sensible heating of the atmosphere is shown in Figure 11-5 as a function of latitude. Curves for the ocean and the land area of the northern hemisphere are shown on a logarithmic scale. As we would expect, the ratio for the ocean decreases markedly as we go from the equator to the poles. For the land, on the

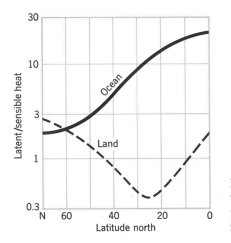

Figure **11-5** Ratio of the yearly average latent to sensible heating of the atmosphere by the ocean and the land for the northern hemisphere. (After Sellers, 1965)

other hand, we find a broad minimum between 20 and 30°N, the generally arid land areas. In high latitudes, the ratio actually becomes greater for the land than the ocean. This is because of the humid conditions combined with higher temperatures during the summer.

The high heat of vaporization of water and the rapid increase in water-vapor content with temperature limit the temperature to which air over the ocean can be heated. An increase in the amount of heat received by the tropical ocean would increase the latent heat added to the air in the form of water vapor without significantly raising the air temperature. Therefore, the presence of the water surface effectively limits the maximum air temperature. If the earth were a dry planet, the temperatures in the tropics would be much higher than they are, since the heat transfer to the air would be limited to sensible heat.

Effect on the Surface Water of the Ocean

The ocean transfers sensible and latent heat to the atmosphere. A transfer of latent heat results in a change of water temperature, while the evaporation of water vapor leads to an increase in the salinity of the sea. The increase in salinity by evaporation is offset by the dilution of seawater through precipitation. Since the capacity of the atmosphere for water is much smaller than the amount of water evaporated per year (Fig. 10-10, p. 150), evaporation must equal precipitation when averaged over the globe. In fact, the oceans lose slightly more water by evaporation than they gain by precipitation, the excess water being returned to the sea by rivers. The continents, on the other hand, have an excess of precipitation over evaporation.

Although there must be a global balance between evaporation and precipitation, the two are not equal if averaged along a parallel of latitude, for moisture is carried between latitudes by the atmosphere. The north-south transport of water vapor shown in Figure 11-6 is derived from atmospheric observations during the International Geophysical Year (1958) for the northern hemisphere. The curve shows the flux of water vapor across latitude circles in units of 10^{11} g sec^{-1}. Between the equator and 23°N, the average flux was to the south; north of this latitude, on the average, the water vapor was moving toward the North Pole.

To determine the water balance of a given latitude belt, one must cal-

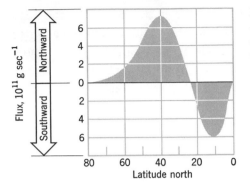

Figure **11-6** Flux of water vapor across parallels of latitude for the northern hemisphere during the International Geophysical Year. Units are in 10^{11} g sec^{-1}. (After Starr, Peixoto, and Crisi, 1965)

culate the difference between the inflow and the outflow. For example, between 25 and 30°N we have the following:

At 30°N 4.19×10^{11} g sec^{-1} flowing north (out of the belt)
At 25°N 1.23×10^{11} g sec^{-1} flowing north (into the belt)
Difference 2.96×10^{11} g sec^{-1} loss from the belt

Thus there is a net loss of water vapor from the 5° zone between 25 and 30°N. This water must have been supplied by excess evaporation, mainly from the ocean. The gain or loss of water by the earth's surface per 5° latitude belt is shown in Figure 11-7 in units of 10^{11} g sec^{-1}.

There are two ways to study the water balance of the earth. One is to measure the rates of evaporation and precipitation separately and determine the net change in water from the difference between the rates. The average results obtained for the northern hemisphere in this way are shown in Figure 10-9 (p. 149). Evaporation exceeds precipitation between 12 and 38°N, resulting in a water loss for this portion of the earth's surface; the rest of the northern hemisphere has an excess of precipitation over evaporation. Alternately, one can estimate the water balance from the north-south transport of water vapor in the atmosphere. These results, shown in Figure 11-7, agree closely with estimates based on the differences between evaporation and precipitation.

Let us consider the effect of evaporation and precipitation on the surface salinity of the ocean. Between 25 and 30°N, there is an excess of about 3×10^{11} g sec^{-1} of evaporation over precipitation. The area of the ocean between these parallels is 1.34×10^7 km². If all the excess evaporation in this 5° belt is derived from the ocean in this belt, then the

Figure **11-7** Loss or gain of water vapor by the earth's surface per 5° latitude band for the northern hemisphere during the International Geophysical Year. Units are in 10^{-11} g sec^{-1}. (After Starr, Peixoto, and Crisi, 1965)

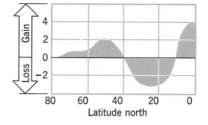

ocean must lose about 10^{19} g of water per year from an area of 1.34×10^{17} cm^2, giving a net evaporation rate of $10^{19}/1.34 \times 10^{17} = 75$ g cm^{-2} yr^{-1}. If the ocean were well mixed to a depth of 75 m, the excess evaporation at this latitude would lead to a 1 percent increase in salinity per year.

This increase in salinity is counteracted by the inflow of water to replace the water lost by evaporation. Otherwise the level of the sea between 25 and 30°N would drop. In addition, because of ocean currents, there is a flow of surface water across the parallel which leads to an exchange of water between this and other latitude belts. The exchange of water by currents tends to mix the surface layer of the sea and thus results in a more uniform salinity distribution. At the same time, the difference between evaporation and precipitation enhances the salinity contrast between the regions of water loss and gain. The net surface salinity results from the competition between these two processes. Let us consider a simple model as an illustration.

Figure 11-8 shows two well-mixed ocean areas, A and B, having salinities S_A and S_B, respectively. For simplicity, we will express the salinity in grams of salt per cubic centimeter. (This is equivalent to the

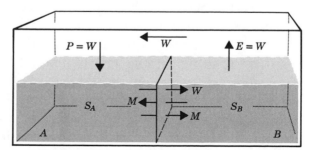

Figure **11-8** Evaporation-precipitation model.

usual designation of salinity in grams of salt per kilogram of seawater if we neglect the slight difference in density between fresh water and seawater.) Over ocean area A there is an excess of precipitation amounting to W cm³ of water per unit time. Since our model is a closed box, there must be the identical rate of excess evaporation in area B and an atmospheric water transport from B to A amounting to W cm³ in unit time.

To maintain a uniform water level in the two areas, there must be a flow of seawater from A to B having a salinity S_A and a flow rate of W. In addition, there will be a certain amount of mixing, M cm³ per unit time, between the two ocean areas. In unit time, this will carry M cm³ of water with salinity S_A from A to B, and M cm³ of water with salinity S_B from B to A. After we start the model, the salinities in A and B will adjust themselves until there is a balance so that as much salt flows into each reservoir as flows out of it.

For the salt fluxes to be equal:

$$S_B M = S_A(M + W)$$

Therefore,

$$\frac{S_B}{S_A} = 1 + \frac{W}{M}$$

If there is no mixing ($M = 0$), the salinity in B becomes infinite; or, rather, all the salt will be transferred to B. If the volume of A is much larger than that of B, this will lead to the formation of a bed of solid sea salt in B. In this manner, isolated portions of the ocean in the geologic past have given rise to thick salt deposits.

If there is exchange of water between A and B, the final salinity ratio will depend on the ratio of the net evaporation rate to the mixing rate. In the North Pacific, the subtropical surface waters have a salinity of 35‰ while the subarctic surface waters have a salinity of 33‰. If we consider this ocean as an isolated system, the ratio W/M is $(35/33) - 1 = 0.06$. Thus the mixing rate is about 17 times the average net evaporation rate in the subtropical area.

The latitudinal variation in surface salinity for the Atlantic and Pacific oceans is shown in Figure 11-9. The data are for summer in the northern hemisphere. The Atlantic values are taken at a longitude of 20°W while the values for the Pacific are from 160°W. In both oceans there is

Figure **11-9** Surface salinity distribution during the summer of the northern hemisphere for the Atlantic Ocean at 20°W and the Pacific Ocean at 160°W.

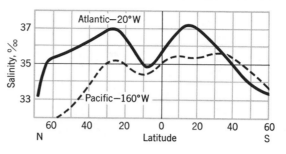

a region of maximum salinity on either side of the equator with slightly lower salinity on the equator itself. The lowest salinities occur in high latitudes, particularly in the north. Note that the Atlantic is generally saltier than the Pacific, with the greatest salinity difference between the North Atlantic and the North Pacific.

The Effect of the Wind on the Sea Surface—Waves

We have seen that heat and water are exchanged between the ocean and the atmosphere across the sea surface. In addition, there is a transfer of motion between the atmosphere and the ocean; as the wind blows over the sea, it produces waves. But the motion of the water differs significantly from the motion of the air that excites it, just as the motion of a violin string differs from that of the bow that sets it in movement. As the bow slides over the string, it excites transverse vibrations in the string. In similar fashion, as the wind blows over the sea, it produces an oscillatory motion in the water.

If we drop a stone into a quiet pond, waves are produced that travel outward from the disturbance. The water, however, does not travel with the wave. This becomes apparent if we place a cork in the water. The cork does not travel with the wave but bobs up and down in place as the wave passes it. If we watch the cork carefully as the wave moves by, we find that it moves in a vertical circle, reaching its highest position as the crest of the wave passes and its lowest position in the trough of the wave. The cork makes a complete revolution for each wave that passes. The time required for the wave to pass is the *wave period T;* the distance from one crest to the next is the *wavelength L.*

In order to make a detailed study of the motion of the water as a wave passes, let us mark a series of spots or particles in the water in the direc-

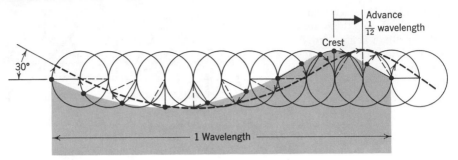

Figure **11-10**   Water motion during wave advance.

tion of the wave advance (Fig. 11-10). Each spot will undergo a circular motion. However, the revolution of each spot in the direction of the wave's advance will lag slightly behind that of the preceding spot. The shaded area in Fig. 11-10 shows the initial appearance of the wave. If it takes 12 sec for the wave to pass a given point, then after 1 sec each water particle will have undergone $\frac{1}{12}$ of a complete rotation; that is, each particle will have rotated by 30° in its orbit. The positions of the spots after 1 sec are shown by the dashed line. During that second, of course, the crest of the wave has advanced $\frac{1}{12}$ wavelength in the direction of rotation. During a complete period, the wave will advance by one wavelength. Therefore, the speed of wave advance, V, is one wavelength, L, per period, T. Thus

$$V = L/T$$

While the wave advances a distance L, each water particle returns to the point it started from. The wave height H, the difference in height between the crest and the trough, is equal to the diameter of the circular orbit of the water particle. Since each water particle describes one full circle per period, it travels a distance πH in time T. Therefore the speed of the water particle, S, is:

$$S = \pi H/T$$

Thus while the speed of the wave advance depends on the wavelength and the period, the speed of the water particles, hence the kinetic energy of the water, depends on the wave height and the period.

Figure 11-10 shows the motion of the water at the sea surface only. If we examine the motion of the underlying water, we find that it is similar to that of the surface layer; that is, each water molecule undergoes a circular motion. However, the amplitude of the motions—that is, the radius of the circles—gradually decreases with depth. At a depth corresponding to approximately $\frac{1}{2}$ wavelength, the motion of the water essentially ceases. Thus the movement of a submarine, provided that it is at sufficient depth, is not affected by heavy waves that cause surface ships to pitch and roll.

What happens if the wind blows over the sea surface? If waves are present, the surface of the sea is undulating and so will disturb the flow of the air over it. Consider a wave moving in the direction of the wind. The air flow over the sea will be as indicated in Figure 11-11. At the crest of the wave, the motion of the water particles is horizontal in the direction of wave advance. The motion of the air at the crest will therefore be roughly in the same direction as that of the water. The wave crest deflects the air upward, and thus a region of relatively low pressure will develop in the wind shadow of the crest. In this region, the water is moving upward so that the low pressure will reinforce this motion. As we approach the next crest, the air is moving downward, and so the air pressure at the water surface will be increased. Note that here the water is moving down, hence the pressure reinforces the motion of the water. The pressure difference at the sea surface produces a small eddy of air moving in a direction opposite to the wind. This, again, reinforces the motion of the water.

Therefore, a wind blowing in the direction of wave advance will rein-

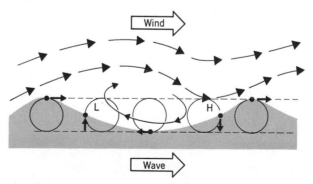

Figure **11-11**   Reinforcement of the wave by the wind. L, low pressure; H, high pressure.

force the waves. Waves moving in the opposite direction will be attenuated, since the water motions would be exactly opposite to those shown
in Figure 11-11. The wind thus selectively reinforces the waves that
move in the same direction. The longer the wind blows, the more the
waves are reinforced, hence the higher the waves build up. Since the
waves propagate in the direction of the wind, the wave height also depends on the horizontal distance that the wind has blown over the sea
in a given direction.

Recently, oceanographers from the University of California at San
Diego observed the propagation of waves across the entire Pacific
Ocean. The waves were generated by winter storms in the Antarctic
region. Using sophisticated measuring equipment and electronic com-
puters to analyze the data, the scientists were able to identify wave
trains from specific storms as they spread from the Antarctic across the
equator to the coast of Alaska. The storms generate a mixture of waves
of varying wavelengths. While the short waves were dissipated rapidly
as they left the storm area, the long waves traveled virtually without attenuation halfway around the world before they spent their energy on
distant shores.

The waves traveling over the surface of the ocean represent a form of
energy. This energy is dissipated when the waves pound against the
shore. Small waves wash sand back and forth on the beach; large waves
can move boulders, automobiles, and even railroad engines. Waves contain potential energy due to the vertical displacement of the water and
kinetic energy due to the circular motion. On the average, the kinetic and
potential energies per unit area of the wave are equal, and they are proportional to the square of the wave height (H). The total energy per unit
area is

$$E = \frac{\rho a_g H^2}{8} \approx 125 \; H^2 \; \text{ergs cm}^{-2}$$

Thus as the heights of the waves build up, their energy content grows at
a still faster rate. Waves of twice the height contain four times the energy
per unit of sea surface. It is not surprising, therefore, that when abnormally large waves impinge on a coast, they can cause a great deal of
destruction.

Let us examine again what happens when the wind blows over a wave

Figure **11-12** Motion of the water in the direction of the wind.

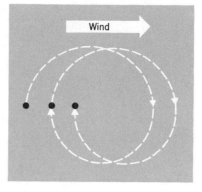

(Fig. 11-11). We have noted that the pressure on the windward side of the wave is higher than the pressure in the wind shadow of the wave. Thus there will be a tendency to push the water in the direction of the wind. As the wind moves over the water surface, there is a tendency to drag the water along. As the waves build up, they tend to break, throwing sea spray in the air; the water droplets in the spray are then blown downwind. When they reenter the sea, they will transmit the forward motion to the water. As a result of these various factors, the water particles will not oscillate in a perfect circle but will be dragged slightly in the direction of the wind (Fig. 11-12).

If the wind stops, the internal friction in the water will gradually dissipate the wave motion and so reduce the wave height. While the wind is blowing, there will be competition between the dissipation of the wave energy by friction and the addition of new energy from the wind. The wave height will increase until the rate of energy loss is equal to the rate of input of energy from the wind. Thus the faster the wind, the more energy will be imparted to the water, hence the greater will be the ultimate wave height.

The actual sea is not so simple as the single wave illustrated in Figure 11-10. At any one time, there is a great profusion of waves of different wavelengths traveling in different directions. Superimposed on the major waves are small ripples. These little ripples look insignificant on top of the larger waves, but they are in fact responsible for most of the forward motion that the wind imparts to the sea. Since the winds drag the water along, one would expect the circulation of the ocean to be patterned after the general circulation of the atmosphere (Figs. 10-11a and b, p. 151). However, while winds blow almost indifferently over land and sea, the disposition of the continents drastically restricts the flow of ocean currents. We will consider the wind-driven circulation of the ocean in the next chapter.

The wind produces waves on the sea surface and tends to drag the water along, giving rise to ocean currents. In addition to producing horizontal drift, the waves also cause vertical mixing. As the circular motion of the water particles in the wave is attenuated by internal friction, it stirs the surface layer of the sea and tends to eliminate vertical variations in temperature and salinity.

The Temperature-Salinity Structure Near the Sea Surface

To study the time and space variations of properties near the sea surface, we could use a number of Nansen bottles spaced closely together. In this region, however, the temperature changes so rapidly with depth that we do not need the great accuracy of the reversing thermometers. Rather, we require a more continuous record of the vertical variation. To measure the temperature variation in the upper 200 m of the sea, we use an instrument called a *bathythermograph* (Fig. 11-13).

The bathythermograph consists of a temperature-sensing element to which a stylus is attached. In order to obtain a fast response, xylene is used as the thermometer fluid instead of mercury. As the xylene, enclosed in a long, thin copper tube, expands and contracts, it moves a stylus which scratches a trace on a special glass slide. One side of the glass slide is attached to a pressure-sensitive element. As the bathythermograph sinks in the sea, the pressure increases, and the slide is moved in one direction while the stylus records the temperature in the other direction. In this way, we obtain a curve of temperature versus pressure.

The bathythermograph is shaped like a torpedo so that it can be lowered while the ship is in motion. During the cruise, therefore, we can take frequent measurements of the temperature near the sea surface without having to stop. To interpret the temperature-depth record, we place the glass slide in a special reader. We calibrate the record by taking a surface-temperature reading with an ordinary thermometer. From bathythermograph observations obtained by many ships at various times of the year, a picture is formed of the seasonal temperature variation of the

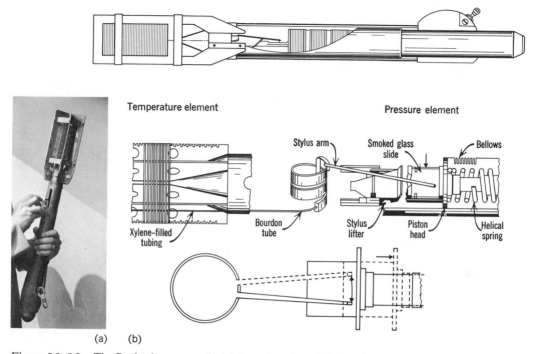

Figure **11-13** The Bathythermograph. (*a*) Inserting glass slide into bathythermograph. (Photograph: ESSA) (*b*) Diagram showing mode of operation.

surface layer of the sea. A series of records from ocean station P (50°N, 145°W), taken in 1956, is shown in Figure 11-14.

In March, the upper 100 m have a uniform temperature of 4.6°C. By April, the sea surface has been warmed slightly. Note that the upper 25 m have a uniform temperature owing to surface mixing and that there is some warming down to 100 m. As we progress into summer, the upper water gets warmer. The depth of the layer of uniform temperature (H), the *mixed layer,* varies somewhat depending on wind conditions. By August, the upper 23 m have reached a temperature just under 14°C. The temperature change at the surface is quite similar to the data for 1957 (Fig. 11-3, p. 158). In August, the mixed layer extends to 23 m. Between 20 and 70 m, the temperature decreases rapidly. This region is called the *thermocline.* Below 70 m the temperature changes more gradually.

The record of surface temperatures (Fig. 11-3) tells us that the sea-

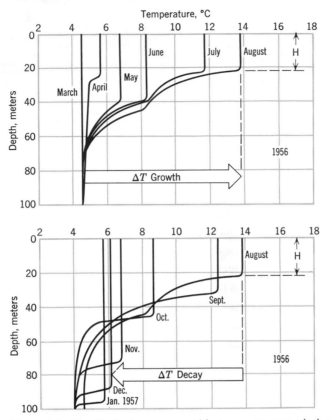

Figure **11-14** Vertical mean monthly temperature variation for the upper 100 m at Ocean Station *P* from March 1956 to January 1957. (After Dodimead, Favorite, and Hirano, 1962)

water was warmed between April and August, but it does not permit us to calculate the amount of heat that was added to the sea. To determine the heat budget, we must know the variation of temperature with depth, for the same amount of heat can produce a large temperature increase in a thin layer or a small increase in a thick layer. Between March and August 1956, there was a 9° temperature increase (ΔT) in the upper 23 m and a variable increase over the next 50 m. The total heat amounts to about 3.4×10^4 cal cm^{-2}. At 50°N, the ground receives about 3×10^{-3} cal cm^{-2} sec^{-1} (Fig. 7-7, p. 88), which amounts to 10^5 cal cm^{-2} yr^{-1}. Thus about 34 percent of the yearly energy received at this latitude is used to warm the ocean at ocean station *P*.

From August to September the sea surface cools about 1.5°. As the surface water cools, it becomes denser and so mixes downward to a depth where the temperature is equal to the new surface temperature. Actually, however, the mixed layer in September is considerably deeper than this; water between 24 and 37 m has actually been warmed by mixing with the overlying water. The depth of the mixed layer depends on both the amount of surface cooling and the stirring by waves. As the season progresses into winter, the surface temperature decreases and the depth of the mixed layer increases. By January 1957, the surface temperature has cooled by 5.7°C and the depth of the mixed layer is 95 m.

Thus far we have considered the effect of temperature and have neglected variations in salinity. At ocean station P, in the northeast Pacific, there is an excess of precipitation over evaporation amounting to about 35 cm of water per year (Fig. 10-9, p. 149). As a result, the surface water has a relatively low salinity and is underlaid at about 150 m by saltier water, with a steep salinity increase, the *halocline*, between 100 and 150 m. There tends to be slightly more excess precipitation in summer than in winter. Thus the density of the surface water in summer is decreased not only by an increase in temperature but also, to a lesser extent, by a slight decrease in salinity.

At ocean station P, the seasonal variations of temperature and salinity are additive and so produce a greater variation in density. When we look at the variations of temperature and salinity in the surface layer of the oceans, however, we find that the two often act in opposition and so reduce the variation in surface density. In high latitudes, where the temperature is low, the excess of precipitation over evaporation reduces the surface salinity. In the regions of high evaporation excess, the temperature is relatively high. While the high salinity increases the density, the high temperature decreases it.

If we average the oceans by latitude bands, we find that the temperature varies from about 0 to 27°C and the salinity from about 30 to 37‰. The temperature variation alone, at a constant salinity of 35‰, would lead to a density variation from 1.0281 to 1.0227. This is a range in σ_T from 28.1 to 22.7, with a difference of 5.4 σ_T units. The salinity variation at a constant temperature of 15°C would lead to a σ_T variation from 22.2 to 27.5, a range of 5.3 units. The combined effect of the temperature and salinity in the ocean leads to a σ_T range from 21.9 to 27.1, or a range of

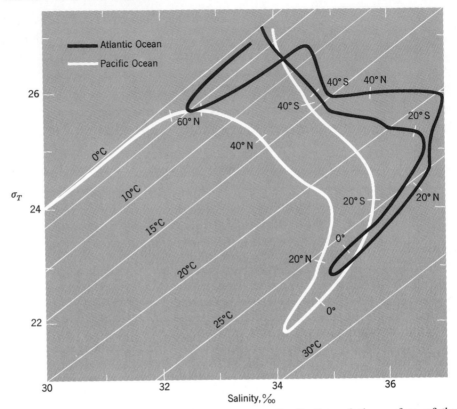

Figure **11-15** Density (σ_T)-salinity-temperature distribution of the surface of the Atlantic and Pacific oceans (5° latitude averages). (After Wüst, 1964)

5.2 units. Thus the combined effect of temperature and salinity on the density, rather than being additive, is slightly less than the effects of these factors considered separately.

To examine the actual distributions in the Atlantic and the Pacific, it is convenient to plot the density (σ_T) against the salinity, as is done in Figure 11-15. On such a plot, the constant temperature curves are approximately straight lines. Owing to the increase of thermal expansion with temperature, the *isotherms* (lines of constant temperature) become farther apart as the temperature increases.

The highest surface densities in both oceans are found near the Antarctic. As we move north, the density decreases to a minimum slightly north of the equator. The highest salinities occur near 20° S. From there to the temperature maximum, the salinity decrease and the increase in temperature both act to reduce the density. While the curves for the Atlantic and Pacific are similar in the southern hemisphere, they diverge markedly in the northern hemisphere. Here the Atlantic has a significantly higher salinity than the Pacific (Fig. 11-9, p. 165).

Since denser water will tend to sink beneath water of lower density, the bottom of the ocean consists of the cold water from the Antarctic region, and the warm equatorial water is underlaid by cooler water of higher salinity derived from the region of the salinity maximum. Thus the exchanges of heat and salt at the sea surface, since they control the density of the surface water, are responsible for the deep circulation of the ocean. Once water has sunk beneath the surface, its characteristics can be altered only by mixing with other water. The characteristics of seawater are therefore determined almost exclusively by the air-sea interaction.

The Formation of Sea Ice

If seawater is cooled below its freezing point, sea ice forms. As the temperature of the surface water is lowered, its density increases so that it mixes downward, bringing warmer water to the surface. As cooling continues, the depth of the mixed layer increases (Fig. 11-14). The rate of heat loss from the Arctic atmosphere amounts to about .005 cal cm^{-2} sec^{-1}, or about 1.5×10^5 cal cm^{-2} yr^{-1}. Thus, if the water starts with a temperature of $0°C$, less than 1 km of seawater can be cooled to the freezing point $(-1.9°C)$ during one year.

For seawater to freeze, either the depth must be shallow or the surface water must be underlaid at shallow depth by water of higher salinity. If a shallow halocline exists, then the surface water, even if cooled to the freezing point, will still be less dense than the warmer, saltier water underneath. Because of the low thermal expansion of seawater at low temperatures, the salinity increase does not have to be very large to counteract the density increase produced by the lowered temperature.

Figures 11-16 and 11-17 show the vertical variation of salinity and temperature, respectively, with depth in the Arctic Ocean. Note that the salinity increases sharply between 5 and 150 m. Because of this steep salinity increase, surface cooling to the freezing point will produce a mixed layer of only about 20 m depth. Thus, as "seen" by the atmos-

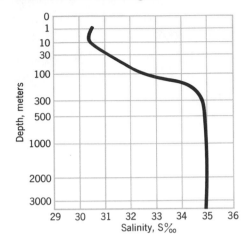

Figure **11-16**   Vertical salinity distribution in the Arctic Ocean at 80°N, 180° (*A* of Figure 11-22). (After Zubov, 1943, p. 413)

phere, the Arctic Ocean acts like a shallow pond. Note that the cold surface water is underlaid by a layer of warmer, saltier water between 200 and 1000 m. If the Arctic Ocean had a uniform salinity from top to bottom, the entire water column would have to be cooled to the freezing point before ice could form. In this case, the heat loss by the atmosphere would probably be insufficient to produce freezing, for the cooled water would tend to exchange with warmer water from lower latitudes. The fossil evidence suggests that during Tertiary times (Fig. 3-2), the climate around the Arctic Ocean was relatively mild. This might have been due to a more uniform vertical salinity structure of the Arctic Ocean.

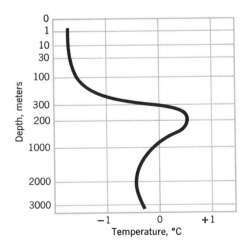

Figure **11-17**   Vertical temperature distribution in the Arctic Ocean at 80°N, 180° (*A* of Figure 11-22). (After Zubov, 1943, p. 413)

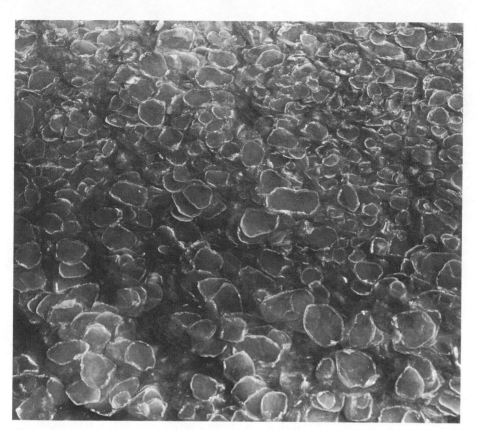

Figure **11-18**   Pancake ice. (Official U.S. Navy photograph)

Once the surface water has cooled to the freezing point and the mixed layer cannot deepen, either because it has reached bottom or because of a halocline, freezing at the surface will commence. At first the sea surface will look oily, with a peculiar leaden tint. As individual ice crystals become larger, they become visible and form a slush. The ice crystals, or spicules, then freeze together, and a very thin layer of ice forms. This layer bends easily and moves up and down with the waves. As the ice thickens it becomes more brittle, cracking and breaking into pieces which drift about. Because of wave motion, these pieces of ice develop a rounded shape through collision with other pieces. The circular pieces of ice from 50 cm to 1 m in diameter are called *pancake ice* (Fig. 11-18). With further freezing, the pieces of pancake ice coalesce to form a continuous sheet of *pack ice* (Fig. 11-19).

Owing to the action of waves and tides, the ice sheets break up, form-

1 Km

Figure **11-19**   Arctic pack ice: (*a*) winter ice pack; (*b*) summer ice pack. (Official U.S. Navy photograph)

ing ridges and hummocks of ice many times as thick as the original ice sheets (Fig. 11-20). Occasionally polynyas, open areas in the ice sheet, form. These polynyas permit nuclear submarines to surface even in the center of the Arctic pack ice (Fig. 11-21).

The formation of sea ice on the surface of the ocean greatly reduces the interaction between the ocean and the atmosphere by preventing the

Figure **11-20**   Various ice topographies caused by pressure. (U.S.N.O.O.)

Figure **11-21** The nuclear submarine Skate surfaced close to the North Pole. (Official U.S. Navy photograph)

convective transfer of heat between the sea surface and the mixed layer of the ocean beneath. Heat then has to be conducted through the ice, and sea ice is a very poor conductor. The conductivity of sea ice is variable, amounting to about 4×10^{-3} cal deg^{-1} cm^{-1} sec^{-1} for older ice. This means that if we have a layer of ice 1 cm thick, across which the temperature differs by 1°C, then each square centimeter of the layer will transmit heat at the rate of 4×10^{-3} cal sec^{-1}.

The Arctic ice is about 2 m thick, and in winter the air temperature drops to about -40°C at the North Pole. Thus the temperature gradient across the ice is about 0.2°C cm^{-1}. The ice will therefore conduct heat from the ocean to the atmosphere at a rate of 0.8×10^{-3} cal sec^{-1} cm^{-2}. This must be compared to a rate of heat loss by the polar atmosphere of 5×10^{-3} cal sec^{-1} cm^{-2}. The ice acts as a good insulator, effectively preventing the ocean from warming the atmosphere. Over the winter, a total of about 10^4 cal cm^{-2} are transferred from the sea to the atmosphere through 2 m of ice. Since the heat of fusion of water is 80 cal g^{-1}, the amount transferred is enough heat to freeze 1.25 m of ice. The added ice melts again during the summer.

Sea ice also has a marked effect on the energy budget of the ocean. Water is a good absorber of the radiant energy from the sun. On the other hand ice, particularly fresh ice or snow, is a very good reflector. The open ocean absorbs about 80 percent of the incident sunlight; sea ice may reflect as much as 80 percent. As a result, the presence of sea ice greatly reduces the heating of the earth's surface by solar radiation.

At the turn of the century, the German research ship *Gauss* was frozen fast into the Antarctic sea ice and spent the winter there. When it looked as if the heat of the summer would not be adequate to free the ship, the crew of the coal-burning *Gauss* took all the ashes and garbage they had on board and spread it on the ice toward an open channel in the pack ice. The surface thus blackened absorbed most of the incident solar radiation and the ice melted, forming a channel to open water and freeing the ship.

The extent of sea ice in the Arctic Ocean in the month of March, when the ice covers the maximum area, is shown in Figure 11-22. Location *A* is where we plotted the vertical salinity and temperature profiles in Figures 11-16 and 11-17. Note that the ice extends farther south in the Pacific Ocean than in the Atlantic. As we move southeast from Greenland

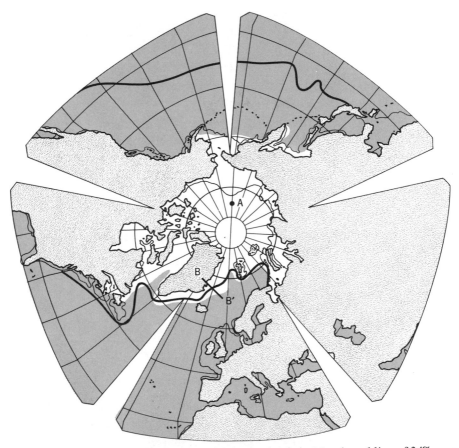

Figure **11-22**   The distribution of sea ice in the Arctic in March and line of 34‰ surface salinity (heavy lines). *A* is the location for Figures 11-16 and 11-17; *B-B'* is the location for Figure 11-23.

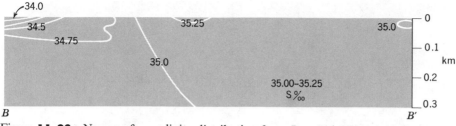

Figure **11-23**   Near surface salinity distribution from B to B' in Figure 11-22.

along the line from B to B', we go from the ice-covered ocean to an open area. When we look at a vertical salinity profile along this line (Fig. 11-23) during the summer, the reason for the contrast becomes apparent. Near Greenland there is a shallow layer of water of lower salinity; to the southwest, the salinity is uniform with depth.

The surface 34‰ constant salinity line, indicated in Figure 11-22, is at about 40°N in the Pacific, so that the North Pacific has a layer of low-salinity water at the surface which permits the southward extension of sea ice. In the North Atlantic, on the other hand, the high-salinity water reaches far north, and the March ice limit tends to parallel this line. Because of the high surface salinity, hence the great depth of the mixed layer, ice cannot form here. The low salinity along the east coasts of Greenland and Canada is due to the East Greenland and Labrador currents, which carry low-salinity, cold Arctic water to the south. These are the currents that bring icebergs south to threaten the North Atlantic shipping lanes.

Before we consider icebergs, let us first look at the distribution of sea ice around the Antarctic Continent. The limits of pack ice around this continent are shown in Figure 11-24. The seasons here are six months out of phase with those in the Arctic, so that the maximum extent of ice is found in September. A unique feature of the Antarctic is the Ross Ice Shelf, which consists of a thick layer of floating ice derived from the continental ice cap. From it are spawned the great Antarctic icebergs.

Icebergs

During the night of April 14–15, 1912, the superliner *Titanic,* on her maiden transatlantic voyage to America, ran into a small iceberg at 41°46′N, 50° 14′W and sank with a loss of 1517 lives. This maritime dis-

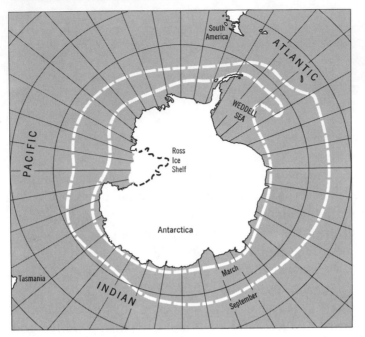

Figure **11-24**   Mean limits of pack ice around the Antarctic continent during March and September. (U.S.N.O.O.)

aster to the supposedly unsinkable ship clearly demonstrated the threat that icebergs pose to navigation. Icebergs are found well south of the sea-ice limit and are not derived from the freezing of seawater. Rather, they have their origin on land. Although ice is a solid, it tends to flow very slowly, like hard tar. As snow builds up in Greenland, on the Antarctic Continent, and on mountains in high latitudes, it flows downhill, forming a glacier. When the glacier meets the shoreline, huge blocks of ice break off; thus the glacier calves an iceberg. Since ice has only about 90 percent of the density of seawater the broken-off ice floats in the water. Depending on the amount of air it contains the iceberg may float with 80 to 90 percent of its volume beneath the sea surface. The iceberg, once calved, drifts with the ocean currents to lower latitudes and gradually melts.

Most of the icebergs that endanger the North Atlantic ship traffic originate on the west coast of Greenland, north of 68°30′. Here, one hundred glaciers give rise to about 15,000 icebergs a year. They first drift north with the West Greenland Current, then south through the Davis Straits to the Labrador Current. During an average year, about 400 ice-

Figure **11-25**   Plane of United States Coast Guard scouting for icebergs as part of the International Ice Patrol. (Official U.S. Coast Guard photograph)

bergs reach south of the 48th parallel. To prevent disasters such as the one that befell the *Titanic,* the international ice patrol now keeps track of icebergs and warns shipping (Fig. 11-25).

The most spectacular icebergs are calved by the Ross Ice Shelf in the Antarctic. These bergs are much more massive than their northern cousins; however, since they are not near major shipping lanes, they have received much less attention. They pose special problems to the yearly convoys, led by icebreakers, that supply the Antarctic research programs (Fig. 11-26).

Figure **11-26**   The icebreakers *Burton Island, Atka,* and *Glacier* jointly push a tabular iceberg from the channel to McMurdo station, Antarctica. (Official U.S. Navy photograph)

Summary

Energy, water, and motion are exchanged between the ocean and the atmosphere at the sea surface. Light penetrates only the upper 100 m of the sea; the depth of penetration depends on the clarity of the water. The ocean and the atmosphere exchange both sensible heat and latent heat of vaporization. The salinity variations of the surface of the ocean result from the interaction between mixing and the net transfer of water from the ocean. As the wind blows over the water, it produces waves and causes a slow drift of surface water in the direction of the wind.

The density of the surface water depends on both the temperature and the salinity, and the two factors generally act in opposition. Temperature tends to be more important in low latitudes, but in high latitudes, owing to the smaller thermal expansion, salinity has more effect on density. The cooling of surface water in high latitudes, particularly in the presence of a shallow halocline, leads to the formation of sea ice. Icebergs consist of continental ice that has broken off from glaciers and floats in the sea.

Study Questions

1. If you were to design a light source for use in underwater photography, over what part of the spectrum should this source have most of its emission, and why?
2. A scheme to produce high temperatures from solar energy consists of building a pond, filling it with brine, and then placing a relatively thin (perhaps 20 cm thick) layer of fresh water over it. When sunlight shines on this composite pond, the pond becomes very much hotter than a similar pond filled with a single liquid. Why?
3. Gypsum (calcium sulphate) precipitates from seawater when its salinity reaches 70‰; ordinary salt (sodium chloride) precipitates only at a salinity of 350‰. Using the model depicted in Fig. 11-8 with $S_A = 35$‰, determine over what range of ratios of evaporation (W) to mixing (M) gypsum will precipitate in reservoir B without the precipitation of salt.
4. Draw a diagram similar to Figure 11-11 for a case in which the wind blows in a direction opposite to the direction of wave advance, and explain why the wind will decrease the wave height.
5. At what latitudes is the surface water in the Atlantic and Pacific more and less dense than the surface water at latitudes to the north and south? At what latitudes would you expect the surface water to sink, and where would it form a stable surface layer. (Use the data in Fig. 11-15.)
6. If one chops a hole in the Arctic sea ice, it will freeze relatively rapidly at first, and then the rate of freezing will drop. Why? If the air temperature is

$-42°C$ and the thermal conductivity of the ice is 4×10^{-3} cal deg^{-1} cm^{-1} sec^{-1}, and if the water has uniform salinity and is at the freezing point, how fast will ice freeze if the ice is (a) 1 cm, (b) 100 cm thick? (Remember that the heat of fusion of ice is 80 cal g^{-1}.) Give the answer in centimeters of ice per day.

Supplementary Reading

(Starred items require little or no scientific background.)

* Bascom, W. (1959). "Ocean Waves," *Scientific American,* August.
Groen, P. (1967). *The Waters of the Sea.* London: D. Van Nostrand Co. Chap. 4.
Kinsman, Blair (1965). *Wind Waves: Their Generation and Propagation on the Ocean Surface.* Englewood Cliffs, N. J.: Prentice-Hall.
* Martin, O. L., Jr. (1966). "The Titanic—50 Years Later," *Science and the Sea,* U. S. Navy Oceanographic Office, pp. 35–44.
* *Bathythermograph Observations* (1963). U. S. Navy Oceanographic Office, *Observers Manual,* Pub. No. 606-c.
* *Ice Observations* (1963). U. S. Navy Oceanographic Office, *Observers Manual,* Pub. No. 606-d.
* *Oceanographic Atlas of the Polar Seas,* U. S. Navy Oceanographic Office, Pub. No. 705.
Williams, J. (1962). *Oceanography,* Boston: Little, Brown & Co. Chaps. 14, 15.

12 The Wind-Driven Circulation of the Ocean

As the wind blows over the ocean, it produces waves and causes the water to drift in the direction of the wind. If the earth were entirely covered by water, the circulation of the ocean would be similar to that of the atmosphere. The ocean currents would circle the earth from east to west in the trade-wind belts and polar regions, and from west to east at intermediate latitudes.

A Circulation Model

The two major oceans, the Atlantic and the Pacific, extend from the Arctic to the Antarctic and are bounded east and west by the American and the Afro-Eurasian supercontinents. The Pacific is bounded in the southwest by Australia and the islands of the Indonesian archipelago. To get a general idea of the oceanic circulation, let us replace the actual complex shorelines of the major oceans by a simple elliptical boundary extending from pole to pole (Fig. 12-1). Let this model ocean be acted on by the average east-west component of the wind (Fig. 10-7, p. 146).

Starting at the North Pole, the wind blows from the east. As we move southward, we come to a region where the wind is predominately from the west. South of the region of westerlies we meet the horse latitudes, and then the region of the trade winds, where the wind blows strongly and consistently from the east. As we approach the equator, the east winds decline in the doldrums, picking up again in the southern trade-wind belt. Next comes the region of the westerlies and finally easterly winds again as we approach the South Pole.

As the system of winds acts on the model ocean, it will produce an east-to-west motion near the poles, a west-to-east motion in the region of the westerlies, and a strong east-to-west motion in the trade-wind belts. If the earth were completely covered by water, these ocean cur-

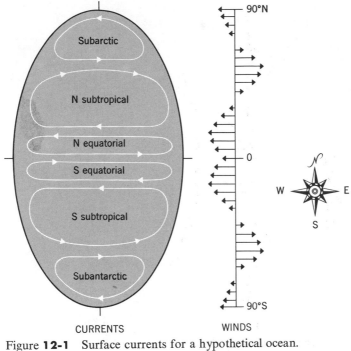

Figure **12-1**   Surface currents for a hypothetical ocean.

rents would simply circle the globe along parallels of latitude. The elliptical boundary of the model ocean, however, deflects the currents, resulting in a series of circular currents, or gyres.

The largest current, the subtropical gyres, are formed by water being pushed to the west by the trade winds, to be returned eastward in the region of the westerlies. Poleward of these gyres are the subpolar gyres, between the polar easterlies and the westerlies. Finally, on either side of the equator, narrow equatorial gyres are formed by the partial return to the east, in the doldrums, of water that has been pushed to the west by the trade winds.

Our simple model ocean thus circulates in six gyres, three on each side of the equator. At the boundaries between the gyres, there will be some mixing of the surface water between one gyre and its neighbor. Superimposed on the circulation pattern is a latitudinal temperature gradient due to the variation in the solar energy received, and a latitudinal variation in salinity due to variations in evaporation and precipitation. The surface distributions of temperature and salinity of the model ocean result from a combination of the circulation and the air-sea interaction. First let us investigate the temperature distribution.

The Temperature Distribution of the Model

Owing to the variation, with latitude, of the energy received from the sun (Fig. 6-3, p. 72), the average temperature increases from each pole to the equator (Fig. 7-3, p. 82). The circulation pattern superimposes an east-west asymetry on the latitudinal temperature variation. Consider the subtropical gyre in the northern hemisphere. The westward-flowing water in the south of the gyre will be heated, both because of the higher insolation and by mixing with water from the tropical gyre. On its northern, eastward-flowing leg, on the other hand, it will be cooled because of the lower insolation and by mixing with water from the subpolar gyre. As a result, the temperature of the north-flowing western part will be warmer than that of the south-flowing eastern part of the gyre. The opposite situation holds for the subpolar gyres of the northern hemisphere. Here the north-flowing eastern branch will be warmer than the south-flowing western branch. The warmest temperatures in the ocean will occur at the eastern margin of the ocean on the equator. Here the eastward-flowing waters have been warmed by flowing along the

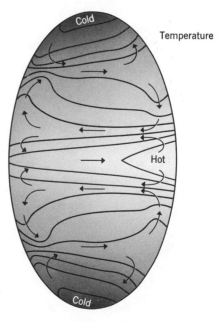

Figure **12-2** Surface temperature distribution for the hypothetical ocean.

equator. The expected distribution of temperature is shown in Figure 12-2.

The distribution of air temperature over the oceans (Fig. 7-1, p. 78) is as expected. Near 60°N the air over the Atlantic and Pacific is significantly warmer in the east than in the west. Near 20°N, on the other hand, the western side is warmer than the eastern side. At 20°S the situation is similar to that at 20°N. At 60°S the temperature is relatively uniform around the globe. Here, however, our model breaks down. Since the continents in the southern hemisphere do not block the ocean currents, the eastward-flowing currents are able to circle the Antarctic continent.

The Salinity Distribution of the Model

Now let us look at the effect of the circulation on the surface salinity of the ocean. The latitudinal variation of the ocean's net loss and gain of water, taken from Figure 11-7, is shown to the right of the model ocean in Figure 12-3. A plus sign indicates a net gain of water, hence a reduc-

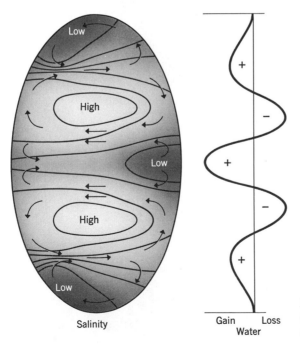

Salinity

Gain   Loss
Water

Figure **12-3** Surface salinity distribution for the hypothetical ocean.

tion in salinity, due to an excess of precipitation over evaporation, which dilutes the water of the sea. Loss of water, hence increased salinity, is the result of an excess of evaporation over precipitation, indicated by a minus sign.

When we combine the effects of the atmospheric water-vapor transport with the circulation of the model ocean, we obtain the relative surface-salinity distribution depicted in Figure 12-3. The highest salinities occur near the centers of the subtropical gyres; the lowest salinities occur at the western margins of the subpolar gyres. The salinity is reduced along the equator, with the lowest values occurring in the east.

The actual distribution of surface salinity in the World Ocean is shown in Figure 12-4. We see that the predictions of our simple model are generally confirmed. Along the equator, the lowest salinities are found at the eastern margins of the oceans. At high latitudes, the western margins tend to be lower in salinity. There are regions of high salinity symmetrically located on either side of the equator. Note that the Atlantic is significantly more saline than the Pacific. This, of course, could not have been predicted by our simple circulation model.

The Currents of the World Ocean

Having seen that our simple circulation model predicts the distributions of temperature and salinity of the ocean surface rather well, let us now look at the distribution of surface currents during the winter of the northern hemisphere (Fig. 12-5a). Strong currents, greater than 50 cm sec^{-1} (1 knot), are indicated by heavy arrows; light arrows indicate a slower drift of the surface water. Because of differences in water characteristics, the surface currents are sometimes visible in satellite photographs (Fig. 12-5b).

While the simple model would predict a current pattern that is symmetrical about the north-south axis of the oceans, we find that all the strong currents are on the western ocean margins. We have the strong Gulf Stream and Kuroshio in the western North Atlantic and Pacific, but the currents at the eastern margins are much weaker. The winds that produce the currents are equally diffuse across the ocean. The currents, on the other hand, flow in a narrow, rapid stream on the western side of the ocean while the return flow in the east is slow and diffuse.

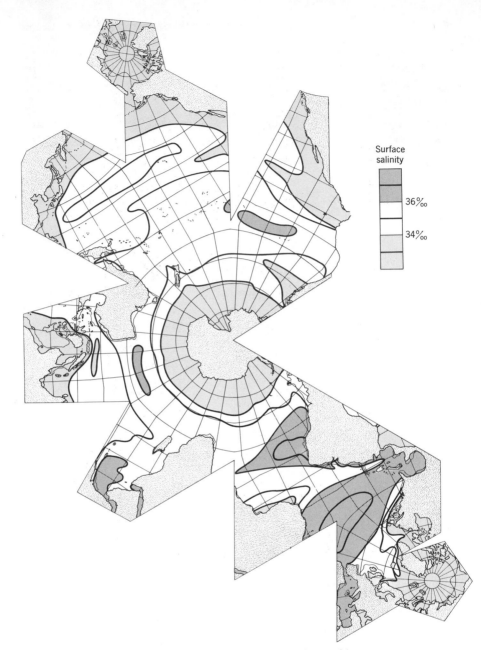

Figure **12-4** Actual surface salinity distribution in the world ocean.

The East-West Asymmetry of the Ocean Currents

What produces the marked east-west asymmetry in the circulation of ocean currents? The answer to this puzzle was provided by the American oceanographer Henry Stommel in 1948. To understand the asymmetry we must consider the forces that cause the surface water to rotate in a gyre. As an example, let us consider the subtropical gyre in the North Atlantic. The water rotates in a clockwise sense. The motive power for this clockwise rotation is provided by the clockwise sense of the wind. To prevent the water from increasing its rate of rotation, there must be an equal counterclockwise tendency. This is provided by friction, acting in a direction opposite to the motion of the water. If the winds stopped, the rotation would gradually be dissipated by frictional drag. Since the motion of the water is clockwise, friction therefore has a counterclockwise sense.

If the wind and water were the only forces producing rotation, the currents would be symmetrical. But Stommel pointed out that there is a third force that produces rotation. This force is due to the rotation of the earth. In Chapter 9 we saw that the rotation introduces the Coriolis acceleration. We also saw that this force per unit mass is proportional to the sine of the latitude and so increases from zero at the equator to $1.5 \times 10^{-4} v$ at the pole. It is the latitudinal variation of the Coriolis acceleration that is responsible for the asymmetry.

To see this, let us consider the three forces separately (Fig. 12-6). The wind blows from west to east in the north and from east to west in the south. It thus produces a clockwise rotation of the water which we will call positive. Friction is roughly proportional to the square of the velocity of the water and acts in a sense opposite to the motion of the currents. It is a counterclockwise or negative force throughout the gyre. Finally, we must consider the effect of the Coriolis acceleration. In the northern hemisphere this acts to the right of the direction of motion of the water. Thus, the northward-flowing western boundary current will be deflected to the east. If the velocity of the stream is constant, the Coriolis effect will increase from south to north as the latitude increases. This increasing deflection will impart a clockwise or positive rotation to the water.

Now let us consider the Coriolis effect on the southward-moving

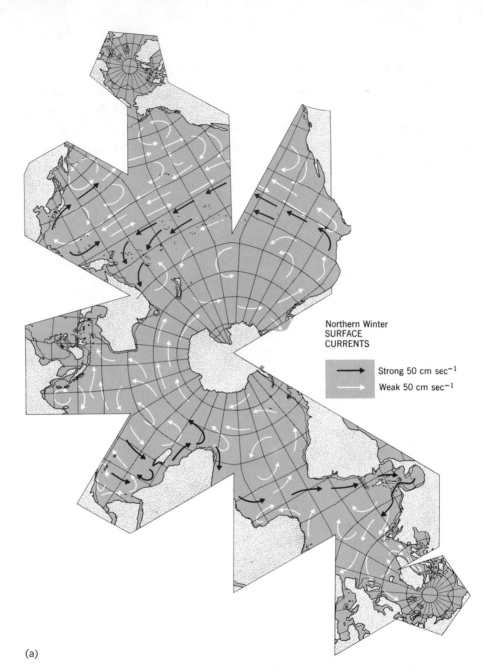

Northern Winter
SURFACE
CURRENTS

→ Strong 50 cm sec^{-1}
→ Weak 50 cm sec^{-1}

(a)

Figure **12-5** (*a*) Surface currents of the world ocean during the northern winter; (*b*) The Peru and South Pacific Equatorial currents are clearly visible in the satellite

(b)
photograph made by NASA's Application Technology Satellite II at 1400 h EST, 10
April, 1967. (La Violette and Chabot, 1967)

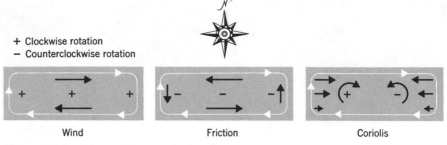

+ Clockwise rotation
− Counterclockwise rotation

Wind   Friction   Coriolis

Figure **12-6**   The rotation-producing forces.

water at the eastern boundary. Here the deflection to the right will tend to move the water to the west. As the water moves from north to south, the Coriolis effect will decrease as the sine of the latitude. As a result of the latitudinal variation of the Coriolis effect, it will impart a counter-clockwise or negative rotation to the gyre. If the rate of rotation is to re-main constant, the positive and negative forces must be in balance.

Let us now assume that the currents are equally strong at both sides of the ocean. Stommel then makes the following estimates for the rela-tive intensities of the three rotation-producing forces:

Rotation-Producing Force	North-Flowing Western Current	South-Flowing Eastern Current
Effect of wind	+1.0	+1.0
Frictional drag	−0.1	−0.1
Coriolis effect	+1.0	−1.0
Net effect	+1.9	−0.1

Thus we see that the forces are not and cannot be in balance so long as the currents at the two sides have the same velocity. The effect of the imbalance is to accelerate the motion at the western boundary and slightly retard the motion at the eastern margin. As the motion in the west accelerates, the effect of the wind will remain constant, since the ocean current still is much slower than the wind, and so the water cannot accelerate the wind. The frictional and Coriolis forces, on the other hand, will increase with the velocity. Now let the current in the east re-main about the same while the western current increases by a factor of about 10. We obtain the following balance:

Rotation-Producing Force	North-Flowing Western Current	South-Flowing Eastern Current
Effect of wind	+1.0	+1.0
Frictional drag	−10.0	−0.1
Coriolis effect	+9.0	−0.9
Net effect	0.0	0.0

Thus we see that the rotation-producing forces can be in balance only if the western boundary currents are much stronger than the flow in the east.

The asymmetry of the currents is a direct result of the latitudinal variation of the Coriolis effect. As a result, the strong Gulf Stream flows along the Atlantic coast of the United States while the currents on our west coast are weak. To find the strong North Pacific current, we have to travel westward to the east coast of Japan, where we find the Kuroshio, the Gulf Stream's counterpart.

Circulation Due to the Wind and Density Differences

In Chapter 9 (Fig. 9-14), we explained the Kuroshio and the Gulf Stream as resulting from differences in density between the subarctic and subtropical waters. These density differences produced a higher sea level for the subtropical waters than for the denser waters to the north. Now we explain the currents as resulting from the action of the wind on the surface of the ocean. Actually, the two factors act in unison. Their inter-action will become important in Chapter 29, where we discuss the sub-surface circulation of the ocean.

The General Circulation of the Ocean

Having seen the reason for the intensification of the ocean currents at their western margins, we must now return to a consideration of the general circulation (Fig. 12-5). We see that the subtropical gyres pre-dicted by our model are in fact developed in each ocean. The equatorial gyres manifest themselves mainly by an equatorial countercurrent run-ning from west to east. In the Atlantic, during the northern winter, this current is developed only in the eastern portion and is north of the equator. During the summer of the northern hemisphere this current intensifies and extends much further west. It is centered at a latitude of about 7°N.

In the Pacific, the equatorial countercurrent also occurs at about 7°N, but extends from the western to the eastern shore. Its width expands from about 3° in winter to over 5° during the summer. The subarctic gyres, instead of consisting of a single circulation cell, tend to be broken up into a number of separate gyres. In the south, as we pointed out earlier, since there are no continental barriers, the subantarctic gyre surrounds the continent of Antarctica.

The Indian Ocean extends only to 20°N, and here the wind regime changes drastically from summer to winter (Fig. 10-11, p. 151). During the summer, the winds blow from the west Indian Ocean onto the Indian subcontinent. As the moist oceanic air rises, by being heated on land, it gives rise to the monsoon rains. In winter, cold, dry air blows from the cold Asian continent over the Indian Ocean. As a result, the circulation of the North Indian Ocean also changes. In the northern winter the currents in the north are primarily from east to west, while in the summer they reverse to flow from west to east (Fig. 12-7). In the south, the circulation is more like that of the other oceans.

The fact that the ocean circulates in a series of gyres retards the mixing

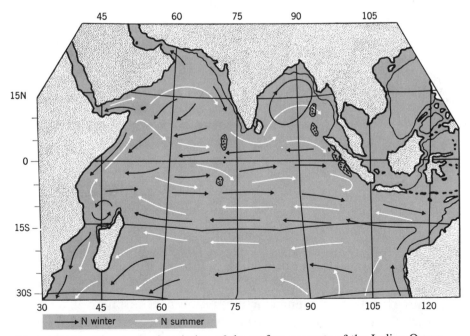

Figure **12-7**   The seasonal variation of the surface currents of the Indian Ocean.

of the surface waters and thus increases the salinity and temperature differences in the ocean. Since the oceans are interconnected only in the far south, the mixing between the North Pacific and the North Atlantic is particularly inhibited. Thus these oceans are able to maintain a significant difference in salinity. In Tertiary times, when there was a seaway across the Isthmus of Panama and when the Atlantic was connected to the Indian Ocean by an extension of the Mediterranean, the differences in surface salinity must have been less than they are at present.

Summary

We have seen that the surface circulation of the ocean (Fig. 12-5) is a direct result of the circulation of the atmosphere. The pattern of gyres results from the winds and the geography of the continents. While the oceans are separated in the north, the free passage around Antarctica permits a great current to flow from west to east around the globe. The following table gives a rough idea of the rate of transport of water by the major ocean currents:

Current	Transport Rate $(10^6 \ m^3 \ sec^{-1})$
Antarctic current	150–200
Gulf Stream—Florida Straits	25
Gulf Stream—Cape Hatteras	100
Kuroshio	50
North Pacific to Arctic Ocean	0.7
All the world's rivers	1

For comparison, the maximum northward transport of atmospheric moisture across a parallel of latitude is 0.7 in these units (Fig. 11-6, p. 162), and the combined flow of all the world's rivers is about 1 unit. Thus the ocean currents transport a tremendous amount of water and keep the surface water of the sea relatively well mixed.

Study Questions

1. If you wanted to cross the model ocean of Figure 12-1 in a sailboat east to west and back again at latitude 30°N, what would be your best route?
2. Where would you expect the density of the surface water due to the temperature and salinity distribution of Figures 12-2 and 12-3 to be at maxi-

mum and at minimum? (Remember that the thermal expansion of sea-water is low, near the freezing point.)

3. Would arctic sea ice extend further south to the east or to the west in the model ocean, and why?

4. The summer monsoon winds over India are accompanied by heavy rains, while the winter winds blowing from the Asian continent over the Indian Ocean contain little moisture and so result in evaporation over the ocean. Combining these facts with the seasonal reversals in the currents of the North Indian Ocean (Fig. 12-7), in which portion of the North Indian Ocean would you expect the highest salinities and why? Compare your results with Figure 12-4. The salinity distribution is affected by the Persian Gulf and the Red Sea. Why are these landlocked seas highly saline?

Supplementary Reading

(Starred item requires little or no scientific background.)

* Munk, W. (1955). "The Circulation of the Oceans," *Scientific American,* September.

Stommel, H. (1965). *The Gulf Stream,* 2nd ed. University of California Press.

Part III
The Earth beneath the Sea

Oceanography is not limited to a study of seawater and the physical, chemical, and biological processes that take place in it. The oceanographer is also interested in the land that lies below the sea. His interest in describing the configuration of the ocean floor naturally leads him to ask further questions: How are the continents supported above the level of the ocean floor? What causes the topographic features beneath the sea? How fast is material eroded from the continents, and how is it deposited in the sea? Since continents and oceans have existed for billions of years, why were the continents not drowned long ago?

In Part II we saw that the fluid envelope of the earth is dynamic. Differential heating of the atmosphere powers the winds and the currents of the ocean. Here we shall see that the solid crust of the earth, from the perspective of geologic time, is also in motion. While the winds and the eroding action of running water wear the continents away, radioactive heating within the earth causes new mountains to rise. It not only maintains the relief of the earth but also changes the geography by causing the continents to drift.

Photograph of the sea floor. (Woods Hole Oceanographic Institute.)

Weights are being added to a bottom corer before it is lowered to the sea floor. The weights push the corer into the soft sediments on the ocean floor to extract a sample. (Official U. S. Coast Guard photograph.)

13 The Relief of the Ocean

It has been said that we know more about the moon's backside than we do about the bottom of the ocean. While the lunar landscape is revealed in all its detail to a moon-orbiting, camera-carrying spacecraft, the floor of the ocean is hidden by a curtain of seawater.

Measuring the Depth of the Ocean

Since ancient times, sailors have found the depth of the sea by lowering a weight on the end of a line, the *sounding line,* into the water. Although this procedure works well in coastal waters, it becomes time-consuming in deep water. Here the hand-held line was replaced by a motor-driven reel of wire (Fig. 13-1). But this method also has its drawbacks. As more and more wire is let out, the wire eventually becomes much heavier than the weight at its end, and it is difficult to tell when the weight has reached bottom. Furthermore, mapping the sea by taking individual soundings is like mapping the surface of the land from a dirigible floating at a constant altitude above sea level. Every 100 km or so, a line is lowered to the surface from the windowless cabin. In this way the elevation is obtained at points that are far apart. The sampling is random, for mountaintops and valleys cannot be seen from the dirigible. Although one can obtain a good idea of the topography of broad plains, mountain ranges are very poorly mapped by this method (Fig. 13-2). Similarly, the mountains and valleys of the ocean floor cannot be mapped properly by using the sounding line.

Because seawater absorbs light, visual methods cannot be used to explore the depth of the ocean from the surface. Since seawater is a fairly good conductor of electricity, it absorbs radio and other electromagnetic radiation, so that radar cannot be used either. Sound waves, however, are readily transmitted by seawater. These waves are utilized by the echo sounder, which emits a pulse of sound and determines the time

Figure **13-1** Dredging and sounding arrangements aboard H.M.S. *Challenger* (from Challenger office report, Great Britain, 1895).

required for the echo from this pulse to return from the sea floor (Fig. 13-3).

One of the first deep-sea echo sounders was employed by the British on the *Discovery I* expedition to the Antarctic in 1925. The German *Meteor* expedition of the same year was the first to make an extensive bathymetric survey with the new instrument. Echo sounders have been

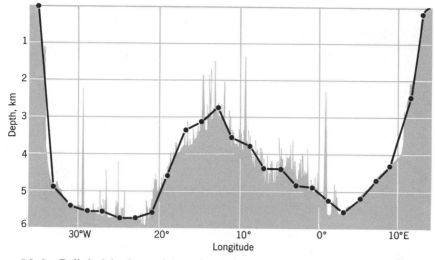

Figure **13-2**   Relief of the floor of the Atlantic Ocean at 8°S as revealed by continuous echo soundings (after Fuglister 1960) and wire soundings (400 × vertical exaggeration).

much improved since then, but the basic principle of their operation is still the same.

The detector of the echo sounder is connected to a recorder. This consists of a stylus which travels across the special paper from left to right (Fig. 13-4) while the paper is slowly advanced. The stylus is charged electrically with the amplified output from the receiver. As the stylus

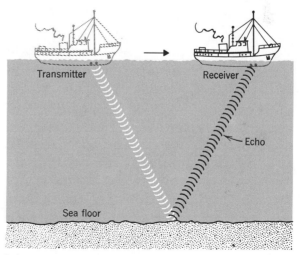

Figure **13-3**   The principle of the echo sounder.

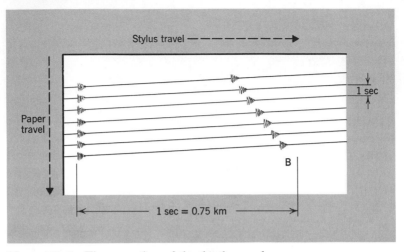

Figure **13-4**   The operation of the depth recorder.

passes the left margin of the paper, a sound pulse is sent out by the transmitter, which the stylus receives and indicates by making an impression on the paper at A. From A to B, the receiver hears only background noise, and the stylus therefore makes only a faint trace on the paper. At B the echo is received, and this louder sound causes the stylus to make a heavier impression. One second after the first signal has been transmitted, another signal is sent out. By repeating this process continuously, we obtain a record of depth along the ship's track (Fig. 13-5).

The velocity of sound in seawater is about 1.5 km sec^{-1}. If the sea floor has a depth of 0.5 km, the sound has to travel 1 km (to the sea floor and then back to the ship); therefore the echo will return after 0.67 sec. The sound velocity varies slightly with the pressure, temperature, and salinity of the seawater, as follows:

	Change in sound velocity (%)
Per km depth increase	+1.1
Per °C temperature increase	+0.3
Per ‰ salinity increase	+0.09

At 1 atm pressure, 0°C, and 35‰, the sound velocity is 1.4455 km sec^{-1}. To convert the travel time of the sound into an accurate depth, one must know the variation in temperature and salinity with depth at the particular location.

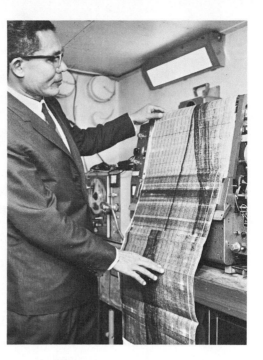

Figure **13-5** Typical record produced by the Precision Depth Recorder. (Photograph: ESSA)

With a 1-sec sweep speed of the stylus, depths can be recorded to about 0.75 km. Over deeper water, the sweep speed of the stylus is slowed down. Periodically, the time is marked on the recorder paper. Later, from the ship's log, the positions of the ship are entered on the depth record. Thus a continuous record of depth is obtained along the track of the ship. We can read the approximate depths from the vertical divisions on the chart paper. These have to be corrected to true depths, using Nansen-cast data of the temperature and salinity distribution.

With the echo sounder, the separate soundings taken by wire or line are replaced by a continuous depth record taken along the track of the ship while it is under way. This represents an enormous improvement in accuracy, but the survey of the ocean is still not so complete as the survey of the topography on land, for the ship's track is an arbitrary line. Since we cannot see the topography of the bottom, we cannot orient the track to pass over the highest points or across the deepest parts of a depression. A peak on the record can indicate an isolated underwater mountain or a ridge. An accurate map of the true topography requires many parallel tracks.

The Distribution of Elevation on Earth

From numerous echo-sounding tracks, more or less accurate maps of the sea floor have been prepared. In areas where the tracks are closely spaced, these maps are very accurate. In other areas, with few tracks, our knowledge of the topography is fragmentary. Using the best charts available, Menard and Smith (1966) analyzed the depth distribution in the World Ocean statistically. They determined the percentage of the ocean area that fell into each depth interval. When we add the distribution of elevations on land to these data, we obtain the curve shown in Figure 13-6.

The distribution of elevations on the earth's surface has two peaks, a narrow peak about 100 m above sea level and a broader peak at about 4.5 km below sea level. If the surface of the earth had been assembled more or less at random, one would expect a single peak in the distribution representing the most probable elevation, with the fractional area decreasing as we get above or below this level. Instead, the surface of the earth appears to be constructed from two different materials, so that we obtain two peaks in the distribution, one representing the continents and the other the ocean basins. The natural division between these two realms—that is, the minimum between the peaks—does not occur at sea level but, rather, at a depth of about 1.5 km.

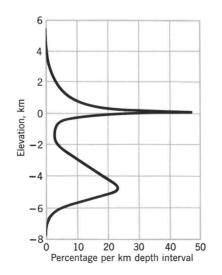

Figure **13-6** The distribution of elevations on earth. (After Menard and Smith, 1966)

Figure **13-7** The hypsometric curve and the simplified earth model.

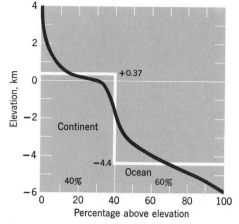

The data of Figure 13-6 can be presented in a different way, by plotting the percentage of the earth's surface whose elevation is above various levels. The result (shown in Fig. 13-7) is called the *hypsometric curve.* According to this curve, 29 percent of the earth, the land area, is above sea level (0), and virtually all of the earth is above −6 km.

If the earth's surface were in a state of minimum potential energy, the hypsometric curve would be a horizontal line. If the contrasts in elevation were evened out, the sea floor would be at a uniform depth of 2.44 km below present sea level. Covering this even sea floor, an area of 510×10^6 km², with 1350×10^6 km³ of seawater produces a water depth of 2.64 km. Thus evening out the topographic relief would cause the sea to stand 200 m above its present level.

In Figure 13-6 we saw that there are two most probable elevations corresponding to the land surface and the sea floor. We can obtain a simplified model of the earth's surface by separately leveling the land and the sea floor. The dividing line between the two realms, as we have noted, is not the present shoreline but rather the 1.5-km depth contour. When we level the land and the sea floor separately, our model earth consists of 40 percent continents with an elevation of +.37 km and 60 percent ocean with a depth of −4.4 km. This simple earth model is indicated by the white line in Figure 13-7.

The hypsometric curve suggests a number of questions that are basic to the earth sciences. What causes the contrast in elevation between the oceans and the continents? How fast is this contrast disappearing as a result of land erosion? In view of the antiquity of the earth, why were the continents not drowned long ago, resulting in an ocean-covered earth? With the aid of recent and current research, we shall attempt to answer these questions in the chapters that follow. First it is necessary to investigate the strength of the earth.

The Strength of the Earth

To get a feeling for the relief on earth and the problem of supporting it, let us make a model, reducing the earth by a factor of 10^8 so that the 6400 km of the earth's radius become 6.4 cm and the 5-km average difference in elevation between the ocean bottom and the land becomes 0.05 mm, or about half the thickness of this page. Thus reduced, the earth appears to be a very smooth ball. It might seem that there should be no problem in supporting this slight relief.

When we make a scale model of the earth, however, different parameters are reduced by different factors. Using material of the same density as the earth, if we reduce the radius by a factor of 10^8, the surface area is reduced by a factor of 10^{16} and the mass by a factor of 10^{24}. How must we scale the strength of the material of our model earth? Rock, for example, has a compressive strength which is expressed as the maximum force per unit area that can be sustained without crushing the material. The forces in the earth arise from the pressures produced by the overlying rock. Therefore, at a given level, the total force depends on the volume of the overlying material and increases as the volume, or the cube of the length. The area to support this weight, on the other hand, increases only as the square of the length. Therefore, if we wish to scale the strength of our model earth properly, the strength of the model must be 10^{-8} times that of the materials making up the actual earth.

Reducing the strength of rocks by 10^8 results in a very weak material which would not be able to support itself or the slight differences in elevation between the mountains and the sea. That this is indeed so is demonstrated by recent geologic history. During the ice ages, northern North America and Scandinavia were covered by ice sheets about 2 km thick. When this ice melted, about 12,000 years ago, a huge load was removed, equivalent to about one-fifth of the relief difference between the oceans and the continents. As a result of this unloading, the land that had been under the ice rose up to compensate for the loss in weight.

The history of this vertical motion has been well documented by determining the present elevation of ancient dated beaches.

As we increase the size of a structure, the relative strength of the material used varies inversely with the increase in the linear dimension, L:

Mass proportional to L^3
Area proportional to L^2
Mass per unit area proportional to $L^3/L^2 = L$

Once we reach dimensions of the size of the earth, the materials become very weak and behave more like liquids than solids. Therefore the difference in elevation between the continents and the ocean basins cannot be supported by materials with the strength of earth materials. A more reasonable mechanism for supporting the differences in elevation is the so-called principle of *isostasy,* illustrated by icebergs. Since water is not able to support a static difference in elevation, the surface of the ocean is level. And since ice is less dense than water, icebergs float. Their tops protrude above the level surface of the sea, supported by the buoyancy of the ice immersed in the water. The deeper the ice extends below the water surface, the higher the top of the iceberg will reach.

The same principle is illustrated by the problem of building a skyscraper in the Gulf Coast region, where the surface soil consists of weak clay. To build a house on bedrock would require a foundation many kilo-

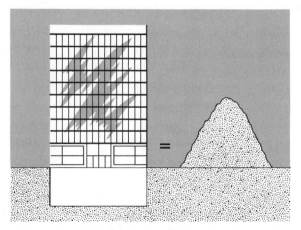

Figure **13-8** Building construction on the Gulf Coast. The weight of soil excavated must equal the weight of the building.

meters deep. In order to support a building, one must therefore make it float in the weak soil. A deep basement is excavated so that the weight of the soil removed is just equal to the weight of the building to be constructed (Fig. 13-8). If the building is heavier than the soil removed, it will slowly sink; on the other hand, a swimming pool, if not kept full of water, will gradually be pushed upward out of the ground. To maintain their elevation above the ocean bottom, then, the continents must have an underlayer of rocks of lower density than the oceans. Under mountain ranges, the "roots" of the mountains either must have a lower density than the surrounding continents or must extend deeper.

The Crust of the Earth

In order to see whether the density under the continents is indeed less than that under the ocean basins, we must study the outer layer of the earth, the earth's crust. How are we to do this? As long as we restrict ourselves to the shallowest layers, we can measure the density of material recovered while drilling oil wells. In this way, we can obtain information to a depth of some 8 km in oil-producing areas. To explore the crust fully, however, we must go much deeper, and we must investigate areas where there are no wells. We are therefore forced to use indirect methods to study the crust of the earth.

One approach is to study the acceleration of gravity. The acceleration of gravity is about 980 cm sec^{-2}. The unit of acceleration, 1 cm sec^{-2}, is also known as the *gal*, in honor of Galileo. The acceleration of gravity results from the gravitational attraction between the earth and a test mass. If the density of the crust is less in one area than another, the lesser density will result in a slight decrease in the acceleration of gravity, since there is less mass nearby to attract the test body.

To measure the acceleration of gravity, we can use a simple pendulum. The period T of such a pendulum is related to its length L and the acceleration of gravity a_g by the following equation:

$$T = 2\pi \sqrt{L/a_g}$$

By carefully timing the swings of a pendulum against a very accurate clock, it is possible to determine the acceleration of gravity to within a few milligals (1 milligal $= 10^{-3}$ gal, or about 10^{-6} of the acceleration of gravity). A gravity meter can also be used to make the measurements.

This consists of a quartz spring from which a weight is suspended. By measuring the relative elongation of the spring, one can determine differences in the acceleration of gravity.

Because of the flattening of the earth at the poles and the centrifugal effect of the earth's rotation, the acceleration of gravity varies with latitude, from 978.04 gals at the equator to 983.22 gals at the poles. Gravitational acceleration also varies with elevation, decreasing at a rate of about 300 milligals per kilometer as the distance between the center of the earth and the test mass increases. Finally, the acceleration of gravity varies with topography, since the nearby mass distribution on a mountain is different from the distribution over the sea. To estimate the variation of density in the lower parts of the earth's crust, we must make allowances for all these variations. What we would like to know is the acceleration of gravity at the surface if the earth were leveled off to sea level and the ocean were filled to this level with rocks having the average density of surface rocks. Furthermore, we must remove the mean latitudinal variation of gravity.

When we make these corrections to the observed acceleration of gravity, we obtain what is called the *Bouguer anomaly*, named after the French mathematician and hydrographer Pierre Bouguer (1698–1758), who made some of the first measurements on the variations of gravity. The Bouguer gravity anomaly for southern California is shown in Figure 13-9. Note that it varies from − 100 milligals on the continent to

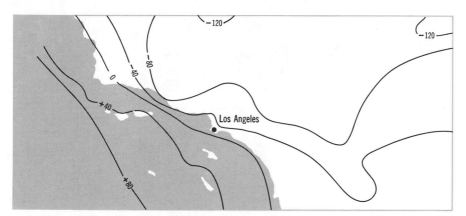

Figure **13-9**   Generalized map of the gravity anomaly of parts of Southern California. (After Emery, 1960)

+100 milligals over the Pacific, with a sharp increase as we approach the coast and move offshore into deep water. This demonstrates that after we level off the topography, the density is significantly higher under the ocean than under the continent. It confirms the isostatic principle that the continents stand above the ocean floor because they are buoyed up by having less dense material under them.

Although the gravity data tell us that there is a deficiency of density under the continents, they do not give us a detailed picture of the density structure. The Bouguer anomaly could be due to a large difference in density over a limited depth interval, or to a small difference over a deeper interval. To explore the structure of the crust, one can use sound waves. Just as the echo sounder determines the depth to the sediment-sea water interface by bouncing sound waves off this interface, it is possible to detect the reflections of more energetic sound waves from interfaces *within* the crust.

Sound waves are partially reflected when they go from one medium to another in which the sound velocity is different. In general, the sound velocity increases with the density of the material. Thus, as sound travels from one layer, *A*, to another layer of greater density, hence of greater sound velocity, *B*, a portion of the sound signal is reflected (Fig. 13-10). If we replace the electronically generated sound pulse of the echo

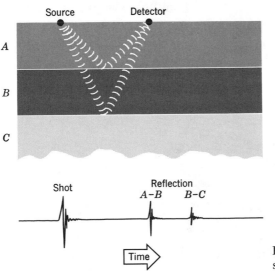

Figure **13-10**   Reflection shooting.

Figure **13-11** Refraction of
light and sound by a plate of
glass.

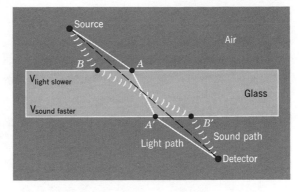

sounder by the noise produced by the explosion of a stick of dynamite,
we obtain much more sonic energy, hence are able to detect the echo
from deeper interfaces than the water-sediment interface—for example,
from *B-C* in Figure 13-10.

In order to calculate the depth of the interface from the travel time of
the sound, we must know the sound velocity within the various layers.
To determine these velocities, we study the travel times of *refracted*
sound waves over various horizontal distances. Sound waves, like the
waves of light, travel from source to detector by the path that takes the
least time. In a material of uniform velocity, the minimum time path is
the straight line between the source and the detector.

As an example, consider the trajectory of a wave from a source to a
detector, both in air, across a thick plate of glass (Fig. 13-11). The shortest distance from source to detector is the straight line joining them, *S-D*.
For a beam of light, the velocity in the glass is slower than that in air, so
the light beam will be refracted to reduce the path in the glass (which
requires an increase in the path length in the air). The resulting light
beam will be refracted to travel via *A-A'*. Sound waves, on the other
hand, travel faster in glass than in air. Therefore, for the minimum travel
time, the path in the glass will be lengthened so as to shorten the path in
air, resulting in the path from *B* to *B'*.

In the ground, the sound velocity increases with depth; going from
one medium, *A*, to an underlying medium, *B* (Fig. 13-12), results in large
increases in sound velocity. For short distances, such as to 1 and 2, the
path will be entirely in layer *A*; for longer distances (3, 4) the shortest
time path will be through layer *B*. By measuring the differences in travel

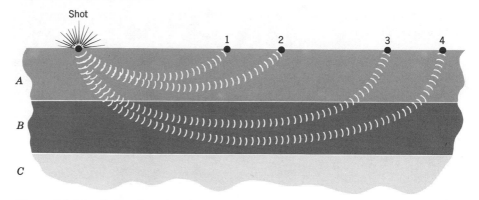

Figure **13-12**   Refraction shooting.

time between shot-1 and shot-2, and between shot-3 and shot-4, we can calculate the sound velocity in layers *A* and *B* if the distances 1 to 2 and 3 to 4 are known. By combining the velocities thus obtained with the travel times to the interfaces *A*-*B* and *B*-*C*, we can determine the actual depth of the boundaries.

From sound waves and gravity measurements, one can infer the structure of the crust of the earth. In particular, one can find out how the crust under the continents differs from that under the ocean. The general picture that emerges is shown in Figure 13-13, for an area across the New Jersey coast. On the continent, the left part of the figure, we find that there is a marked discontinuity in sound velocity at a depth of about 34 km. Here the sound velocity jumps from about 6 to 8 km sec^{-1}. This discontinuity, called the Mohorovicic discontinuity after the Yugoslavian geophysicist who discovered it, represents the boundary between the crust of the earth and the deeper layer, the *mantle*. However, when we

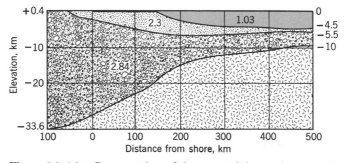

Figure **13-13**   Cross section of the crust of the earth perpendicular to the New Jersey coast. Numbers are average densities in g cm^{-3}. (After Worzel and Shurbet, 1955)

look at the crust under the oceans (at the right of the figure), we find that the Moho is at a depth of only 10 km below sea level.

As we leave the continent, the Moho gradually rises from its continental depth of -34 km to about -10 km once we are a few hundred kilometers offshore. An elevation of -10 km is typical for the oceanic boundary between the crust and the mantle. Thus, the less dense crust is much deeper under the continents than under the oceans. Let us see if the two are in isostatic equilibrium.

On land (the left margin of Fig. 13-13), we go from an elevation of $+0.4$ km to the Moho at -33.6 km, with an average density of 2.84 g cm^{-3}. The total mass of the crust is therefore

$$34 \times 10^5 \times 2.84 = 96.2 \times 10^5 \text{ g cm}^{-2}$$

Under the ocean, we first have about 4.5 km of seawater with an average density of 1.03; next, 1 km of sediments of density 2.3; and finally, 4.5 km of rocks having an average density of 2.84. We must add the weight of the mantle to a depth corresponding to the Moho on the continents. This amounts to 23.6 km with an average density of 3.27. Thus we have

water	4.5 km $\times$ 1.03 =	4.6
sediments	1.0 km $\times$ 2.30 =	2.3
rocks	4.5 km $\times$ 2.84 =	12.8
mantle	23.6 km $\times$ 3.27 =	77.1
	33.6 km	96.8 $\times$ 10^5 g cm^{-2}

compared to 96.2×10^5 for the continents.

The data show that the continents and the ocean basins are approximately in balance. Since the density is inferred from the gravity readings and the sound velocity, it is somewhat uncertain, and so we can not expect to obtain an accurate balance. We see from the cross section that the difference in elevation between the ocean basins and the continents is due primarily to a difference in the depth to the Moho. While the Moho on the continents is overlaid by a thick (almost 34 km) blanket of crustal rocks, the crust over the oceans is much thinner (4.5 km). Under high mountains the Moho is pushed even farther down, enabling the mountains to "float" high in the air.

Because of the horizontal variation in density, the upper layers of the earth are not in a state of equilibrium. Just as in the atmosphere and the ocean, horizontal pressure gradients also exist in the solid surface of the earth. Although it is weak, the material of the solid earth is much more viscous than either the water or the air. The weather varies in a few days, and the ocean water circles a gyre in tens of years; the motions of the solid earth require tens and hundreds of millions of years. In addition to the interior, or *tectonic,* motion of the solid earth, there is another motion of solids at the earth's surface. This is the removal of continental material to the ocean by the action of wind and water. Let us look next at the denudation of the continents.

Summary

The distribution of elevations on earth has two peaks, corresponding to the oceanic and the continental crust. Because of its size, the earth would not be sufficiently strong to maintain this difference in elevation if the density distributions under the ocean and the continents were the same. Gravity measurements and seismic data show that the lighter crust is deeper under the continents than under the ocean basins. The difference in depth of the crust-mantle boundary accounts for the difference in elevation between the continents and the floor of the ocean.

Study Questions

1. The deepest depressions in the ocean are about 10 km deep. How long will it take a sound pulse to travel to the bottom and return? How far will the ship have moved during this time if it has a speed of 500 cm sec^{-1}?
2. What is the average elevation of the solid earth, the modal (most frequent) elevation, and the elevation that divides the earth into equal halves (the 50th percentile)?
3. A typical rock has a density of 2.8 g cm^{-3} and a strength of 5000 kg cm^{-2}. What is the maximum height a slab of this material can have before the rock at the bottom of the slab will fail?
4. A mountain has a density of 2.84 g cm^{-3} and is 5 km high. How deep must the Moho be under the mountain so that the mountain is in static equilibrium with the average continental crust?

Supplementary Reading

(Starred item requires little or no scientific background.)

* Hubbert, M. K. (1945). "Strength of the Earth," *Bulletin of the American Association of Petroleum Geologists,* Vol. 29, No. 11, November.
Poldervaart, Arie, ed. (1963). *Crust of the Earth,* Special Paper 62, New York: Geological Society of America.

14 The Denudation of the Continents

We have seen that the continents stand an average of about 5 km above the floor of the ocean and are supported by a crust thicker than the one under the ocean. Although the continents are in isostatic balance, the earth's surface is not in a state of equilibrium, for the surface of the earth would have a lower potential energy if the thickness of the crust were uniform and if the ocean covered the earth to a constant depth. There are processes at work that constantly tend to reduce the topographic relief by eroding the land and carrying it to the sea. At the same time there must be other processes that maintain the topographic differences or the continents would have drowned long ago. First let us look at the mechanisms that tend to reduce the relief on earth. The major transport of continental material to the ocean results from the flow of water.

The Water Balance of the Continents

The continents receive more water by precipitation than they lose by evaporation. The excess water, about 10^{12} g sec^{-1}, is returned to the oceans by the rivers of the world and as ice in the form of icebergs. While the oceans offer an unlimited supply of water for the atmosphere, the land cannot lose more water by evaporation than it receives from the atmosphere as rain and snow. Interior continental depressions that have no outlets to the sea will gather the surface runoff as lakes, whose areas will adjust themselves until the water loss by evaporation equals the gain by precipitation. If the climate gets wetter, the area of the lake increases, leading to an increase in evaporation. Evaporation is not limited to lakes and rivers but also takes place from the soil and from vegetation. Not only the free water but the entire flora adjusts itself to keep the continental water budget in balance.

219

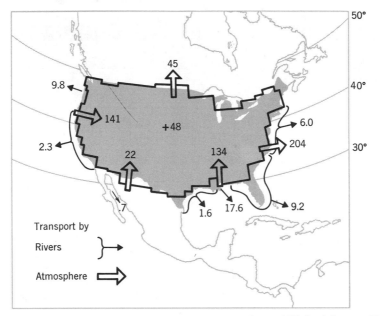

Figure **14-1**   The water balance of the continental United States. Units are in 10^9 g sec^{-1}. (After Rasmussen 1967, Judson and Ritter 1964)

Of the land area of the world, an average of 33 percent has no drainage to the sea, ranging from a low of 10 percent interior drainage for North America to a high of 64 percent for Australia. On the average, the continental United States receives about 76 cm of precipitation per year. Of this, 22 cm (29%) is returned to the oceans as river water, and the remaining 54 cm (71%) is returned to the atmosphere by evaporation from the soil, vegetation, and interior lakes and streams. Since the area of the continental United States is 7.8×10^6 km^2, the annual runoff amounts to 1.7×10^{18} g yr^{-1}, or 55×10^9 g sec^{-1}. Let us see where this excess water comes from.

Rasmussen (1967) estimated the atmospheric water transport over North America between about 30 and 50N latitude (Fig. 14-1). In this region, the winds are predominantly from the west, so that the moisture is transported from the Pacific to the Atlantic. In addition, there is a large influx of water vapor from the Gulf of Mexico. For the period May 1961 to April 1963 he obtained the following fluxes into the area:

Water Flux on Continent	Flux (10^9 g sec^{-1})
From Pacific	163
From Gulf of Mexico	134
From Atlantic	-204
From North	-45
Net water gain	48

Thus the area receives an excess of about 5×10^{10} g sec^{-1} of water from the atmosphere, which must be returned to the sea by rivers. The runoff for the continental United States is summarized in Table 14-1. The two areas do not correspond exactly, since the northern runoff is not included in the table.

It should be clear that we have three independent methods of estimating the runoff from the continental United States:

From estimates of precipitation-evaporation	55×10^9 g sec^{-1}
From estimates of the water-vapor flux	48×10^9 g sec^{-1}
From measurements of river discharge	47×10^9 g sec^{-1}

Considering the nature of the data on which these estimates are based, the fact that they are in rough agreement is encouraging.

TABLE **14-1** RATES OF REGIONAL DENUDATION IN THE UNITED STATES*

Drainage Region	Area 10^6 km^2	Runoff 10^9 g sec^{-1}	Runoff g cm^{-2} yr^{-1}	Concentration ppm Solution	Concentration ppm Solid	Load 10^{-3} g cm^{-2} yr^{-1} Solution	Load 10^{-3} g cm^{-2} yr^{-1} Solid	Load 10^{-3} g cm^{-2} yr^{-1} Total
Columbia	0.673	9.77	46	120	30	5.7	1.4	7.1
Pacific Slopes	0.302	2.27	24	150	870	3.6	20.9	24.5
Colorado	0.637	0.65	3	760	14,000	2.3	41.7	44.0
Western Gulf	0.828	1.56	6	680	1,700	4.1	10.1	14.2
Mississippi	3.24	17.56	17	230	560	3.9	9.4	13.2
South Atlantic and Eastern Gulf	0.735	9.20	40	150	120	6.1	4.9	11.0
North Atlantic	0.383	5.95	49	120	140	5.7	6.9	12.6
Total U. S.†	6.800	46.95	22			4.2	11.6	15.8

* (Modified from Judson and Ritter 1964).
† St. Lawrence and Hudson Bay drainage are not included.

The Rate of Denudation

So far we have considered only the water balance. Although precipitation is almost pure water, river water also contains mineral matter that is transported to the ocean. As rainwater and melted snow flow over and through the soil, they dissolve some of the rock material and carry small mineral particles in suspension. Judson and Ritter (1964) have estimated the amount of dissolved and suspended matter that is carried to the sea by rivers. Their results are summarized in Table 14-1.

As an example, let us consider the Columbia River. It has a drainage area of 0.673×10^6 km^2 and at its mouth has an average flow of 9.77×10^9 g sec^{-1} of water. This amounts to an annual runoff of 46 cm of water averaged over the drainage area. The annual precipitation ranges from over 350 cm per year in the coastal rainforest of the Olympic Peninsula to less than 25 cm in the rain shadow of the Cascade Range. The Columbia River contains about 120 parts per million (ppm) of dissolved mineral salts and 30 ppm of suspended mineral matter. Per square centimeter of its drainage area, it carries 5.7×10^{-3} g dissolved and 1.4×10^{-3} g solid mineral matter to the Pacific Ocean per year, removing, on the average, about 7×10^{-3} g yr^{-1}. Using a density of surface rock of 2.6 g cm^{-3}, we find that the land is being removed at a rate of 2.7 cm per 1000 years.

Comparing the various areas in Table 14-1, we note that the greater the runoff per unit area, the purer the river water. The Colorado River, which drains the arid south-central region of the continent, has the highest concentration of dissolved and suspended matter. While the concentration of mineral matter in the rivers varies by a factor of 100, the rate of denudation varies only by a factor of 6.

The average elevation of the United States is about 700 m. Taking 2.6 g cm^{-3} as the average density of the surface rock, this amounts to about 1.8×10^5 g cm^2 of mass above sea level. On the average, this is removed at a rate of 16 g cm^{-2} per 1000 years. Thus it would require only 11 million years at the current rate to erode the United States to sea level. Since the beginning of the Cambrian, the United States could have been completely eroded 60 times. Therefore, there must be forces that counteract erosion and permit the continents to hold their surface above the water.

Isostatic Adjustment

Let us return to our simple model of the oceanic and continental crust (Fig. 13-7, p. 209). The continents comprise 40 percent of the earth, with an average elevation of about 0.40 km above sea level, while the oceans occupy 60 percent of the area, with an average depth of 4.4 km. Under the continents, the Moho is at a depth of 33.6 km below sea level, while under the oceans it is at a depth of 10 km (Fig. 13-13, p. 216). Suppose that we could erode the upper 0.37 km of the continents and spread this rock uniformly over the sea floor (Fig. 14-2). Since the area of the ocean is 3/2 of the area of the land, we would obtain a layer 0.25 km thick on the floor of the ocean.

Assume that before erosion the continents and the oceans were in isostatic equilibrium (Fig. 14-2a). By removing a load from the continents and placing it on the ocean floor (Fig. 14-2b), we have disturbed this equilibrium. As a result, the continental section will tend to rise relative to the oceanic one until the two are once more in isostatic equilibrium (Fig. 14-2c). If the continental rocks have a density of 2.6 g cm^{-3}, we have removed a load of 9.6×10^4 g cm^{-2} from the continents and added a load of 6.4×10^4 g cm^{-2} to the oceans. The total imbalance, therefore, amounts to 1.6×10^5 g cm^{-2}. We saw (Fig. 13-12) that the rocks below the Moho have a average density of 3.27 g cm^{-3}. Thus if we

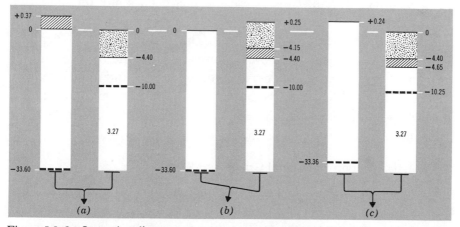

Figure **14-2**   Isostatic adjustment in response to denudation: (*a*) initial state; (*b*) after the removal of 370 m from the land surface; (*c*) after isostatic adjustment. (Not to scale.)

lower the oceanic section by 1 km relative to the continental section by transferring mantle material from under the oceans to under the continents, we reduce the loading of the ocean section relative to the continental section by 3.27×10^5 g cm^{-2}. Because we have an imbalance of 1.6×10^5 g cm^{-2}, we have to shift the oceanic section down by 0.49 km relative to the continental section.

Since our earth is shifting vertically, we have no fixed level that we can use to measure absolute displacements. The best we can do is to consider sea level as our reference plane. Since we have added 0.25 km of sediments to the ocean, the oceanic Moho will now be at a depth of 10.25 km below sea level (Fig. 14-2c). To maintain isostatic equilibrium, we have to lower the oceanic Moho 0.49 km relative to the continental Moho. So far we have lowered it 0.25 km. Therefore we must raise the continental Moho $0.49 - 0.25 = 0.24$ km so that it will lie 33.36 km below sea level and the surface of the continents will stand 0.24 km above sea level. By removing 0.37 km of continental material, after isostatic adjustment, we have lowered the level of the land by only $0.37 - 0.24 = 0.13$ km.

Because of isostatic adjustment, about 3.5 km of the continents must be eroded in order to lower the land surface by 1 km relative to sea level. To lower the United States to sea level, therefore, would require the erosion of 2.45 km. At the current rate of erosion, this would take about 40 million years. Thus while isostatic adjustment reduces the rate at which the continents are drowned, it does not explain their preservation since the Precambrian. Also, if erosion has been going on since that time, there should be about 30 km of material derived from the continents under the floor of the ocean. This is three times the depth of the oceanic Moho below sea level. Therefore, by some mechanism, the earth must enable the continents to continue to rise, and it must remove sediments from the floor of the ocean. In Chapter 18 we shall examine in detail how this is accomplished. First, however, let us see how the rivers are able to carry mineral matter to the sea.

The Relative Motion of Mineral Grains and Water

Moving water has the capacity to transport fine-grained solids. Thus the sand on a beach is moved back and forth by each wave, and rivers carry mineral grains to the sea. The dynamics of the relative motion of solid grains and water are virtually the same for fresh and seawater, for the slight (2%) difference in density between the waters is insignificant relative to the density difference between rocks (2.5 − 3.5) and water (1.0). The laws that govern the movement of sand in a river are therefore the same as the laws that control the motion of sand in the sea.

The denudation of the continents depends on the ability of running water to carry solid matter and to dissolve minerals partially. We shall discuss the fate of the dissolved mineral matter in Part IV. Here let us examine the interaction of water with solid particles.

A solid object at the surface of the earth experiences a downward force equal to its mass times the acceleration of gravity. If the object is placed in water, this downward force is reduced by the buoyancy of the water. According to Archimedes' principle, the downward force is reduced by the force acting on an equal volume of water. The object in water there-fore acts as if its density d_o were reduced by the density of the water d_w:

$$\text{effective density} = d_o - d_w$$

If the grain has a volume V, then the downward force on it in water is

$$\text{force} = a_g V(d_o - d_w)$$

If we drop a stone in water, the downward force will accelerate it so that it will drop at an ever-faster rate. As the stone moves through the water, the water must move around the stone to permit it to sink, at the same time exerting an upward force on the stone. This is the drag of the water, which impedes the downward motion of the stone. In order to move a ship through water, the engines have to do work to overcome the drag of the water. The drag increases as the relative motion between the ob-ject and the water is increased. As the stone starts to sink, it accelerates until the force of the drag is equal to the downward force. Once it reaches this velocity, the two forces are equal and opposite, and so the stone will continue to fall at a uniform rate.

In water the period of acceleration is very short, so that only the steady velocity of the mineral particle as it sinks is of importance. If the grain sinks rapidly, it will not be carried far by the moving water before it reaches bottom. A slowly sinking object, on the other hand, can be transported a long distance before coming to rest on the bottom.

To find the sinking rate v, we must equate the downward force to the drag force and then solve for the velocity. For objects of the same density, the downward force is proportional to the mass of the object. If the shapes of the objects are similar, therefore, the force is proportional to the cube of the diameter, D:

$$\text{downward force proportional to } D^3$$

For similar shapes, the drag force depends on the size of the object and its velocity through the water, v. For small, slowly moving grains, the water moves around each grain as a smooth stream, in so-called *laminar flow*. In this case, the drag is proportional to the diameter of the grain times the velocity:

$$\text{laminar: drag force proportional to } Dv$$

For uniform motion:

$$\text{drag force} = \text{downward force}$$

Therefore:

$$Dv \text{ proportional to } D^3$$

and so the velocity, v, under conditions of laminar flow, is proportional to the square of the diameter:

$$\text{laminar: velocity proportional to } D^2$$

If the object is larger and so moves more rapidly through the water, the smooth laminar flow of the water becomes disturbed and the flow becomes *turbulent*. Under turbulent conditions, the drag increases markedly and is proportional to the square of the velocity times the area of the object. Since the area varies as D^2, we obtain

$$\text{turbulent: drag force proportional to } D^2v^2$$

Again equating the turbulent drag force to the downward force we obtain

$$D^2v^2 \text{ proportional to } D^3$$

Therefore

$$\text{turbulent: velocity proportional to } \sqrt{D}$$

Thus, while the sinking rate for small grains is proportional to the square of the diameter, for large grains the sinking rate increases only as the square root of the diameter.

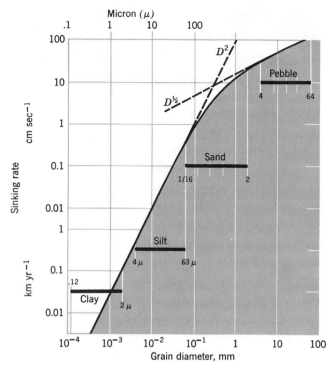

Figure **14-3**   Sinking rate of average shape quartz grains in pure water. (After Bagnold, 1962)

Between these two simple cases is a range in size where the flow gradually changes from laminar to turbulent. The variation in sinking rate in water for grains of a density of 2.65 (quartz), having average grain shapes, is shown in Figure 14-3. The logarithm of the sinking rate is plotted against the logarithm of the grain diameter. On such a plot, a simple power law, such as $v = AD^n$, will appear as a straight line of slope n, for if we take the logarithm of both sides of the equation, we obtain

$$\log v = \log A + n \log D$$

For small grain sizes, the velocity increases as the square of the diameter; for large diameters, the velocity increases as the square root of the diameter. The transition between laminar and turbulent flow occurs between 0.1 and 2 mm. Grains of this size are designated as sand-size.

Depending on their size, rock particles are given different names, ranging from pebble to clay. The names and their size ranges are indicated in Figure 14-3. While sand-size and larger particles fall relatively rapidly, silt and clay take a long time to settle. Thus fine silt takes about a year to settle 1 km, while fine clay particles require over 100 years.

Transportation, Deposition, and Erosion

Depending on their size, rock particles settle rapidly or drift downward extremely slowly. The settling rates summarized in Figure 14-3 apply to still water. If the water of a river enters a reservoir, the sand and coarser silt-size particles will settle out of the water and gradually fill up the reservoir. Depending on the depth of the reservoir and the time the water spends in it before flowing out, only particles below a certain size will continue the trip downstream.

In a flowing stream, however, the situation is different. If the water in the stream had a smooth motion parallel to the stream bed, then the particles would settle out relative to the motion of the water. Thus the finer particles would be carried further downstream than the coarse particles.

Actually, however, the motion of the stream is turbulent. As a result, the water not only flows parallel to the stream bed but also moves up and down in irregular eddies. This irregular motion prevents the particles from settling out, for the vertical stirring produced by the irregular flow counteracts the settling of the particles to the bottom. Whether a particle will settle out depends on the efficiency of the vertical stirring relative to the rate of settling. The rate of stirring depends on the velocity of the stream, while the rate of settling depends on the grain size. For a given stream velocity, particles larger than a particular size will settle out while smaller particles will be carried by the stream. The size limit cannot be defined exactly, for it depends on the shapes of the particles and on the details of the stream flow. However, we can establish a rough boundary between transportation and deposition as a function of grain size and stream velocity. This is shown in Figure 14-4.

A comparison of Figure 14-4 and 14-3 shows that the stream velocity at the boundary between transportation and deposition roughly corresponds to the settling rate of the grain in the water. If the grain sinks at a rate greater than the mean flow velocity of the stream, it will settle out and be deposited, while grains having a slower settling rate will be carried by the stream.

Figure **14-4**   The relationship between current velocity and the fate of sediment particles. (After Hjulström, 1939)

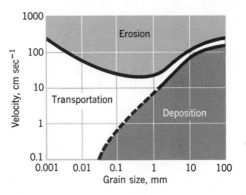

Moving water is able to transport particles in two ways. The body of the moving fluid carries the suspended grains with it. River water is often brown in color because it contains fine silt and clay particles in suspension. If we allow the water to stand, these fine particles gradually settle out, and the water becomes clear. During floods, the clay-laden water will breach the river's banks and inundate fields. There the motion of the water is greatly reduced, and the clay and silt settle out to coat the land.

Particles that are too large to remain in suspension can be transported along the bottom of a river. The turbulent motion of the water at the stream bed imparts a drag force on the grains sitting on the bottom. When this force becomes sufficiently great, the grains will start moving. The force required to lift a grain increases as the cube of its diameter. The turbulent drag of the flowing water on the grain, however, increases as D^2v^2. Thus coarse grains that produce turbulence are dragged along at the bottom of the stream bed, and the diameter of the grains so carried increases with the square root of the stream velocity. Meanwhile, the finer particles are carried in suspension in the body of the flowing water.

So far we have considered how flowing water is able to transport solid material in suspension and along the bottom of its channel. How does this material get into the stream in the first place? What happens when pure water flows over the ground? If the turbulent water encounters grains, it will exert a drag force and cause the grains to move. With increased velocity, the size of the grains that can be moved increases so that the square root of the grain size is proportional to the velocity. It is more difficult to initiate motion than to maintain it, since collisions between moving grains will aid the general motion. To *erode* the ground—that is, to initiate a downslope motion—requires somewhat higher velocities than to maintain this motion by transportation.

Erosion requires turbulent drag. As the size of the particles becomes smaller, the bottom becomes more nearly regular and the flow at the solid-fluid interface becomes laminar. Without turbulence, the water cannot act on individual grains but, rather, acts on the bottom as if it were a continuous solid. In Figure 14-3 we saw that the transition from turbulent to laminar flow around a sinking grain occurs at a grain size between 1 and 0.1 mm. The easiest grain size to erode should be the smallest grain that results in turbulent motion. The stream velocity required to initiate erosion is also indicated in Figure 14-4 as a function of grain size. We see that erosion takes place most easily for grain sizes between 0.1 and 1 mm, which is just the size range where turbulence first appears. To erode grains of sand size takes a lower water velocity than to erode any other size. Silt and clay are more resistant to erosion because of the smoothness of the surface. Once eroded, however, the finer materials are much easier to transport (Fig. 14-5).

The continents are eroded mainly by the action of flowing water. In addition to water, air is also able to erode and transport fine particles. Since water is more viscous than air, the drag forces in air at the same velocity and grain size are much smaller. Therefore higher velocities are required, and the dust particles that can be carried in air are smaller than the suspended matter in water. The ease with which the wind can pick up dust also depends on the moisture content of the soil. Great dust storms develop only in desert regions, where there is no moisture to bind the soil together. The suspended dust gradually falls out as the wind speed decreases. The finer haze particles, however, may be carried into the upper atmosphere, from which they gradually fall back to earth.

The wind thus contributes to the erosion of the continents. It can pick up dust only in dry continental regions, and it drops its suspended load indiscriminately over the continents and the oceans. The very fine material will be distributed worldwide, while the coarse dust particles will be deposited primarily where winds blow off desert regions. Dustfalls at sea are frequently observed off the west coast of Africa, near 20°N, where the trade winds blow from the Sahara Desert. The frequency of haze over the oceans is shown in Figure 14-6. The airblown

Figure **14-5** Erosion of the landscape. (Photograph courtesy of Union Pacific Railroad)

dust is a major component of the deep-sea sediments in regions where the contribution of water-borne sediments from the continents is negligible.

The Effect of Man

We have seen that the action of water and, to a lesser extent, of air erodes the continents and transports mineral matter to the oceans. The continents are continually wasting away and the oceans are being filled in, and yet the ocean basins and the continents have endured. During the last instant of geologic time, man has been altering the natural processes of continental denudation. By agricultural practices we try to reduce surface erosion. We build dams on major rivers to slow the downstream flow of the water; therefore the water drops its load of sediment behind the dam, gradually filling in the man-made lake. Sometimes this process is very rapid. In southern New Mexico, for example, a small dam was built a few years ago to provide water for a new agricultural community called Hope. The sediments carried by the river soon filled in the reservoir. Today Hope, New Mexico, is a ghost town, a reminder that while we can temporarily interrupt the flow of sediments to the sea, the capacity of man-made reservoirs is limited.

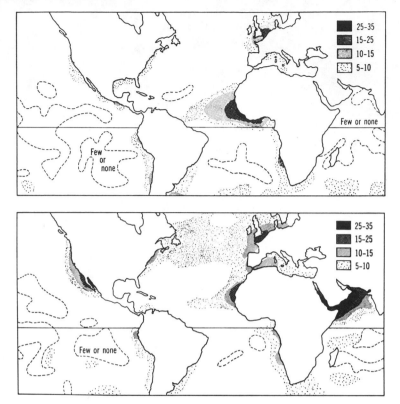

Figure **14-6.**   Average frequency of haze over the ocean during the northern winter (upper map) and summer (lower map). Frequencies are given in per cent of total number of observations. (From Arrhenius, 1959, *Researches in Geochemistry,* John Wiley.)

By measuring the rate of sediment transport, it is possible to determine how rapidly the land behind a proposed dam will be filled in and so ascertain whether the life expectancy of the reservoir is sufficient to provide an economic advantage. Sooner or later, all existing and proposed man-made reservoirs will be filled in by the denudation of the land above the level of the dam. Meanwhile, the interruption of the flow of sediment to the sea will produce changes at the coast. To understand these changes, we must next study what happens to the sediments when they are carried to the sea.

Another effect of man's agricultural effort is to alter the water balance of the continent (Fig. 14-1). By irrigating fields, we increase the rate of evaporation over the continent and so reduce the amount of runoff.

Could we utilize all the excess precipitation in this way until there is no runoff from the land? If we attempt to do this, the concentration of dissolved matter in the water will increase until the water is too salty to be used for agriculture. The lower Colorado is already close to this limit. While the runoff from the Columbia and the Mississippi could be reduced, this reduction cannot go on indefinitely, for we must permit the water to carry the dissolved salts to the sea lest we convert all inland waters to salt lakes.

Judson (1968) has investigated the effect of man on the rate of erosion. In order to obtain a long historical record, he has studied archeological sites near Rome, to determine the level of the soil surface at the time the site was constructed and compare this with the present soil level. Erosion rates during the last 2000 years, so obtained for agricultural areas, amount to about 40 cm per 1000 years. Sediment cores from lakes in the same area indicate an increase in denudation rate with the introduction of agriculture. Initially, the denudation rate over the watershed of the lake was 2 to 3 cm per 1000 years. After 200 B.C. when the area began to be farmed extensively, the erosion rate increased to 20 cm per 1000 years. Thus exploitation by man in central Italy has increased the rate of erosion by a factor of 10.

The increase in evaporation due to irrigation also alters the heat budget of the land, since heat is expended in evaporation. In this way agriculture has an effect on the climate. Because of water's important role in the heat budget, both as vapor and as clouds, it is difficult to make reliable predictions. Man is interfering with the natural denudation of the continents in various ways and thus is altering the environment. He is making these changes for short-term effects. Our current knowledge of the delicate interactions that control the environment is inadequate for reliable long-range predictions of the consequences of man's action.

Summary

The runoff from the continents carries mineral matter to the ocean and thus reduces the elevation of the land surface. This process is rapid geologically. The preservation of the continents therefore requires a means of removing sediments from the ocean floor and returning them to the continents. Isostatic adjustment in response to erosion is not adequate to explain the continued existence of the continents.

The ability of moving water to carry mineral grains depends on the velocity of the water, the size of the grains, and the density of the grains in the water. The flow around a grain changes from laminar to turbulent as the grain size increases from about 0.1 to 1 mm, and these sand-size grains are the easiest to erode. By agricultural practices and by building dams, man is altering the natural processes of denudation.

Study Questions

1. Three ways of estimating the water balance of the continental United States are given in the text; the results differ by about 15 percent. Discuss some of the difficulties that are likely to be encountered in making these estimates.
2. What is the net transfer of water from the Pacific to the Atlantic and the Gulf of Mexico across the United States? (The net transfer is the atmospheric transport minus the water returned to the ocean by rivers.)
3. What is the total annual rate of addition of mineral matter, dissolved and particulate, from the continental United States to the Pacific Ocean?
4. Contrast the behavior of sand and clay with respect to erosion and transportation. Why do beaches contain sand but no clay?
5. Discuss ways in which man increases and decreases the rate of erosion.

Supplementary Reading

(Starred items require little or no scientific background)

Bagnold, R. A. (1962). "Mechanics of Marine Sedimentation," *The Sea*. New York: Interscience Publishers. Vol. 3, pp. 507–528.

* Judson, S. (1968). "Erosion of the Land—Or What's Happening to Our Continents," *American Scientist*, Vol. 56, pp. 356–374.

Kuenen, Ph. H. (1950). *Marine Geology*. New York: John Wiley and Sons. Chap. 4.

15 The Moving Shoreline

The continents are gradually eroded away by rivers which carry the mineral debris into the sea. To understand what happens to this material once it reaches the ocean, we must look carefully at the boundary between the ocean and the land. On a map the shoreline appears to be static. When we stand on the beach, however, we can see that the dividing line between land and water is in continuous motion.

Every wave pushes water up the beach. The water then flows back down before the advance of the next wave. If we observe the shore for many hours, we find that, in addition to the fast pulse of the waves, there is a much slower rhythmic advance and retreat of the water. This is the tidal oscillation which causes the sea to advance and retreat approximately twice a day. If we could have observed the shoreline for the last 100,000 years, we would have noted a third cyclic movement of the shoreline. This cycle occurs because we live in a period of glaciations, when the land in high latitudes is periodically covered by great sheets of ice. The removal of water from the ocean to form the ice sheets has caused the level of the sea to drop about 100 m. As the ice sheets melt, the water is returned to the ocean, and the level of the sea rises again.

Thus the division between the land and the ocean moves back and forth with each wave, with the tides, and in response to glaciation. These three motions control the fate of the material the rivers carry to the sea.

Waves

The wind blowing over the ocean produces waves. As we saw in Chapter 11, the water particles move in circular orbits whose radii decrease with depth until the motion ceases, at a depth corresponding to about $\frac{1}{2}$ wavelength. What happens as a wave approaches the shore? As long as the water is deeper than $\frac{1}{2}$ wavelength, the wave is not affected by the ocean bottom. When the water becomes shallower, how-

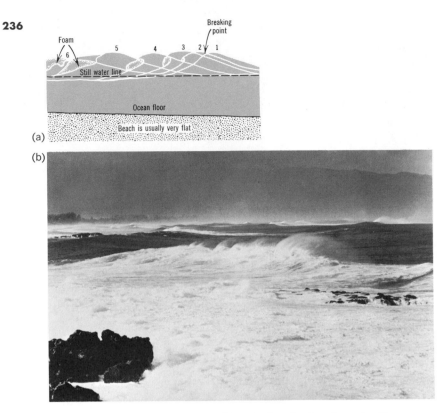

(a)

(b)

Figure **15-1** Spilling waves, (a, b); and plunging breakers, (c, d).

(c)

(d)

ever, the sea floor interferes with the motion of the water particles, reducing the wavelength and therefore the speed of wave advance.

While the speed of the wave's advance is slowed down, the water particles at the surface are still rotating with their original kinetic energy. They advance more rapidly than the wave and so tend to overtake it. This causes the wave to break. The manner in which the wave breaks depends on the slope of the sea bottom. If the slope is gentle, the wave is slowed down gradually, and the water at the top of the wave spills over, producing whitecaps (Figs. 15-1*a* and *b*). If the bottom shoals rapidly, the slowing down is more sudden, and the wave tends to curl over and descend in one great splash (Figs. 15-1*c* and *d*).

The plunging breaker enables the surfer to ride the wave. In the curl of the wave, the water is moving upward in a circle while gravity tends to pull the surfer on his board downslope. Relative to the water, the surfer is continually sliding downhill (Fig. 15-2). The upward motion of the water, however, causes him to remain at the same elevation. Thus the plunging wave carries him in to the beach.

Lines of breaking waves warn the mariner that he is approaching shallow water. Islands and coasts are often surrounded by reefs or bars. A *reef* is an offshore, consolidated rock hazard to navigation, less than 20 m deep; a *bar* is a hazard consisting of unconsolidated sediment. As the waves from the open ocean approach shallow offshore areas, they form a line of breakers. On low coastlines or islands the line of breakers is often the first indication that land is near.

Figure **15-2**   A surfer in the curl of a big wave. (Photograph: Dr. Don James)

As the waves break, the rotary motion of the water is converted into a back-and-forth motion. The water rushes up the beach and then reverses to flow back to the sea. This back-and-forth movement keeps any loose material at the water-sediment interface in constant motion. The size of the material the waves are able to move depends on the energy of the waves. The back-and-forth motion of the sediments impelled by the waves is modulated by the lower-frequency tidal oscillation.

Tides and the Tide-Generating Force

On most shores, the sea advances and retreats with a regular rhythm about once or twice a day. The amplitude of this oscillation varies from fractions of a meter to as much as 15 m (in the Bay of Fundy). In the Mediterranean the tides are almost imperceptible; the Greek and Roman navigators were therefore not concerned with this phenomenon until they ventured through the Strait of Gibraltar into the Atlantic. Other maritime people, however, were at the mercy of the tides, for many harbors can be entered only at high tide.

Ignorance of the tides has had an impact on history. Caesar's war galleys were devastated on the British shore because he failed to pull them high enough out of the water to avoid the returning tide. In October 1216, King John of England (1167–1216) was caught by the high tide near the coast of England and lost all his baggage and treasure and part of his army. This so enraged the King that he fell ill; he died a week later.

In his geography, written in A.D. 902, Ibn Al Fakih retold some of the legends used to explain the tides. According to one legend, the tide rises and falls because an angel dips his finger in the China Sea and then pulls it out again. Another legend ascribes the tides to the breathing cycle of a giant whale. In a book written slightly later, in A.D. 985, another Arab geographer, Al Mokaddasi, described the use of tidal power to turn mills and pointed out that the timing of the tides is related to the cycles of the moon. In A.D. 1213, the Abbot of St. Alban's published a table from which the time of high tide at London Bridge could be predicted from the number of elapsed days since new moon.

While coastal people developed an empirical knowledge of the tides, a scientific explanation had to await the development of the universal

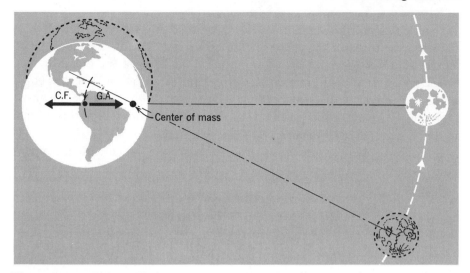

Figure **15-3** The earth-moon system. C.F., centrifugal force; G.A., gravitational attraction. (Not to scale.)

theory of gravitation. In his *Principia Mathematica* of 1687, Newton showed how the tides arise from the gravitational attraction of the moon and the sun on the earth. Of the two, the effect of the moon is stronger.

Let us consider only the earth-moon system. Since the average distance between the center of the earth and the moon does not change with time, the gravitational attraction of the moon on the earth must be exactly balanced by the centrifugal force due to the rotation of the earth about the center of mass of the earth-moon system (Fig. 15-3). While this centrifugal force is the same everywhere on earth, the gravitational attraction of the moon varies from place to place. If we draw a line through the centers of the two bodies, the point on the line nearest the moon will experience a greater gravitational attraction by the moon than the point farthest away. Since the centrifugal force must equal the *average* gravitational attraction, at the point closest to the moon the gravitational attraction will exceed the centrifugal force, resulting in a net force toward the moon. At the point farthest from the moon, on the other hand, the centrifugal force will exceed the gravitational attraction, leading to a net force away from the moon.

We can calculate the strength of the tide-producing force as follows: Let the distance between the center of the earth and that of the moon be R, let the radius of the earth be r, and let m be the mass of the moon divided by the mass of the earth. The average gravitational attraction, which is equal to the centrifugal acceleration, is then proportional to m/R^2, since the gravititational attraction is proportional to the mass and inversely proportional to the square of the distance.

Consider a point on the earth closest to the moon. The distance of this point from the center of the moon is $R - r$ and so the gravitational attraction is proportional to

$$\frac{m}{(R - r)^2}$$

The tide-producing force, which is equal to the difference between the gravitational attraction and the centrifugal force, is proportional to

$$\frac{m}{(R - r)^2} - \frac{m}{R^2} = \frac{mR^2 - m(R - r)^2}{(R - r)^2 R^2}$$

Since r is small compared to R, the equation is approximately equal to $2mr/R^3$. In contrast, in the same units, the gravitational attraction of the earth is $1/r^2$. Thus the ratio of the tide-producing force to the acceleration of gravity is

$$\frac{\text{tidal force}}{a_g} = \frac{2mr^3}{R^3} = 1.176 \times 10^{-7}$$

Since $m = 1/81.45$, $R = 60.34r$.

The point nearest the moon along the line of centers of the earth-moon system experiences an acceleration toward the moon of 0.115×10^{-3} cm sec^{-2}, while the point farthest from the moon experiences a similar acceleration away from the moon. Thus we are dealing once again with a very weak force that produces large effects when it acts on the ocean. For points on the earth's surface away from the line of centers, the tide-producing force is the vector sum of the centrifugal and gravitational forces (Fig. 15-4).

Consider an earth entirely covered with water, which is acted on by the lunar tide-producing forces. These forces will tend to pile up the

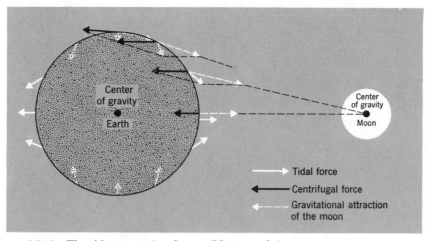

Figure **15-4**   The tide-generating forces. (Not to scale.)

water at the points closest and farthest away from the moon and to depress the water level at right angles to the line between the centers of the earth and moon. As the earth rotates and the moon makes its orbit about the earth, the tidal bulges will rotate with the line of centers, making one complete revolution about the earth in 24 hours 50 minutes. (Since the moon orbits the earth in the same direction as the earth rotates about its axis, it takes somewhat longer than one day for the moon to complete one revolution.)

Since there are two tidal bulges as the earth-moon line makes one complete revolution, two high tides will occur at any one place. The period of the principal lunar tide is therefore 12 hours 25 minutes. If we consider the detailed motions of the sun, moon, and earth, other tide-producing forces appear with different frequencies. Some of the important components of these forces, their periods, and their relative intensities are as follows:

	Period in Hours	*Intensity*
Semidiurnal		
Principal lunar	12.42	0.454
Principal solar	12.00	0.212
Diurnal		
Lunisolar	23.93	0.266
Principal lunar	25.82	0.189

The principal components of the tide-producing forces have periods that are close to either once a day (diurnal) or twice a day (semidiurnal). In addition, there are longer period components, semimonthly, monthly, and semiannual.

So far we have considered only the forces that produce tides. We have yet to examine the motion of the water in the sea in response to the tide-producing forces. To see how the water of the ocean responds to external disturbances, let us first investigate the misnamed tidal waves. Since these waves have nothing to do with the tide-producing forces, we shall call them by their Japanese name, *tsunamis*.

Tsunamis

Periodically, great waves have inundated coastal areas, causing much damage and the loss of many lives. These waves originate from seismic disturbances, such as volcanic explosions or earthquakes. For example, the collapse of the volcano Krakatoa in 1833 gave rise to waves that killed over 36,000 people in the East Indies. Japan lost 27,000 lives as a result of a seismic sea wave in 1896, and 1000 to a wave in 1933.

Tsunamis are most frequent in the Pacific. To minimize the loss of life, the Coast and Geodetic Survey of the U. S. Department of Commerce has established a Seismic Sea-Wave Warning System. To learn about the tsunami, let us see how this system operated in response to the tsunami generated by the Good Friday (March 28) earthquake in Alaska in 1964. The following excerpts are taken from the communications log of the Seismic Sea-Wave Warning System. All times are Greenwich Mean Time (GMT), and the decimal code 0344 means 3 hours 44 minutes GMT.

0502 Honolulu Observatory issues first advisory via FAA and Defense Communications network: THIS IS A TIDAL-WAVE ADVISORY. A SEVERE EARTHQUAKE HAS OCCURRED AT LAT. 61 N., LONG. 147.5 W., VICINITY OF SEWARD, ALASKA, AT 0336 GMT, 28 MAR. IT IS NOT KNOWN, REPEAT NOT KNOWN, AT THIS TIME THAT A SEA WAVE HAS BEEN GENERATED. YOU WILL BE KEPT INFORMED AS FURTHER INFORMATION IS AVAILABLE. IF A WAVE HAS BEEN GENERATED, ITS ETA [expected time of arrival] FOR THE HAWAIIAN ISLANDS (HONOLULU) IS 0900 GMT, 28 MAR. . . .

0530 Communications inoperative on Alaskan mainland. Honolulu Observatory issues second bulletin, stating that it is not yet known whether a seismic sea wave has been generated.

0555 Kodiak replies: EXPERIENCE SEISMIC SEA WAVE AT 0435 GMT. WATER LEVEL 10–12 FT ABOVE MEAN SEA LEVEL. WILL ADVISE.

0630 Kodiak: WATER LEVEL STARTED RISING AT 0435 GMT ROSE 15–20 FT ABOVE MEAN SEA LEVEL BY 0445 GMT. EBB TIDE STARTED 0445 WATER LEVEL 15–18 FT BELOW MEAN SEA LEVEL AT 0507.

0637 Honolulu Observatory issues third bulletin: THIS IS A SEISMIC SEA-WAVE WARNING. A SEVERE EARTHQUAKE HAS OCCURRED AT LAT. 61 N., LONG. 147.5 W., VICINITY OF SEWARD, ALASKA, AT 0336 GMT, 28 MAR. A SEA WAVE HAS BEEN GENERATED WHICH IS SPREADING OVER THE PACIFIC OCEAN. THE ETA OF THE FIRST WAVE AT OAHU IS 0.900 GMT, 28 MAR. THE INTENSITY CANNOT, CANNOT BE PREDICTED. HOWEVER, THIS WAVE COULD CAUSE GREAT DAMAGE IN THE HAWAIIAN ISLANDS AND ELSEWHERE IN THE PACIFIC AREA. THE DANGER MAY LAST FOR SEVERAL HOURS. . . .

0708 Kodiak reports a series of waves: SEA WAVES AT 0435 GMT; 32 FT AT 0540 GMT; 35 FT AT 0630 GMT; 30 FT SEAS DIMINISHING, WATER RECEDING. EXPECT 6 MORE WAVES.

0711 Sitka: UNUSUAL TIDAL ACTIVITY BEGAN 0510 GMT. FLUCTUATIONS LARGE. STILL OCCURRING. [Fig. 15-5a].

0739 First wave arrives at Crescent City, California. . . . Some evacuees return to danger area. [Fig. 15-5b].

0750+ Four persons drown at Depoe Bay, Oregon.

0900 Tsunami reaches Hawaiian Islands. Damage slight at Hilo, with three restaurants and a house inundated; at Kahului, a shopping center is flooded. [Fig. 15-5c.]

0920 A 12-foot wave—probably the fourth—sweeps into Crescent City, California. This wave, and its successors, destroy or displace more than 300 buildings; five bulk gasoline-storage tanks explode; 27 blocks are substantially destroyed; there are casualties.

1020 Tsunami reaches east coast of Hokkaido, Japan.

1038 Tsunami reaches northeast coast of Honshu, Japan.

1100 Honolulu Observatory: THIS IS A TIDAL-WAVE INFORMATION BULLETIN. THE LARGER WAVES HAVE APPARENTLY

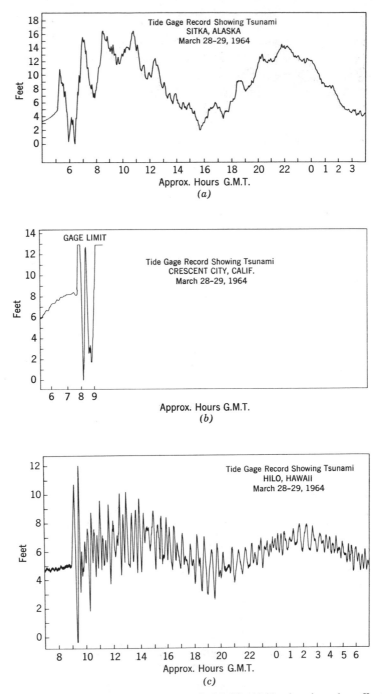

Figure **15-5** Tide-gage records for March 28–29, 1964, showing the effect of the tsunami: (*a*) Sitka, Alaska; (*b*) Crescent City, California; (*c*) Hilo, Hawaii.

PASSED HAWAII. AN ALL-CLEAR STATUS FOR HAWAII CAN BE ASSUMED AT 1100 GMT. ALL PARTICIPANTS IN THE SEISMIC SEA-WAVE WARNING SYSTEM SHOULD ASSUME THE ALL-CLEAR STATUS 2 HOURS AFTER THEIR PARTICU-LAR EXPECTED TIME OF ARRIVAL UNLESS LOCAL CONDI-TIONS WARRANT CONTINUANCE OF THE ALERT STATUS. MAXIMUM WAVE HEIGHTS REPORTED BY VARIOUS MEDIA ARE: OAHU 8 FT., HAWAII 6 FT., KODIAK 30 FT., MIDWAY 1.5 FT., KAUAI 3 FT., CRESCENT CITY 12 FT., CORDOVA 30 FT., TOFINO 8 FT.

The record of the tsunami of March 28, 1964 illustrates a number of points. The Seismic Sea-Wave Warning System was able to make an ac-curate forecast of the time of arrival of the disturbance in Hawaii. On the other hand, the amplitude of the wave at various points around the Pa-cific could not be predicted. While Crescent City, California experi-enced a 12-foot wave, places closer to the source of the disturbance were less affected. From Seward, Alaska to Crescent City, California, a dis-tance of about 3000 km, the wave took only 4 hours, so that its average velocity was 750 km hr^{-1} (2.1×10^4 cm sec^{-1}).

The speed of a wave is equal to the wavelength divided by the period. The wavelength is proportional to the square of the period, so that the speed is proportional to the square root of the wavelength. This relation-ship holds for waves in deep water—that is, waves whose wavelength is shorter than the depth of the water. The square of the wave velocity in centimeters per second is 156 times the wavelength in centimeters. A velocity of 2×10^4 cm sec^{-1}—the speed between the source of the March 28 tsunami and Crescent City—therefore requires a wavelength of 25 km. This is over five times the average depth of the ocean; the ocean, then, is not deep relative to a wave with the speed of the tsunami.

Waves in "shallow water" obey different rules. For these waves, the velocity v depends only on the depth of the water D and the acceleration of gravity a_g:

$$v = \sqrt{a_g D}$$

For a depth of 4 km we obtain

$$v = (10^3 \times 4 \times 10^5)^{\frac{1}{2}} = 2 \times 10^4 \text{ cm sec}^{-1}$$

To predict the time of arrival of a tsunami wave, we need to know only the depth of the ocean between the source of the disturbance and the place of arrival.

As the tsunami approaches the coast, the depth decreases, and so the speed of propagation is reduced. This slowing down of the leading edge of the wave causes the water to pile up, and therefore the height of the wave increases markedly. At sea, in deep water, the wave that hit Crescent City would hardly have been noticed. The height to which the wave rises on the shore depends on how it interacts with the offshore topography. In some areas the energy of the wave may be funneled into a small section of the coast, while in other areas it may be dispersed. Thus, while it is easy to predict the time of the wave's arrival, it is difficult to predict its effect on the coast.

The Response to the Tide-Generating Force

The tsunami is generated by a sudden displacement at the sea floor. This disturbance sends out a wave that travels with a speed of about 750 km hr^{-1} in the open ocean. The tide-generating force, acting in the direction of the moon, circles the earth once every 25 hours. Since the radius of the earth is 6.4×10^3 km and its circumference is 4.0×10^4 km, the tidal force circles the equator with a velocity of 1700 km hr^{-1}. Thus the tide-generating force moves at the equator about twice as fast as any waves it could generate. If the ocean were about 20 km deep, the generating force and the wave response would have the same frequency, resulting in a large tidal excursion.

But the world ocean is not tuned in to the frequency of the tide-generating forces. In high latitudes, where the two frequencies do become equal, we have either the Antarctic continent or the broad, shallow shelves of the Arctic Ocean. The ocean acts like a grotesque violin designed by an abstract impressionist. The strings are tuned to the frequencies of the tide-generating forces, but the sounding box is assembled without rhyme or reason from a collection of boxes of various shapes and sizes. While the total sound of the instrument will hardly rival that of a symphony orchestra, some parts of the sound box will resonate strongly with some of the strings. As a result, the tides will differ markedly in amplitude and frequency characteristics from place to place. Typical tidal curves for a number of ports over one lunar cycle are shown in Figure 15-6. While New York Harbor is mildly in tune with the

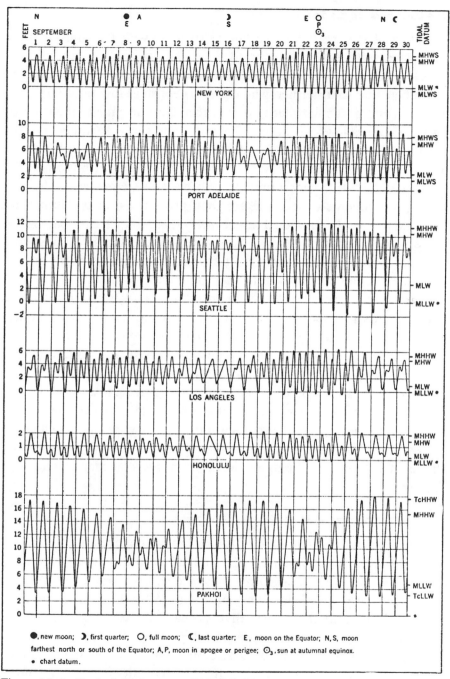

Figure **15-6** Typical tidal curves. (U.S.N.O.O.)

semidiurnal strings, Pakhoi resonates strongly with the diurnal string, modulated with a monthly vibrato.

A knowledge of the tides is important to navigators sailing in and out of harbors and to fishermen who collect shellfish at low tide. To predict the tides at a particular place, we first install a tide gage to measure the variation of sea level with time. (Records from such instruments, showing the effect of the March 28, 1964 tsunami were shown in Fig. 15-5.) The tide record is then analyzed to determine the contributions to the total tide of the tide-generating forces of various frequencies. The individual components of the tide are then projected into the future and summed to give the predicted tide. Finally tide tables are prepared from these computations. Actual sea level is also affected by atmospheric pressure, wind, and occasionally by tsunamis, leading to slight and occasionally large deviations from the tide tables.

The variations in sea level are the result of a movement of water, the tidal currents, in response to the tide-generating forces. Over deep water, these tidal currents are quite weak. At the entrances to large embayments, however, these currents can become very strong (of the order of 100 cm sec^{-1}). Tidal-current charts and tables exist for many coastal areas, to assist the navigator. The tidal currents reverse at low and high tide during slack water and are at maximum halfway between the extremes of sea level. Tidal currents are the primary means by which the waters of coastal embayments are exchanged with the open sea. We shall discuss tidal currents in more detail when we consider estuaries, in Chapter 28.

Sea-Level Changes in Response to Glaciation

We live in a geologic period of rapid climatic changes. Large areas of North America and northern Eurasia have periodically been covered by glacial ice. The water that forms these ice sheets is removed from the ocean. During a typical glacial period, about 150 m of water are removed from the 70 percent of the earth that is ocean; the resulting ice cap is deposited over a land area comprising 3.5 percent of the earth's surface, leaving 26.5 percent of the area unglaciated. The water removed from the ocean will form a cover of ice 3 km thick. As a result of the redistribution of the load, the crust of the earth will have to readjust itself in order to remain in isostatic balance. Since the density of the subcrustal material

Figure **15-7** Sea level during the last 10^5 years.

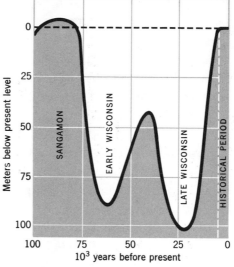

Meters below present level

SANGAMON

EARLY WISCONSIN

LATE WISCONSIN

HISTORICAL PERIOD

10^3 years before present

is 3.27 (Fig. 14-2, p. 223), the floor of the ocean will rise 46 m relative to the unglaciated land area, and the glaciated land surface will subside 910 m. Thus, measured relative to the unglaciated land, sea level will fall 104 m. Because the isostatic adjustment results from plastic flow in the crust and mantle, the adjustment will not be instantaneous but will lag behind the change in load. By locating old shorelines and dating them, we can reconstruct the history of sea-level changes.

At any one place, the change in sea level will depend on the amount of water removed from the ocean and on the changes in elevation of the local land surface. By comparing data from various locations we can estimate the movement of the local land level and obtain a curve of global sea-level changes. When we do this we obtain the curve shown in Figure 15-7. During the last 6000 years sea level has been essentially at its present elevation. Between 17,000 and 6000 years ago (15,000 to 4000 B.C.) sea level rose from about 100 m below its present stand. The average rate of rise was therefore about 1 cm yr^{-1}.

Allowing for isostatic adjustment, about 150 m of seawater were removed from the ocean and deposited as ice on the land during this time. Since the average annual evaporation from the ocean is about 1 m, the rate of glacial melting was about 1.5 percent of the total annual evaporation or precipitation. A sea-level rise of 1 cm yr^{-1} does not appear to be an impressive rate. Just offshore, the slope of the land is about 2 m km^{-1}. Thus the sea encroached on the land at a rate of 5 m yr^{-1}, or 100 m per generation.

People living along the coast or on the lower reaches of rivers were continually forced to abandon their homesites and retreat before the slowly advancing sea. The rise of the ancient civilizations along the lower Nile, Tigris and Euphrates, and Indus rivers had to wait for the stabilization of sea level, about 4000 B.C. Once sea level became static, urban civilization based on an irrigation agriculture became possible. By 3000 B.C. the introduction of writing in Egypt and Sumer marked the opening of the historical period. Our knowledge of the world has increased tremendously since the advance of the sea stopped 6000 years ago.

During the geologic past, there have been long periods (250 × 10^6 years) when the earth was unglaciated, interrupted by periods having many glacial-interglacial oscillations. There is no reason to believe that we are at the end of the period of climatic oscillations; sea level will probably retreat again as we enter the next glacial period. We shall discuss possible causes for these major climatic changes in Chapter 30. At present, our understanding of climatic oscillations is inadequate to predict the onset of the next glacial period.

Summary

The shoreline is not static. Rather, the division between land and ocean moves back and forth at various frequencies. Each wave moves up the beach to retreat again, and the tide raises and lowers sea level with close to a semidiurnal or diurnal rhythm. The amplitude of the tide varies from place to place and depends on the geometry of the sea as well as on the tide-generating forces. Tsunami waves are generated by submarine earthquakes and propagate with a velocity that depends on the depth of the ocean. The largest oscillations of sea level are in response to cyclic glaciations during the Quaternary. Civilization based on irrigation agriculture on the lower reaches of rivers became possible only when sea level stabilized at its present level, some 6000 years ago.

Study Questions

1. On a coast where the water depth increases gradually, how does the distance to the line of breaking waves change as the wavelength of the waves increases, and why?
2. If the distance between the moon and the earth were half as far as it is, by what factor would the strength of the lunar tide-producing forces change?
3. What is the velocity of a tsunami wave in water with a depth of 100 m?
4. From the tide curves in Figure 15-6, list the ports where diurnal and semidiurnal tides predominate and where the tide varies during the month from diurnal to semidiurnal.
5. Why do captains wait for high tide in order to enter a harbor?
6. The teeth of mastodon, Pleistocene elephants, have been found offshore in 70 m of water. How can you explain this fact?
7. Where would you expect to find the habitations of prehistoric fisherman from 15,000 years ago?

Supplementary Reading

(Starred items require little or no scientific background.)

* Bascom, Willard (1964). *Waves and Beaches.* Garden City, N. Y.: Anchor Science Study Series, Doubleday and Co.

Defant, Albert (1958). *Ebb and Flow, the Tides of Earth, Air and Water.* Ann Arbor: University of Michigan Press.

* Fairbridge, R. W. (1960). "The Changing Level of the Sea," *Scientific American,* May.

* U. S. Coast and Geodetic Survey (1965). *Tsunami, The Story of the Seismic Sea-Wave Warning System.*

16 The Margins of the Continents

The continents of the earth are eroded away, and their debris is carried by the rivers into the sea. There, the added mineral matter modifies the topography of the margins of the continents. The sediments interact with the waves and the tides, and the glacial sea level changes. The nature of this interaction depends on the size of the mineral particles. The small clay- and silt-size particles are carried in suspension and gradually settle out as the water velocity diminishes. Once deposited, the clay is difficult to erode (Fig. 14-4, p. 229). Sand-size and larger particles are carried along the bottom. They settle out readily and are easily eroded.

Beaches

To see how sediment is transported by waves and tides, let us perform an imaginary experiment. Suppose that we dump a truckload of sediment having a wide range in size on the intertidal portion of a beach. What will happen to this material? The very large cobbles and boulders will be too big for the waves to move and will remain in place. (The waves produced by a tsunami or a hurricane, however, may be strong enough to move such material high up on the beach.) The fine clay- and silt-size material will be put in suspension by the waves. Because of the constant wave action near shore, this material will be carried to deeper water, where it will gradually settle out.

The sand-size particles, on the other hand, will be carried near the sea bottom. A wave may momentarily carry them in suspension, but they will rapidly settle out again. Thus the sand at the sediment-water interface will be in constant back-and-forth motion as the waves rush up the beach and retreat again. Only when the sand is carried into deep water, where the motion of the waves is too slight to move it, will it come to rest. The size of the sand particles carried back and forth by the waves will depend on the wave energy. If the waves are small, only fine sand will be

253

(a)

Figure **16-1**   Beach at La Jolla, California. (*a*) Appearance in summer. (*b*) The same beach during a winter marked by heavy waves. The sand has been removed to deeper water by the waves exposing cusps of gravel. (Photograph courtesy of Francis Shepard)

(b)

moved. As the wave energy increases, larger and larger particles can be kept in motion. On beaches, which are exposed to large waves from the open ocean, all the finer material is transported to deep water, and only coarse pebbles remain.

In many areas—for example, on the coast of California—the waves are much stronger in winter than in summer. During the summer, we find wide beaches composed of sand. As the wave energy increases in the fall, this material is carried to deeper water and the width of the beach is drastically reduced. As the energy of the waves decreases in the spring, the sand is able to remain in the beach zone again, and so the beach widens (Fig. 16-1).

When waves break on a shallow bar in rapid succession, they can produce rip currents. In this case the water hurled against the beach by the waves does not return seaward over the bar but returns through channels in the bar, producing a strong, continuous, seaward flow (Fig. 16-2). Inexperienced swimmers caught in a rip current generally attempt to swim against the current back to shore. They may become exhausted in their futile attempt to fight the current and drown. The correct procedure is not to swim against the current but to swim parallel to the beach out of the narrow region of strong seaward flow.

Because the waves move the sand, beaches are dynamic. The movement of the sand is not exclusively at right angles to the shoreline. If the wave crests are not exactly parallel to the shoreline, they tend to move the sand along the shore as well as up and down the beach (Fig. 16-3).

When man attempts to modify the coastline by building a harbor, the

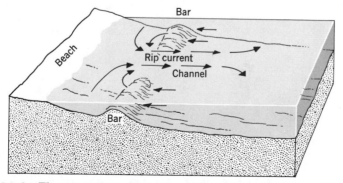

Figure **16-2**   The generation of a rip current by a channel between offshore bars.

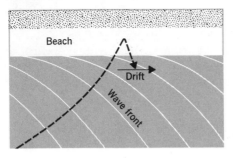

Figure **16-3** Longshore transport of sediment by waves that impinge obliquely to the coast. (After Bascom, 1964)

movement of sand along the shore poses a problem. In California, for example, the coast turns from a N-S to an E-W direction at Point Conception. The waves that come primarily from the west hit perpendicular to the coast north of Point Conception. South of the Point, however, they tend to run along the coast and produce a long-shore drift of sand from west to east (Fig. 16-4). This drift amounts to about 2×10^5 m^3 yr^{-1}. Thus, in one year, the long-shore drift can fill a ditch 1 km long by 100 m wide to a depth of 2 m.

To protect their boats from the waves of the open Pacific, the people of Santa Barbara built a breakwater (Fig. 16-5). By reducing the wave energy, the breakwater caused the sand to pile up just inside the harbor and so fill it in. At the same time, the beaches to the east were deprived of their supply of sand. The long-shore drift caused the beach to disappear, and the land behind the former beach started to erode rapidly. In

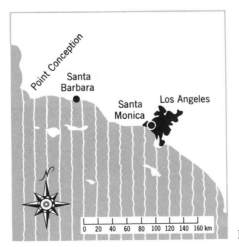

Figure **16-4** Map of southern California.

order to maintain the harbor and prevent erosion, a dredge has to be used continually to pump sand out of the harbor and deposit it on the beach to the east. In this manner, the long-shore drift of the sand is artificially maintained in spite of the obstruction introduced by the breakwater.

To avoid the problem of Santa Barbara, the town of Santa Monica built its breakwater parallel to the coast, several hundred meters offshore (Fig. 16-6). In this way, they believed, the coast would be protected from heavy waves while the sand would be able to move unhindered to the east. Unfortunately, this clever scheme did not work. Because of the protection of the breakwater, the beach behind it grew seaward and started to form an obstacle to the long-shore drift of the sand. Meanwhile, the beaches to the east of the breakwater were being eroded. Thus Santa Monica also had to employ a dredge to move the sand around the harbor.

The coastline is the result of the dynamic interplay of natural forces. If man attempts to alter the natural equilibrium to suit his convenience, he must pay a price. It is difficult enough to maintain harbors at the present time, when sea level is essentially stationary. During periods of changing sea level, the maintenance of harbors would be much more costly.

Where Rivers Enter the Sea

The waves transport sediments along the shoreline. The sediments are derived either from the erosion of the coast or from the interior of the continents whence they are brought to the shore by rivers. Therefore it is important to examine in detail those portions of the coastline where rivers enter the sea. The mouth of the Nile (Fig. 16-7) is a striking example. Here, the sediments carried by the river during its annual flood have enlarged the land seaward to form a delta. Other rivers do not form a convex coastline but enter the sea at the landward termination of a concave embayment. As examples of these two types of river termination, let us consider the Mississippi and the Hudson.

The Mississippi. We will start our examination of the Mississippi at Cairo, Illinois, where the Ohio River enters the Mississippi. Although we are about 1000 airline kilometers above the mouth of the river, Cairo is only 110 m above sea level. The Mississippi flows down the middle of

Figure **16-5** Aerial photograph of the harbor at Santa Barbara, California. (Photograph: Mark Hurd Aerial Surveys)

a 150-km-wide valley which gradually broadens as we move downstream. At 500 km from the mouth we are barely 2 m above sea level. As we move southeast from New Orleans, we enter the delta of the river. Here the Mississippi has built a 10- to 20-km-wide strip of land seaward, with the river channel in the middle. Finally the river splits into a number of channels and forms the "birdfoot delta" shown in Figure 16-8.

Once we leave South Pass to enter the Gulf of Mexico, the water deepens very rapidly. Only 15 km from the pass the water is already 200 m deep, and at 50 km from the pass the depth is 1 km. If, instead of following the river, we travel south, 150 km east of the river, the bottom configuration is quite different.

Leaving the coast, we first traverse the shallow Mississippi Sound. About 20 km offshore we cross a long, narrow island of sand, Horn Island. The water depth gradually increases until at about 100 km off the mainland we have about 50 m of water below our keel. From here the depth increases more rapidly; once we reach 100 m, the slope increases to 1 in 60. The relatively flat, shallow offshore area, with a seaward slope of about 1 in 1500, is known as the *continental shelf*.

The two sections along the river and away from the river indicate that the sediments carried by the Mississippi have been added to the land. The river has built its banks seaward to the edge of the continental shelf and at the present time is adding sediment at a rate of 3×10^{14} g yr^{-1}

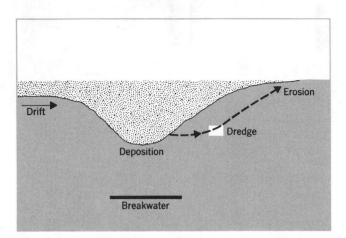

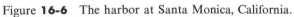

Figure **16-6**   The harbor at Santa Monica, California.

Figure **16-7**   The Nile Delta as photographed by Frank Borman during the flight of Gemini IV. (Photograph: NASA)

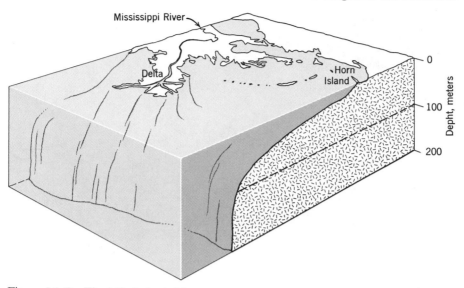

Figure **16-8** The Mississippi delta.

(see Table 14-1, p. 221). The Mississippi is not unique in adding to the land. In Sumerian times, about 3000 B.C., the Tigris and Euphrates entered the Persian Gulf as separate rivers. Since then, sediments carried by these rivers have pushed the sea back 250 km, and the two rivers now enter the Persian Gulf together as the Shatt al Arab.

The Hudson. Let us examine a different type of river, the Hudson. From Albany, New York, to New York City, a distance of 220 km, the river drops less than 2 m. As we head out of New York Harbor past the Sandy Hook light, the depth increases to 60 m rapidly and then remains relatively flat until, after 120 km, we reach a depth of 80 m. Then the sea deepens quickly again with a slope of 1 in 20 (Fig. 16-9).

If, instead of following the Hudson, we cross the continental shelf 80 km to the east, the bottom configuration is quite different. Moving south from Long Island, we first cross the shallow Great South Bay to Fire Island, a long, narrow spit of sand. Continuing south, we find that the water deepens gradually to the edge of the continental shelf. At comparable distances from the shelf edge, the water is significantly shallower than when we moved south from New York Harbor. A detailed survey shows that a channel leads from the shelf edge to the mouth of the Hudson.

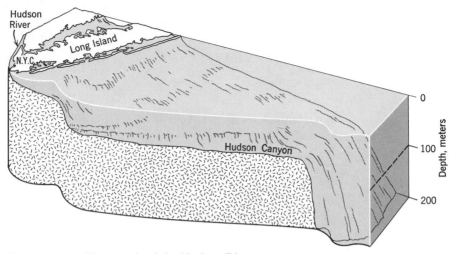

Figure **16-9**   The mouth of the Hudson River.

The existence of the channel can be understood when we remember that only 17,000 years ago the sea stood 100 m below its present level. At that time, the Hudson cut its river channel across the present continental shelf. In the case of the Mississippi, the large amounts of sediments have filled in the ancient channel and covered the continental shelf. The lesser amounts of sediments carried by the Hudson, on the other hand, have not been sufficient to bury the old river channel completely.

We can explain the canyon in the continental shelf as carved by the Hudson during the last ice age. When we follow the canyon seaward, however, we find that it does not stop at the edge of the continental shelf but continues all the way to the floor of the Atlantic at a depth of 4 km. Since the fresh water of the Hudson is less dense than seawater, it is difficult to see how this river could have carved a canyon many kilometers below sea level. We shall examine this problem in the next chapter, when we discuss the ocean floor.

We live in a very dynamic period of earth history. In many regions the coast bears the scars of former river valleys carved when the sea was 100 m below its present stand. Embayments in the coastline, such as Chesapeake Bay, are drowned river valleys. Other inlets, such as the fjords of Scandinavia, were carved by glaciers during the ice age. In other areas, where the rivers bring sediments from vast areas of the con-

Figure **16-10** The Oregon Coast. Cannon Beach forms a crescent between basalt headlands. (Photograph: Ray Atkeson)

tinents, the scars have been covered over and the land is displacing the sea.

While some rivers are pushing back the sea, elsewhere the waves of the ocean are eroding the coastline. In Oregon, for example, the waves from the Pacific are gradually wearing away the steep cliffs along the shore. Here, weak sedimentary rocks are interrupted by intrusions of the much stronger basalt. The sediments wear away more rapidly, forming concave beaches between basalt headlands (Fig. 16-10).

The Continental Shelf

In our trips out to sea we had to cross a relatively flat area, the continental shelf, before the water deepened to typical ocean depths (4 to 5 km). The width of the continental shelf is variable. Off Siberia, in the Arctic Ocean, the shelf extends 800 km from the shore. In other areas, such as off the west coast of South America, the shelf is absent. Worldwide, the average width of the continental shelf is about 70 km, with a slope of 2 m km^{-1}.

Generally, the shelf extends to a depth of 135 m. At that point the slope steepens to 70 m km^{-1} until we reach the ocean floor. This part of the continental margin is known as the *continental slope*. As we move away from the shoreline, it is a considerable distance to the −200-m contour line. Once we pass this depth, however, we rapidly cross the 1-, 2-, 3-, and 4-km depth contours.

As we have seen, coarse sediment particles settle out rapidly, but as the size of the particles decreases they sink more and more slowly. Therefore, we might expect to find coarse sediments near shore grading into finer particle sizes as we move offshore. When we look at maps of sediment-size distribution, however, we find that the pattern is complex. In some areas we find muds near the shore and coarse material at the shelf edge. The distribution of sediments is the result of sea-level changes and variations in wave energy. To understand the pattern, we must study present conditions as well as the rather drastic changes during the last million years of earth history.

In our seaward traverses from the coast, we encounter long, narrow offshore barrier islands. Since these islands protect the shallow inshore waters from the waves of the ocean, silt and clay are able to settle out in the coastal bays. On the seaward side of the islands, the stronger wave action washes out any fine material, and so we find sand and coarser-size deposits.

Off formerly glaciated areas, the continental shelf is very irregular. Here the shelves were sculpted by the moving glacial ice. As the ice melted at its seaward margin, it deposited boulders, pebbles, and smaller rock fragments that had been embedded in the ice. The shelf sediments thus reflect the advancing and retreating margins of the continental glaciers as modified by waves and tides.

When sea level was 100 m below its present stand, the glacial coastline must have displayed many of the features of the present coast. Thus part of the coast must have been protected by offshore islands, while elsewhere there were steep cliffs being eroded by the sea. As sea level rose, many of these features were drowned and partially covered by new sediments. Where large rivers rapidly add sediment, glacial features have been buried. In areas of slow sediment addition, however, the ancient drowned shoreline may be almost unaltered.

Because of its complex history, the continental shelf varies from place to place. To study it, we must map its surface configuration in detail with an echo sounder, and we must sample the sediments on its surface. Using different kinds of grabs (Fig. 16-11a) we can pick up material from the sea floor. To get a deeper look, we drop a heavily weighted, long, thin pipe to the floor. The pipe, known as a *corer,* penetrates the sediments and is then pulled back on board where the *core* of sediment is removed and examined (Fig. 16-12). By using sound waves, we can probe the layering below the sea. By using signals stronger than those of the echo sounder, we can penetrate below the surface sediments to the rock layers beneath.

Slowly we accumulate information to improve our description of the continental shelf and slope. Combining the data derived from many diverse techniques, we obtain a more accurate reconstruction of the geological history of the continental margins.

Figure **16-11** Instruments for sampling the sediment on the ocean floor. (*a*) A Van Veen grab sampler used to collect large samples of the sea floor. (Official U.S. Coast Guard photograph) (*b*) Corer (ESSA)

(a)

(b)

Figure **16-12** Scientists examining cores brought up from the bottom of the Pacific Ocean. (Scripps Institution of Oceanography, U.C.S.D.)

The Ocean Trenches

One might expect to find that the oceans are deepest in their centers and become shallower near the margins. Actually the opposite is true. The centers of the oceans contain huge mountain ranges that rise from the deep sea floor, while the deepest depressions in the oceans are found near some margins of the continents. The greatest depressions in the world ring the Pacific Ocean. Those that extend lower than 10 km are listed below, in order of depth. Their locations are shown in Figure 16-13.

While the deepest trench extends more than 11 km below sea level, the world's highest mountain, Mount Everest, rises only 9.5 km above

Name	Location	Maximum Depth (km)	Length (km)	Width (km)
1. Marianas Trench	NW Pacific	11.022	2550	70
2. Tonga Trench	SW Pacific	10.882	1400	55
3. Kurile-Kamchatka Trench	NW Pacific	10.542	2200	120
4. Philippine Trench	NW Pacific	10.497	1400	60
5. Kermadec Trench	SW Pacific	10.047	1500	60

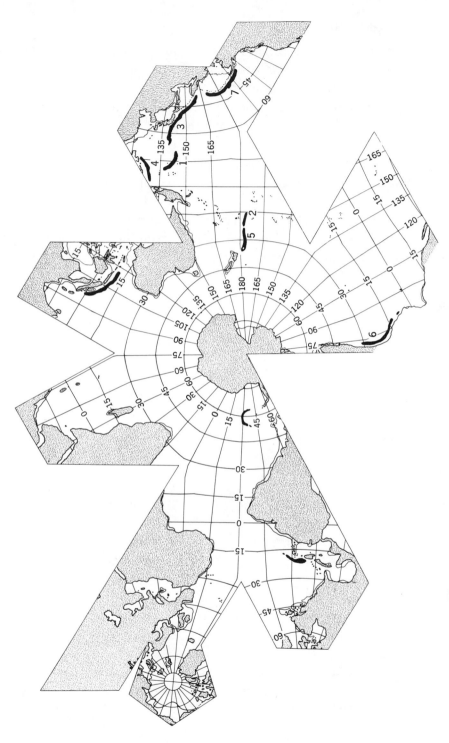

Figure **16-13** The major trenches of the world ocean: (1) Marianas Trench; (2) Tonga Trench; (3) Kurile-Kamchatka Trench; (4) Philippine Trench; (5) Kermadec Trench; (6) Peru-Chile Trench; and (7) Aleutian Trench.

sea level. The great depressions are all very much longer than they are wide, hence they are called the *ocean trenches*. The band of trenches that circles the Pacific Ocean is also known as the "ring of fire," because these trenches are adjacent to active volcanoes. In addition, many earthquakes occur in the region of the trenches. Thus the 1964 Good Friday earthquake, which gave rise to the tsunami discussed in Chapter 15, occurred at the eastern margin of the Aleutian Trench.

A schematic west-east section through the margin of the Eastern Pacific Ocean near 25°S is shown in Figure 16-14. The elevation ranges from −8 km for the Peru-Chile Trench to +6 km for the volcanic peaks of the Andes. The centers of deep earthquakes from this region are also shown. They are shallowest near the trench and become deeper to the east, under the continent.

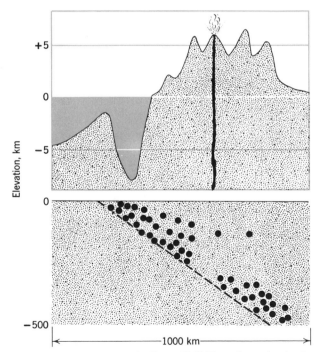

Figure **16-14** W-E section across the Peru-Chili Trench to the Andes near 25°S (*top*, 40 × vertical exaggeration; *bottom*, true scale showing locations of earthquake foci).

Summary

Beaches result when waves carry sand and coarser materials perpendicular to the coastline. Waves that impinge obliquely on the coast produce a long-shore drift of sediments. Where man interferes with this long-shore transport by building harbors, dredges must be employed to bypass the harbor. The present coastline is a result of history. In some areas estuaries now occupy drowned former river valleys; elsewhere the sediments brought by the rivers have filled in the valleys and are building the land seaward in the form of a delta. In most areas the continents are bordered by a broad, shallow continental shelf which merges with the steeper continental slope at a depth of about 135 m. The deepest depressions in the ocean, the ocean trenches, are adjacent to volcanic mountain ranges or island chains. The trenches ring the Pacific Ocean and are associated with earthquake activity.

Study Questions

1. What effect does the damming of rivers have on the future of beaches?
2. How will the shape of a beach be altered if we build a solid damlike structure that reaches above sea level and extends 30 m out to sea, perpendicular to the shore? (Such structures are called *groins.*) Sketch the beach, indicating the direction of the long-shore drift of sand.
3. Determine by looking at an atlas whether the following rivers are forming deltas or terminate in drowned river valleys.
   a. Po River
   b. Niger River
   c. Columbia River
   d. Ganges
   e. St. Lawrence River
   f. Uruguay River
4. Identify the trenches that are not in the Pacific Ocean. Are they associated with volcanism?

Supplementary Reading

(Starred items require little or no scientific background.)

* Bascom, Willard (1960). "Beaches," *Scientific American,* August.
Emery, K. O. (1960). *The Sea off Southern California.* New York: John Wiley and Sons.
* Fisher, R. L., and R. Revelle (1955). "The Trenches of the Pacific," *Scientific American,* November.
* Stetson, H. C. (1955). "The Continental Shelf," *Scientific American,* March.
* Shepard, Francis P. (1959). *The Earth Beneath the Sea.* Baltimore: The Johns Hopkins Press.

17 The Ocean Basins

When we sail over the open ocean we see nothing but water from horizon to horizon. What would the landscape look like if we drained off all the water to reveal the ocean floor? Would the topography be monotonous, or as varied as that of the land? Before the development of the echo sounder, the shape of the ocean floor had to be inferred from widely separated soundings. Only in coastal areas, where there was a close grid of soundings, could a realistic picture of the sea floor be obtained. Because of a lack of data, the charts of the deep ocean suggested a monotonous topography.

Not only were we ignorant of the shape of the ocean floor, but we also knew very little about processes that might alter its topography. In shallow water, the waves and tides move the sediments about. Are there also currents on the ocean floor that are strong enough to move sediments? All of the ocean floor cannot be monotonous, for here and there tiny islands appear above the waves far from land. Occasionally a volcanic eruption produces a new island where there was only water before (Fig. 17-1). Even today, our knowledge of the topography of the ocean floor is only fragmentary, and the layers of sediment that lie below the surface skin are just beginning to be examined.

The Physiographic Diagram

Now that the echo sounder has replaced the sounding line, we can obtain a continuous profile of the sea bottom along the ship's track. However, in order to draw a detailed contour map of the floor of the sea, many parallel profiles, close together, are required. There are very few ocean areas for which we have enough data to construct a detailed contour map. One such area is the Aleutian Islands and the adjacent trench, recently surveyed by the U. S. Coast and Geodetic Survey. In areas for which detailed surveys are not available, we can obtain a general impres-

Figure **17-1** The newly "born" volcanic island, Surtsey, photographed from the air on June 18, 1964. (Photograph: Sigurdur Thorarinsson)

sion of the appearance of the ocean floor from a *physiographic diagram*. Such a diagram indicates the general shape of the sea floor without attempting to map it accurately. It tells us if the bottom is a smooth plane, a series of hills, mountainous terrain, or isolated high peaks standing on a flat sea floor.

Heezen and Tharp, of the Lamont Geological Observatory of Columbia University, have prepared physiographic diagrams of the major oceans, using the method illustrated in Figure 17-2. The cruise tracks along which echo soundings have been obtained are plotted on the chart (lines *A* and *B* in Fig. 17-2*a*). The echo-sounding records along these tracks are shown in Figure 17-2*b*. These records indicate not only the

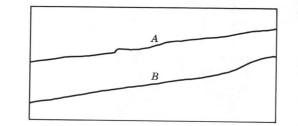

(a)

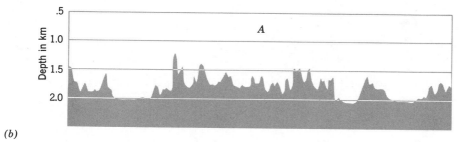

(b)

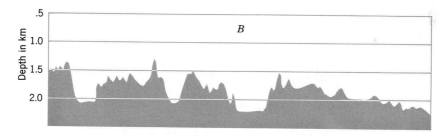

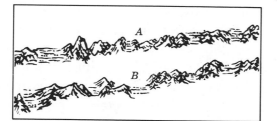

(c)

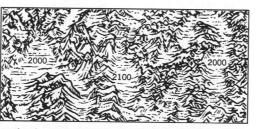

(d)

Figure **17-2** Preparing a physiographic diagram: (*a*) the position of the sounding lines is plotted; (*b*) soundings are plotted with 40:1 vertical exaggeration; (*c*) features of the soundings are sketched in along tracks; (*d*) the area between tracks is filled in by interpolating and extrapolating trends. (Heezen, Tharp, and Ewing, 1959)

Figure **17-3** Physiographic diagram of the tropical Atlantic. (Heezen and Tharp, c. 1961. Reproduced by permission)

variation in relief but also the nature of the bottom, whether it is smooth or undulating. Next, the general nature of the topography as revealed by the echo sounding is sketched in along the cruise tracks (Fig. 17-2*c*). After the relief from all available sounding lines has been sketched, the area between the lines is filled in by extrapolation and interpolation (Fig. 17-2*d*). Obviously, the accuracy of the physiographic diagram depends on the density of the sounding lines.

Figure 17-3 is a physiographic diagram of the tropical Atlantic Ocean. It shows that the topography of the ocean floor is at least as varied as that of the continents. The diagram gives a much more realistic picture of the ocean basins than can be obtained from a chart showing the general distribution of depth in the ocean (Fig. 17-4). The depth chart indicates only that the central portions of the tropical Atlantic are shallower than the basins at either side; the physiographic diagram reveals that the central rise consists of chains of mountain ranges rather than a smooth plateau above the sea floor.

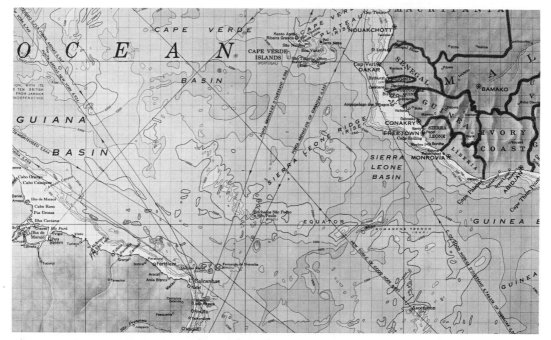

Figure **17-4**   Bathymetric chart of the tropical Atlantic; same area as in Figure 17-3. Depth are in fathoms, 1000 fathoms = 1.83 km. (U.S.N.O.O.)

The Variety of Oceanic Topography

The topography found on the ocean floor can be classified into a number of categories. Adjacent to the continents is the continental shelf, which we considered in Chapter 16. Between the edge of the continental shelf and the floor of the deep ocean lie the *continental slope* and the *continental rise*. The continental slope is very steep, with an average angle of 4°. In many areas this steep boundary extends all the way to the floor of the deep ocean; elsewhere there is a smoother, less steep transition region, the continental rise, between the slope and the ocean basin. The continental slope, on the other hand, is rather irregular and is frequently incised by canyons, such as the Hudson canyon, that run from the edge of the shelf to the bottom of the slope.

Seaward of the slope and rise we often find very flat, smooth areas, called the *abyssal plains*. Often steep, isolated peaks rise from the floor of the deep ocean, some of which extend above sea level to form oceanic islands. The largest features of the ocean floor are the mid-ocean ridges, huge mountain chains that bisect the Atlantic Ocean and traverse the other oceans. Let us now examine these various features of the ocean floor in some detail.

Abyssal Plains

A number of abyssal plains can be seen in Figure 17-3. These are probably the smoothest surfaces on earth. Usually the up-and-down motion of a ship due to waves is much greater than the relief of the abyssal plain. From a few precision depth-recorder curves taken in very calm seas, it appears that the relief on the plains is less than 1 m.

What process has leveled off this portion of the ocean floor? When we replace the weak depth recorder by the more powerful seismic profiler, the sound waves are able to penetrate beneath the sea floor to reveal the structure below.

The abyssal plain is underlaid by a layer with an irregular top surface. Between this layer and the smooth surface of the sea floor, the profiler record is dark, indicating appreciable reflection of sound waves from within the top layer of the abyssal plain (Fig. 17-5a). When we sample the sea floor with a sediment grab or a corer, we find that the abyssal plain consists of sediment. Thus the plain was not formed by leveling off a hard, irregular rock surface but must have resulted from the accumulation of sediments over an irregular surface. If the sedimentation were the result of a slow settling out of fine material from the seawater, one would expect a homogeneous layer of sediment that gradually reduced the relief of the underlying rock surface. Such a homogeneous layer would reflect sound waves only from the water-sediment interface, and possibly from interfaces where the nature of the sediment had changed in the past. Thus we would expect the profiler record to look as it does in Figure 17-5b.

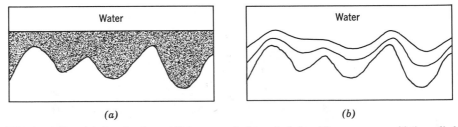

Figure **17-5** (a) A seismic profiler trace of abyssal plain; (b) appearance if the relief had been covered by a slow accumulation of sediment.

Figure **17-6** Graded bedding.

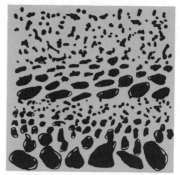

To resolve the mystery, we must examine the sediment. A long corer at the end of a cable is pushed into the sediment by a heavy weight (Fig. 16-11b, p. 265). The tube of the corer removes a cylinder of sediment 10 m long and about 5 cm in diameter. When the core is retrieved and the sediment examined, we find that it consists of a mixture of sand, silt, and clay. Within the sediment are the skeletons of animals that live only on the bottom of the shallow water of the continental shelf. How did this coarse material get all the way out to sea, and why is the plain so level? Only strong bottom currents could have carried the sand-size material from the shelf. Are the same currents responsible for carving canyons in the continental shelf?

When we inspect the core from the abyssal plain in more detail, we find that it consists of a series of layers. Each layer starts with coarse material, grading upward into finer silt and clay (Fig. 17-6). This is known as *graded bedding*. The vertical variation in composition explains why these sediments produce a broad dark band on the seismic profiler. Each graded bed contains variations in sound velocity and thus acts as a separate reflector. How were these layers on the bottom of the deep ocean formed?

Turbidity Currents

The phenomenon responsible for the formation of the abyssal plains was first observed in lakes. If fine sediments, such as clay, become suspended in fresh water, they slightly increase its density. For example, if we add 1‰ of mineral matter with a density of 2 to the water, the resulting mixture will have a density of 1.001. This excess density will cause the water to sink. As it sinks, the potential energy due to the excess density is converted into kinetic energy. The farther the denser water moves downslope, the more rapidly it will travel. In this way the slight increase in density produced by the suspended mineral grains can lead to very high velocities.

If friction is negligible:

$$\text{change in potential energy} = \text{kinetic energy}$$

Let ED be the excess density in g cm^{-3},
$\quad H$ be the change in elevation in cm,
$\quad V$ be the resultant velocity in cm sec^{-1}, and
$\quad a_g$ be the acceleration of gravity.
Per gram of the water then

$$ED \times a_g \times H = \tfrac{1}{2} V^2$$
$$V^2 = 1960 ED \times H$$
$$V = 45 \sqrt{ED \times H} \text{ cm sec}^{-1}$$

Example:
$$ED = 1\%_0 = 10^{-3} \text{ g cm}^{-3}$$
$$H = 2 \text{ km} = 2 \times 10^5 \text{ cm}$$
$$V = 45 \sqrt{200} = 640 \text{ cm sec}^{-1} = 23 \text{ km hr}^{-1}$$

As the velocity of the descending turbid water increases, the frictional forces will increase until the frictional dissipation is equal to the change in potential energy. If the volume of the turbid water is large, the friction per unit mass will be small so that even a slight density increase can lead to velocities of several tens of kilometers per hour.

Do such velocities actually occur on the ocean floor? Heezen and Ewing (1952) pointed out that an earthquake on the Grand Banks off Newfoundland was followed by a series of cable breaks on the adjacent sea floor (Fig. 17-7). They suggested that the earthquake caused slumping on the continental slope. The slumping suspended sediments in the overlying water and thus increased its density. As this denser water moved downslope, it accelerated until the frictional losses equaled the rate of change in potential energy. As the slope of the sea floor decreased, the current slowed down and the suspended sediment settled out.

From the time sequence in the breaks of the submarine cables, it is possible to determine the speed of advance of the turbid water, known

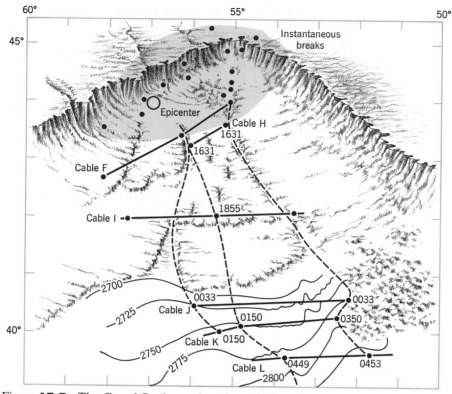

Figure 17-7 The Grand Banks earthquake of 1929 and the subsequent sequence of cable breaks. The times of the breaks are indicated. (From Menard, *Marine Geology of the Pacific,* McGraw-Hill, N. Y., 1964. Used by permission of McGraw-Hill Book Company.)

as a *turbidity current.* Speeds as high as 2000 cm sec^{-1} were obtained, confirming that turbidity currents can reach extremely high velocities. To verify that the disturbance which spread down the ocean floor and destroyed the cables was indeed a turbidity current, the area was cored. The cores revealed a surface layer 1 m thick consisting of graded silt, containing shallow-water microfossils. Thus the turbidity current, set off by an earthquake, transported shallow-water sediments from the edge of the continental shelf into deep water.

The Grand Banks event is not unique. Cable breaks have occurred in many other locations and can be related to turbidity currents from the continental slope. The ocean floor off the continents is evidently not a quiet location devoid of currents. Because of their high speeds, the turbidity currents are able to erode deep canyons in the edges of the continental shelf all the way to the ocean floor.

The sediments carried into the deep ocean by the turbidity current will spread out, depending on the local topography. They will fill in the depression adjacent to the continents to form abyssal plains. All that is required is a supply of sediments from the shelf, a mechanism for putting the sediment in suspension, and a continuous path for the turbidity currents. The currents will be excluded from areas that are shielded from the edge of the continents by intervening ridges.

When the turbidity current reaches the abyssal plain, it slows down and gradually stops. As the velocity is reduced, the coarse particles settle out rapidly. Eventually, the finer components of the suspended load come to rest on the sea floor. Thus we find the coarse material at the bottom of each turbidity-current layer grading upward into finer silt and clay, as shown in Figure 17-6.

Other Sources of Deep-Sea Sediments

We have seen how turbidity currents are able to carry sediments from the continental shelf to the adjacent ocean floor. Farther away from the edge of the continents, the rate of sedimentation is very slow. However, we find that most areas of the ocean floor are covered by sediments. What is the origin of these sediments, and how did they get deposited on the ocean floor?

In high latitudes, ice derived from the land drifts out to sea. As it melts, it releases its load of land-derived mineral matter, which sinks to the ocean floor. Thus, surrounding the Antarctic continent, we find a zone containing large, ice-rafted rock grains. By taking long cores, we can see whether such ice-rafted material extended to lower latitudes during the glacial periods. Conally and Ewing (1965) have shown that, at the present time, near 52°W, ice-rafted material extends only to 65°S. A core from 42°S at this longitude contains three layers of coarse-grained material, indicating that some Antarctic ice reached this latitude before melting during the ice ages.

Coarse material, then, can be carried to the deep ocean by turbidity currents and ice. Fine material, if it is sufficiently fine, can be carried in suspension in the atmosphere or in seawater. Fine dust particles are picked up over land, particularly in desert regions (Fig. 14-6, p. 232).

Gradually, the particles settle out, or they may be washed out by rain. Once they enter the sea, they slowly sink to the bottom. At the same time, fine sediment particles introduced into the sea by rivers may be carried far from shore by ocean currents before they settle to the bottom. Large volcanic eruptions may introduce vast quantities of dust into the atmosphere, producing a layer of volcanic ash on the ocean floor downwind from its source.

So far we have considered marine sediments that were ultimately derived from the land. Sediments can also be produced by submarine volcanism or by precipitation from seawater. Many marine plants and animals have hard skeletons formed from silica and calcium carbonate that was originally dissolved in seawater. As the organisms die or are eaten, their skeletons sink to the sea floor. In their descent, the skeletons may redissolve in the water, or they may reach the sea floor and slowly accumulate there. On the margins of tropical islands, calcium carbonate is deposited rapidly by plants and animals, and the organic skeletal debris can be carried to the deep sea by slumping and turbidity currents. We shall discuss these organically derived sediments further in Part V, where we study the life in the sea.

Marine sediments such as phosphate minerals and manganese nodules can also be produced by direct chemical precipitation from sea-

Figure **17-8** Manganese nodules on the floor of the South Pacific. Location, 45°S, 145°W, depth 5009 m. Research vessel USNS Eltanin. (National Science Foundation photograph)

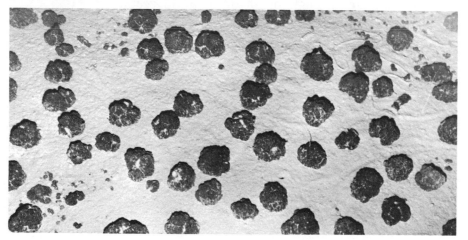

water. These chemical precipitates are found primarily where there is a lack of accumulation of terrigenous and biologic sediments (Fig. 17-8).

The rate of sediment accumulation varies widely. On the continental shelves, where rivers bring in large amounts of sediment, the rate of sediment accumulation locally was able to keep up with a sea-level rise of 1 cm yr^{-1}. Off the coast of Southern California, Emery (1960) estimated the sedimentation rate to vary from 40 to 5 cm per 1000 years. The rates of accumulation of the fine-grained deep-sea sediments range from a few centimeters to fractions of a centimeter per 1000 years.

Isolated Oceanic Peaks

The physiographic diagram of the equatorial Atlantic (Fig. 17-3) shows a number of peaks arranged along linear trends or in a close grouping. Some of these, such as the Cape Verde Islands, extend above the sea surface. When we examine high islands rising from the floor of the deep ocean, we find that they are volcanic in origin. The creation of new volcanic islands is still going on. A recent example is the birth of Surtsey, south of Iceland (Fig. 17-1, p. 272).

On November 13, 1963, a ship in search of herring noticed a local warming of the sea of about 2°C. During the early morning of November 14, a submarine eruption was observed 5 to 6 km WSW of the then-southernmost islands of the Westmann Islands group. The following night the island was born. Two days after the eruption, the island had a height of 40 m and a length of 500 m. By mid-December the island was circular, with a diameter of 1.1 km and a height of 145 m. After one year the island had an area of 2 km^2 and a maximum height of 174 m.

The state of Hawaii affords an example of a volcanic island chain, ranging from the large island of Hawaii to tiny Kure Island, 2500 km to the WNW. On Hawaii, the twin volcanic peaks of Mauna Kea and Mauna Loa rise 4.2 km above the sea. Since the adjacent sea floor has a depth of 5 km, these volcanic peaks, relative to the sea floor, are comparable in height to Mount Everest.

The main islands from Hawaii to Niihau are high and consist of volcanic rocks. The rest of the islands in the chain to Midway and Kure barely reach above sea level and are composed of limestone, which consists of the skeletons of plants, algae, and animals such as corals. When

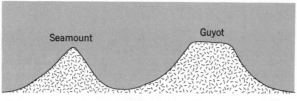

Figure **17-9**   Seamounts and guyots rising from the ocean floor ($5\times$ vertical exaggeration).

we drill into these islands, we find that the limestone is underlaid by volcanic rock. We shall discuss these coral atolls in Chapter 27. Since corals can grow only in shallow water, we know that the volcanic cores of these islands must have sunk. The coral and plant growth, however, produced sufficient limestone to keep the top of the mount above the sea.

Although many oceanic peaks reach above sea level to form islands, others do not. These submarine peaks can be categorized as seamounts or guyots (Fig. 17-9). *Seamounts* are isolated, pointed submarine volcanoes that do not reach the sea surface. During World War II, while on duty with the U. S. Navy, H. H. Hess (1946) of Princeton University charted many new submarine peaks in the Pacific, noting that many of these were flat. He named the flat peaks *guyots,* after Arnold Henry Guyot (1807–1884), a fellow professor of geology and physical geography at Princeton.

Because of their mode of formation, volcanic peaks are pointed. How can we explain the flat-topped guyots? When we dredge sediments from the top of one of these mounds, we find well-rounded volcanic cobbles and a typical reef flora and fauna. The dredgings show that the top of the guyot must once have been close to sea level. There, wave action eroded the peak of the island, producing rounded, coarse sediment. The waves leveled the volcano to slightly below sea level, and this platform became the habitat of corals.

The fossils dredged from the top of the guyot tell us when the volcano sank below about 100 m, the maximum depth at which reef-building corals can flourish. We find that the fossils from the guyots in the Pacific range in age from Middle Cretaceous to Recent. Thus we see that at least during the last 10^8 years, volcanoes have risen from the sea floor, poked their heads above the water, been leveled to slightly below sea level, and then subsided again.

Mid-Ocean Ridges

The most prominent feature of the physiographic diagram of the equatorial Atlantic (Fig. 17-3) is the mountain range running down the center of the ocean. As we head east, we leave the abyssal plain at a depth of about 5 km, and the terrain gradually rises, becoming hilly and then mountainous (Fig. 13-2, p. 205). Near the center, we find a series of high mountains that rise to a depth of 2 km or higher. Next, we encounter a deep rift valley whose floor is at about 3.5 km. The valley has a width of about 10 km. As we continue to the east, we ascend to about 2 km below sea level over mountainous terrain. Gradually the peaks get lower until we arrive at the abyssal plains bordering Africa.

The terrain is roughly symmetrical about the central rift valley. Looking north-south, we see that the rift valley does not run continuously down the center of the ocean. It is offset by a series of faults. One of these fracture zones near the equator has a deep depression, the Romanche Trench, which has a depth of 7865 m. If we look at the rest of the Atlantic, we find that the ridge, called the *Mid-Atlantic Ridge,* traverses the entire ocean, separating the deep waters of the Atlantic into an eastern and western basin. The Romanche fracture zone provides a communicating channel between these two basins.

If we study the source locations of earthquakes in the Atlantic, we find that they occur primarily along the axis of the Mid-Atlantic Ridge and along the faults that offset it. While the earthquakes associated with the oceanic trenches frequently occur at depths greater than 100 km, the earthquakes associated with the ridge occur at relatively shallow depths. The ridge, therefore, is an active part of the earth's surface where relative movement is taking place (Fig. 17-10).

If we examine the topography of the rest of the world ocean, we find that the Mid-Atlantic is a portion of a large ridge that circles the globe (Fig. 17-11). Leaving the South Atlantic, it turns east halfway between Africa and Antarctica. It splits off in the Indian Ocean, one branch heading into the Red Sea, the other heading east to the Pacific, halfway between Australia and Antarctica. The main part of the ridge continues east and then heads north in the eastern Pacific as the East Pacific Rise. It leaves the ocean at the northern end of the Gulf of California and moves offshore again north of Cape Mendocine, California.

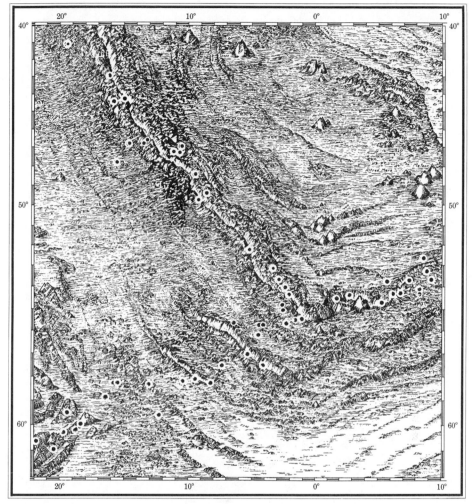

Figure **17-10**   Physiographic diagram showing earthquake epicenters. (Stover, 1968)

Branches of the ridge connect it with New Zealand and the tip of South America. In the North Atlantic, the ridge continues across Iceland and through the Arctic Ocean into Siberia. This ridge is the longest feature on earth. All along its course it is marked by frequent shallow earthquake activity.

The ocean floor also contains other mountain chains that are not seismically active at the present time. One of these is the Ninety East Ridge, which extends from 10°N to 30°S in the Indian Ocean at latitude 90°E,

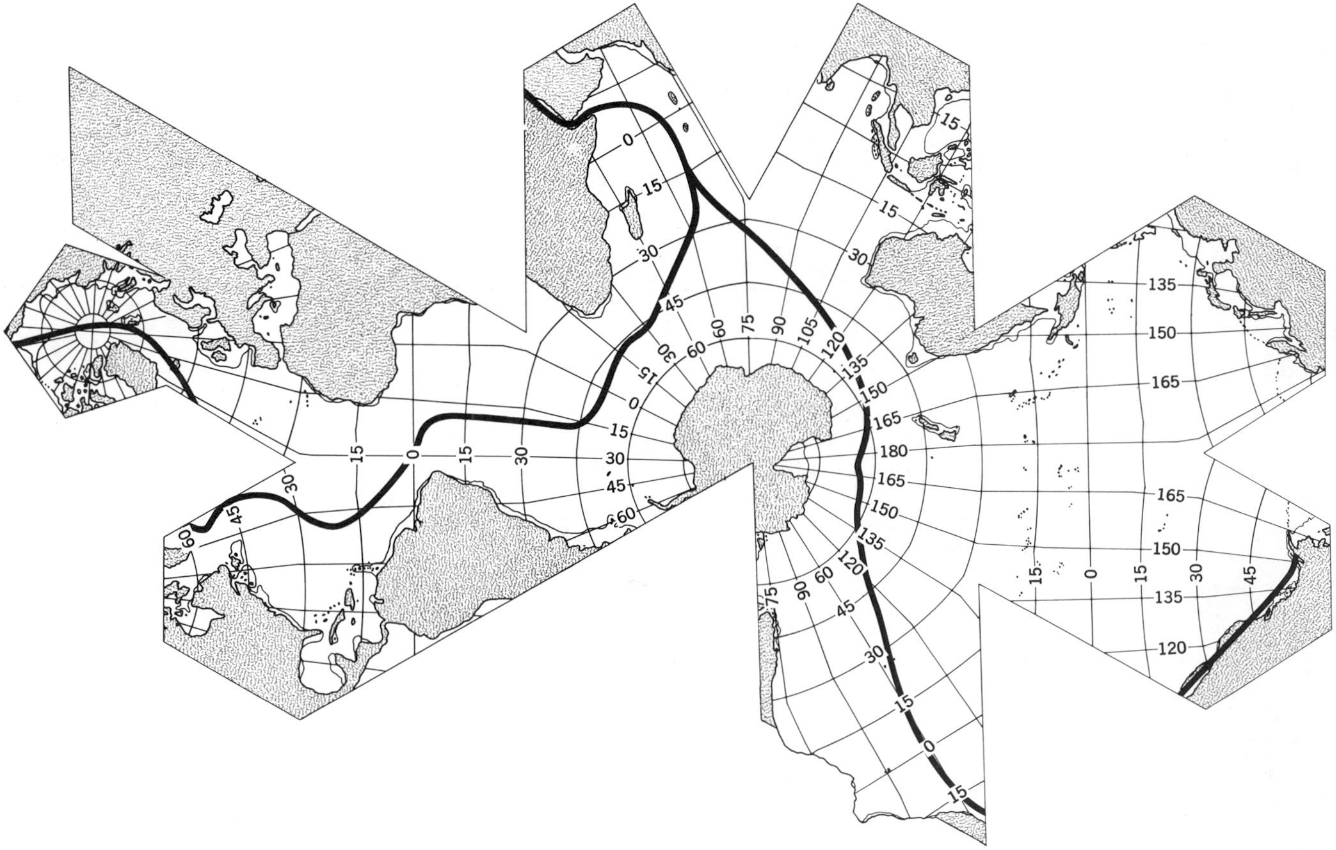

Figure **17-11** A schematic view of the mid-ocean ridges of the world ocean (Actually, the ridges are offset by numerous faults: for example, see Figure 17-3 for the equatorial region of the mid-Atlantic ridge.)

a distance of 4000 km. Thus we see that the ocean floor has a variety of active and inactive features.

Deep-Sea Drilling

So far our examination of the ocean floor has been largely limited to its surface. The seismic profiler gives us information about the layering of the submarine sediments. To determine the nature and geologic ages of these layers, samples are required. We have noted that corers are able to sample soft sediments to a depth of a few tens of meters at most. Where the bottom consists of hard rock, we can collect loose rock fragments or occasionally break off pieces of rock with a dredge. In a few locations, some of the deeper layers become exposed, and it is then possible to sample them. However, to study the history of the ocean basins in detail requires drilling into the bottom and extracting long cores of the sediments.

Long, deep cores of sediments are routinely obtained when exploratory oil wells are drilled on land or over shallow water. To drill such holes in deep water presents new problems. The drilling derrick must be mounted on a vessel (Fig. 17-12), and the vessel must be kept directly

Figure **17-12**   The *Glomar Challenger,* a deep sea drilling ship which has drilled cores in the deep ocean. (Global Marine Inc.)

over the hole that is being drilled. This cannot be accomplished by an-
choring the ship in deep water. Rather, the ship must be held in place
dynamically, with a number of propellers controlling its position. Preliminary tests have shown that drilling over deep water is possible and
that cores up to 300 m deep can be obtained.

An ambitious drilling project to sample the floors of the Atlantic and
Pacific Oceans began in the fall of 1968. The results of this project will
permit us to examine the floor of the ocean in depth for the first time.
By boring through layers of sediments that have slowly accumulated
over geologic time, we shall be able to study the history of the oceans.
The results are likely to revolutionize our understanding of the ocean.
Many parts of this text will undoubtedly require revision when the new
results become available. Keeping in mind our present state of ignorance
about the deeper layers beneath the sea floor, let us now look at an attempt to explain many of the features of the continents and the ocean
floor, the theory of continental drift.

Summary

The topography of the ocean floor is varied. Near the edge of the continents are abyssal plains, produced by the sediments carried to the
ocean floor by turbidity currents. Volcanic peaks are a common feature
of the ocean floor. Some extend above sea level as islands, and some have
flat tops. Deep-sea sediments, in addition to turbidity-current deposits,
consist of fine, terrigenous matter carried to the ocean basins by winds
and ocean currents. The mineral remains of plants and animals form
deposits on the ocean floor, and some deep-sea deposits are directly precipitated from seawater. The largest features of the ocean floor are the
mid-ocean ridges that bisect the Atlantic and extend through the rest
of the World Ocean.

Study Questions

1. Describe the topographic features of the ocean along the shortest path
   between South America and Africa. (Remember that because of the perspective of the rendering, the continental slope off the tip of Brazil is not
   shown in Figure 17-3.)
2. Where is the widest part of the South American continental shelf in Figure
   17-3, and why is the shelf especially wide in these locations?
3. Using tracing paper, make an overlay on Figure 17-3 indicating the prob-

able route by which turbidity currents have added sediments to the abyssal plains.
4. If the average sedimentation rate on the abyssal plain is 5 mm per 1000 years, how often would a turbidity current similar in magnitude to the Grand Banks earthquake have to be repeated?
5. The mid-ocean ridge system was first observed in the Atlantic, where its name is appropriate. Where in Figure 17-11 is the name inappropriate?

Supplementary Reading

(Starred items require little or no scientific background.)

* Dietz, R. S. (1952). "The Pacific Floor," *Scientific American,* April.
* Heezen, Bruce C., (1956). "The Origin of Submarine Canyons," *Scientific American,* August.
* ———— (1960). "The Rift in the Ocean Floor," *Scientific American,* October.
———— , Marie Tharp, and Maurice Ewing (1959). *The Floors of the Ocean,* Vol. I—The North Atlantic. Geological Society of America, Special Paper 65.
Hill, M. N., ed. (1962). *The Sea,* Vol. 3. New York: Interscience Publishers.
Kuenen, Ph. H. (1950) *Marine Geology.* New York: John Wiley and Sons.
* Menard, H. W. (1955). "Fractures in the Pacific Floor," *Scientific American,* July.
* ———— (1961). "The East Pacific Rise," *Scientific American,* December.
———— (1964). *Marine Geology of the Pacific.* New York: McGraw-Hill.
Shepard, Francis P. (1963). *Submarine Geology,* 2nd ed. New York: Harper and Row.

18 Continental Drift

The floor of the ocean has a varied and dynamic topography. Molten rock spews out of submarine fissures to build up volcanic cones. When the volcanic activity ceases, the mountains subside, leaving seamounts and guyots behind. Huge suboceanic mountain ranges circle the globe. Their central rift valleys are sources of earthquakes, and the fresh basalt that floors the rifts indicates recent solidification of new oceanic crust. Other active areas are the deep trenches where earthquakes indicate movement of the crust. These signs of relative motion suggest that the oceans of the world have not always had their present configuration. The geography of the land and the ocean basins is not static. Today we are merely looking at a single frame of the moving picture of the geography of our planet.

The Permanence of the Continents

All the continents contain rocks from the Precambrian Period. Cambrian sediments also are distributed over all the continents. Fossils and rocks from this period, 500 to 600 million years ago, show that there were land and shallow seas on the present continents during the Cambrian. Thus the continents have existed for as long as we have a geologic record. They have been permanent features for at least the last billion years of earth history. However, it does not necessarily follow that the location of the continents has been static.

Former Latitude and Longitude

To locate a position on earth, we measure its latitude and longitude. The latitude has physical significance, for it is measured relative to the earth's axis of rotation. The amount of sunlight received, hence the climate, varies with latitude, and the magnetic field of the earth is roughly aligned with its axis of rotation. Longitude, on the other hand, has only

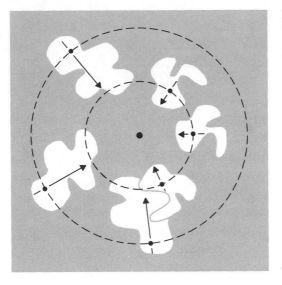

Figure **18-1** Knowing the latitude and the direction to the pole does not uniquely determine the location on earth.

relative meaning. It is arbitrarily defined with respect to a particular spot on the land, the Royal Observatory at Greenwich, England. Placed on a deserted island without any means of communication, we would be able to determine our approximate latitude, but we would have no way of determining longitude.

If we attempt to determine the past locations of the continents, we are in a similar predicament. We can deduce former latitudes, but there is no way by which we can determine the longitudes of the past (Fig. 18-1). To obtain ancient latitudes, we can look at the climatic evidence. Do the fossils suggest a warm or a cold climate? Do we find evidence of glaciation, suggesting high latitudes, or do we see evidence of tropical processes? When we examine the data, we must keep in mind that the climate at a given latitude has not necessarily remained the same. For example, during the last ice ages, the glaciers extended to lower latitudes and then retreated again.

Past Magnetic Directions

There is another way in which we can obtain information about former latitudes. We have pointed out that at the present time, the magnetic field of the earth is roughly aligned with its axis of rotation. This approximate alignment is not coincidental. All theories explaining the magnetic field of the earth involve its rotation and predict that the magnetic poles are aligned with the axis of rotation. If we could find compass needles that had been frozen into ancient rocks, we would be able to determine the location of the ancient pole relative to the rock.

Obviously, we cannot expect to find rusty compass needles that date back millions of years. The first mariner's compass, however, was not magnitized iron but lodestone, a natural magnetic rock consisting of the mineral magnetite. Small grains of magnetite are a common minor constituent of many sediments. When they are suspended in water, the magnetic grains will tend to align their poles with the magnetic field of the earth. As they settle out, they will partially preserve this alignment.

Once the grains are buried, they are no longer able to rotate. Thus the magnetic direction is "frozen" into the sedimentary rock formed by consolidating the sediment. Should the relative position of the earth's pole change after the sediment is formed, the magnetite grains will still point in the direction of the pole at the time they were deposited.

The ancient magnetic field can also be frozen into a rock in a different manner. At high temperatures, the mobility of the atoms prevents matter from being strongly magnetic (ferromagnetic). Some materials, such as iron and magnetite, become strongly magnetic when they cool below a particular temperature, the Curie temperature. This temperature depends on the material but is usually near 500°C. As heated rocks, such as lavas, cool through the Curie temperature, the magnetic minerals they contain become magnetized in the direction of the magnetic field of the earth. Once cooled, they retain this direction of magnetization regardless of later changes in the external magnetic field. In order to take on a new magnetic direction, they would first have to be heated again.

The magnetization locked into rocks either by cooling or by sedimentation is very weak. Using sensitive modern instruments, however, it is possible to read the ancient compass needles. Workers in various parts of the world have made magnetic measurements on many rocks from many locations and from different geological periods. If the continents and the earth's axis had remained fixed throughout time, the results of these measurements would be very uninteresting, for all the magnetic directions would point in the approximate direction of the present magnetic field.

The Paleomagnetic Data

What do the magnetic data indicate? If we look at the North American data, for example, we find that the location of the pole relative to the continent has changed with time. During any one interval of geologic time, the relative location of the pole is approximately the same. As we go back in time, however, there is a gradual movement of the North Pole from its present position to a location in the middle of the Pacific.

We could explain this change as a shift of the earth's axis of rotation relative to the crust. When we look at magnetic data from other continents, however, we find that the pole took a different path relative to these land areas. Thus not only did the pole shift relative to the continents, but the continents must also have moved relative to one another (Fig. 18-2). Because we cannot determine ancient longitudes, the magnetic data alone do not enable us to reconstruct the globe as it appeared during various periods of the geologic past. However, only certain configurations are consistent with the magnetic data. A static geography is specifically excluded.

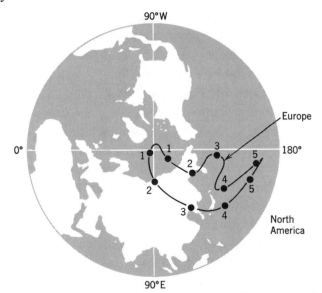

Figure **18-2** Ancient pole positions as determined from Europe and North America. Numbers are time before the present in 10^8 years. (After Deutsch, 1966)

The Jigsaw Puzzle

Rock magnetism suggests that the continents have moved about on the surface of the earth. Long before these magnetic data became available, some geologists suggested that the continents were mobile. In 1910, an American, F. B. Taylor, and, slightly later, a German, Alfred Wegener, advocated a theory of continental drift. In 1937, a South African, A. L. Du Toit, wrote a book entitled *Our Wandering Continents.*

These men started thinking about moving continents because of the similarity they observed between the west coast of Africa and the east coast of South America. If we cut these continents out of a globe, they fit together like pieces of a jigsaw puzzle (Fig. 18-3). The best fit is obtained if we make the cut along the continental slope and eliminate recent river deposits. If we cut the continents from a two-dimensional map, they do not fit very well because of the distortion of shape in making the map projection.

The fact that the coastlines fit together could, of course, be a mere coincidence. If South America and Africa had in fact been joined together at one time, we would expect not only that the coastlines would fit but also that the geologic patterns of the two continents would be similar. The more recent deposits, formed since the Atlantic opened, would be different on the two sides of the ocean. The geologic pattern formed before separation, however, should be continuous.

When we compare the geology of the two continents, we find striking similarities. If we compare the fossil plants and animals of Africa with those of South America, we find differences in fossils younger than Cretaceous times (about 120 million years ago). The fossils prior to Cretaceous times, however, are very similar, indicating free migration between Africa and South America.

Comparing different geologic strata on the two continents, we find a continuous pattern. If we slide the nose of South America into the bend of Africa, the boundaries between different deposits fit perfectly. Just in the bend, for example, there is a strip of 600-million-year-old rocks (indicated as 0.6 in Fig. 18-4) between billion-year-old ones. Although the fit of the coastlines could be a coincidence, the geologic similarities are too great to be the result of chance.

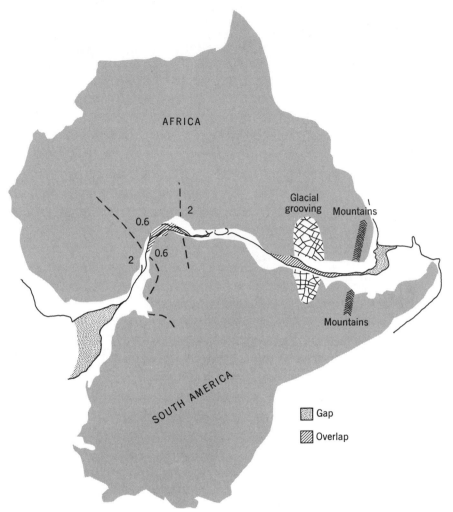

Figure **18-3** The fit of South America and Africa along the 1 km depth curve. (After Bullard et al., 1965) Numbers are ages of rocks in 10^9 years. (After Hurley et al., 1967)

The Floor of the Atlantic Ocean

We have seen that the evidence from the continents suggests that the Atlantic Ocean opened up about 120 million years ago. Now let us look at the evidence from the ocean floor. The most prominent feature of the Atlantic is the Mid-Atlantic Ridge, which bisects the ocean (Fig. 17-11). Could this ridge be the area along which the Atlantic has been opening up? If so, the center of the ridge should be geologically very young, and the floor of the ocean should get progressively older as we move away from the central ridge. If the Atlantic began to be formed in the early Cretaceous period then it should contain no marine sediments older than Cretaceous.

In fact, the age of the basalt of mid-ocean islands gradually increases with the distance from the mid-ocean ridges. In the central Atlantic there are no islands older than 120 million years. Marine sediments increase in thickness away from the ridge, although the rate of increase is not uniform. These data support the idea that the Atlantic Ocean originated about 120 million years ago by the creation of new ocean floor along the Mid-Atlantic Ridge.

Magnetic Growth Stripes

F. J. Vine and D. H. Matthews (1963) have suggested that information on the spreading of the mid-ocean ridges could be obtained from magnetic data. Although the axis of the earth's magnetic field is roughly aligned with the axis of rotation, the sense of the magnetic field periodically reverses. If we examine the magnetization of lavas that have solidified during the last four million years, we find that the pole location corresponds to the present geographic pole. Periodically, however, the magnetic polarity reverses so that the magnetic north pole of a compass becomes a south-pointing pole.

By measuring the magnetization and the time since the solidification of many lavas, Doell and Dalrymple (1966) have determined the recent magnetic history of the earth (Fig. 18-4). If rocks were solidified when the polarity of the earth's field was the same as it is today, their magneti-

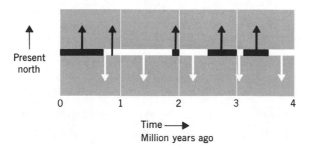

Figure **18-4**   Polarity history of the magnetic field of the earth during the last 4×10^6 years. (After Doell and Dalrymple, 1966)

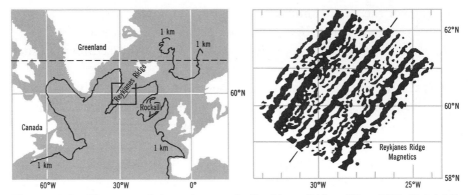

Figure **18-5**   The magnetic intensity over the Reykjanes Ridge. (Vine, 1966, © AAAS)

zation adds to the strength of the present field. If they cooled through the Curie point while the field of the earth was reversed, their magnetization reduces the present field intensity.

As we sail across a mid-ocean ridge, we can measure the earth's field by pulling a magnetometer, a sensitive instrument that measures the magnetic field, behind the ship. When we pass over rocks that have a reverse magnetization, we obtain a slightly lower value, while rocks that were magnetized in the same direction as the present field will give a slightly higher reading. A magnetic map of the Reykjanes Ridge, the portion of the Mid-Atlantic Ridge just south of Iceland, is shown in Figure 18-5. Areas with above-average field intensity are shown in black and areas with less than average field intensity are shown in white. The resulting magnetic map shows a series of stripes that are aligned with the axis of the ridge. Centered over the ridge is a broad stripe of high magnetic intensity, indicating the presence of rocks that have cooled through the Curie point during the last 700,000 years. As we move out from the axis, we find a symmetric arrangement of magnetic bands produced by the spreading and cooling of new basalt. The symmetrical magnetic stripes are like the growth rings on a tree, in that each was formed during a single magnetic polarity period. From the spacing of these bands and the history of pole reversals, we can determine the rate of spreading (Fig. 18-6). The present rate of spreading of the Atlantic, as determined from the magnetic data, ranges from 2 cm yr^{-1} for the Reykjanes Ridge to 3 cm yr^{-1} for the South Atlantic Ridge. Similar data from

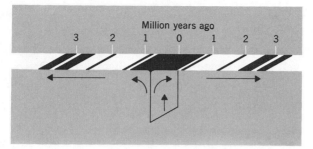

Figure **18-6**   The magnetic pattern produced by lateral spreading of new rock from the axis of a ridge. Numbers are 10^6 years before the present.

other oceans indicate that they are also spreading apart from their mid-ocean ridges.

Motion of the Blocks of the Earth's Surface

The magnetic and other evidence suggests that the floor of the ocean is spreading apart along the mid-ocean ridges. If this is so, then either the total surface area of the earth must be increasing, or else there must be a loss of surface somewhere to compensate for the creation of new ocean floor. If there is no such loss, the radius of the earth would have to be expanding at a rate of the order of 1 cm yr^{-1}, a fantastic rate which could not be maintained for any geologically significant length of time. We must therefore look for places where surface area is being destroyed.

Most of the surface of the earth is quiescent, showing no evidence of horizontal stresses. Earthquakes do not occur uniformly over the surface of the earth but are restricted to certain earthquake belts which include the mid-ocean ridges, the ocean trenches, and active mountain belts, such as the Himalayas (Isacks et al., 1968). Between these active belts, sediments and rocks show no evidence that they are currently being deformed. We are led to the conclusion, therefore, that the surface of the earth acts like a series of rigid plates or blocks which are deformed only at their boundaries.

Le Pichon(1968) after a careful examination of the currently active features of the land and the ocean floor has suggested that the earth now consists of six large blocks which are in motion relative to each other (Fig. 18-8). The six major blocks of the earth's surface are as follows:

The Pacific Block consists of most of the Pacific Ocean and includes small fragments of the west coast of North America.

The American Block consists of the continents of North and South America as well as the western half of the Atlantic Ocean. The Caribbean region forms a small subblock. There may be a slight differential motion between the northern and southern halves of the American Block.

The Eurasian Block is almost entirely continental, consisting of all of Europe and most of Asia, including the East Indies and the Philippines.

The African Block consists of the continent of Africa, Madagascar, the eastern half of the South Atlantic Ocean, and the western half of the Indian Ocean.

The Indian Block stretches from the Arabian peninsula to New Zealand. The block contains the portions of Asia south of the great mountain range from Turkey on the west to the Himalayas and the west coast of Burma on the east. The eastern half of the Indian Ocean, New Guinea, and Australia are included in this block. This long block may consist of eastern and western halves with a slight differential motion.

The Antarctic Block consists of the continent of Antarctica and the Southern Ocean. It includes a strip of the Pacific Ocean between the East Pacific Rise and the west coast of South America.

The differential motions of these six blocks lead to the formation of new ocean floor (indicated as + in Fig. 18-7) where the blocks are moving apart and to a loss of surface (designated as − in Fig. 18-7) where the blocks are forced against each other. The areas where new surface is created correspond to the mid-ocean ridges (compare Fig. 18-7 with Fig. 17-11, p. 286). By studying the magnetic stripes along these ridges, it is possible to reconstruct the recent history of addition of new ocean floor.

Where the blocks are driven into each other (the − areas), the surface expression is more complex. Some of these areas correspond to the ocean trenches (see Fig. 16-13, p. 267). Along other segments where the blocks are pressing against each other we find active mountain belts, such as the Himalayas. While these appear to be compressional features, trenches seem to be the result of tension. It is clear how compres-

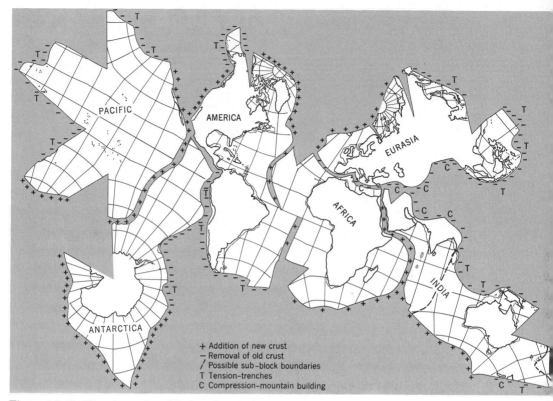

Figure **18-7**   The six surface blocks of earth. (After Le Pichon, 1968)

sion can result from the pressing together of two blocks, but it is difficult to understand how tension can result from the interference of moving blocks.

The Mechanics of Block Motion

It appears that the rigid surface of the earth, the *lithosphere,* which is
tens of kilometers thick, rests on a weaker layer, the *asthenosphere.* The
lithosphere is divided into large blocks which move relative to each
other. The boundaries between blocks are zones of weakness, where
earthquake activity reveals relative motion. The fact that the present
blocks have existed for tens of millions of years indicates that breaks
in the lithosphere tend to be preserved. However, over long periods,
these breaks heal and new breaks develop; in this manner the pattern
of blocks may change. Geologists have yet to reconstruct the history

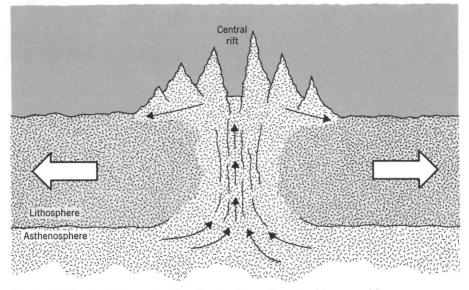

Figure **18-8** Possible mechanism for the formation of mid-ocean ridges.

of the geography and motion of the blocks of the lithosphere. Thus paleogeography not only has the task of studying past shorelines on the present continents but must also chart the dynamic history of the blocks of the lithosphere through geologic time. Such a reconstruction will reveal the physiography of the oceans of the past, their mid-ocean ridges and trenches, as well as ancient regions of mountain building.

Where the blocks of the lithosphere are pulled apart, fluid rock (fresh basalt) is injected into the crack from the asthenosphere, giving rise to the mid-ocean ridges (Fig. 18-8). As this material spreads laterally from the central ridge and cools through the Curie point, it gives rise to the magnetic stripes. The rock recently injected from below is warmer than the old crust at comparable depth. The resulting lower density leads to isostatic uplift and so produces a ridge. As the rock is pushed away from the center of the ridge by the injection of fresh rock, the displaced rock cools, its density increases, and it slowly subsides.

For simplicity, Figure 18-7 shows the rifts between the blocks as smooth curves. Actually the rifts are more irregular, consisting of a series of line segments, offset by normal tears. This can be seen clearly in the physiographic diagram of the equatorial Atlantic (Fig. 17-3). The

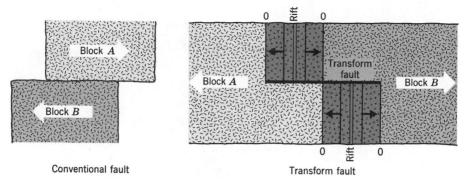

Figure **18-9** Contrast between a conventional fault between two blocks and a transform fault produced by an offset along a mid-ocean ridge. (After Wilson, 1965)

prominent offset at the equator contains the Romanche gap. The discontinuity in the rift gives rise to what J. T. Wilson (1965) has called *transform* faults. These faults are very different from conventional faults between moving blocks, as shown in Figure 18-9.

Where the moving blocks converge, surface area is lost. The mechanism of loss depends on the rate of convergence. At moderate rates, between 5 and 6 cm yr^{-1}, compression results in mountain building. Horizontal sedimentary layers are deformed and the lithosphere increases in thickness, resulting in isostatic uplift (Fig. 18-10).

More rapid convergence, between 6 and 9 cm yr^{-1}, is associated with the formation of ocean trenches (see Fig. 16-13, p. 267). The steeply dipping plane of earthquake centers (Fig. 16-14, p. 268) suggests that

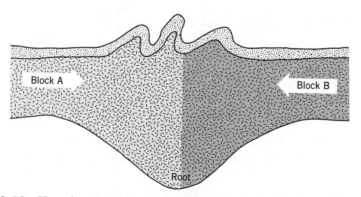

Figure **18-10** How the convergence between two blocks may lead to mountain building (not to scale).

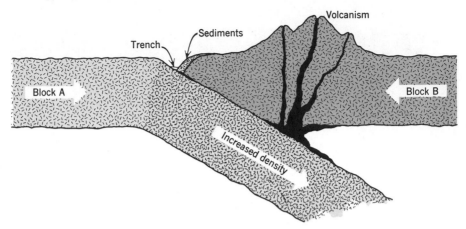

Figure **18-11**   How the convergence between two blocks may lead to the formation of a trench (not to scale). Compare with Figure 16-14, p. 268.

one block of the lithosphere is sliding below the other block. Le Pichon (1968) has proposed that the increased density in the subsiding block may suck it down, resulting in tension. The density is increased not only by compression but also by the loss of low-density volatile components which are extruded to the surface as volcanic lavas (Fig. 18-11). The depth to which a surface block can underride its adjacent block is limited. Therefore this mode of convergence must cease after about 1800 km of ocean floor have been lost. At this point, the nature of the motion between the blocks must change. Thus old blocks may be fused together and new breaks may occur.

The Driving Force for Continental Drift

Many of the surface features of the earth can be explained by assuming that the lithosphere consists of a number of blocks that move relative to one another. The existence of a more plastic asthenosphere under the relatively rigid lithosphere facilitates the motion. Nevertheless, frictional forces at the bottom of the blocks and at the points where the blocks are converging would tend to stop the drift in the absence of a driving force. The kinetic energy of the blocks would gradually decrease until the motion ceased.

Let us calculate the kinetic energy of the blocks per unit area. Assuming a depth of 50 km, a block weighs about 1.5×10^7 g cm^{-2}. A drift velocity of 3 cm yr^{-1}, or 10^{-7} cm sec^{-1}, gives a kinetic energy per unit area of

$$KE = \tfrac{1}{2} mv^2 = \frac{1.5 \times 10^7}{2} (10^{-7})^2 = 7.5 \times 10^{-8} \text{ erg cm}^{-2}$$

The forces to move the blocks about must be applied at the bottom of the lithosphere. The main force within the earth is thermal energy, which results from the radioactive decay of uranium, thorium, and an isotope of potassium. The outward flux of heat amounts to about 1.5×10^{-6} cal sec^{-1} cm^{-2}. This is very much less heat than is received from the sun, so that the heat flux from the interior does not directly affect the climate on earth. The radioactive heat flux, however, is 60 ergs cm^{-2} sec^{-1}, which is 10^9 times the kinetic energy of the lithosphere.

The heat from the interior can be transferred by conduction, radiation, and convection. Of these three modes of heat transfer, only convection is associated with motion and so could be the driving force for continental drift.

Convection currents result if a fluid is heated from below. The density difference between the heated fluid and the cool fluid results in vertical motion, since the heated fluid rises while the cooled fluid sinks. The vertical motions are not random but become spatially organized and result in a pattern of vertical convection cells (Fig. 18-12). The horizontal motions at the top of the asthenosphere exert forces on the blocks of the lithosphere. The direction of the net force will depend on the disposition of the blocks relative to the convection cells. The total force on the bottom of a block is the vector sum of the forces produced by the

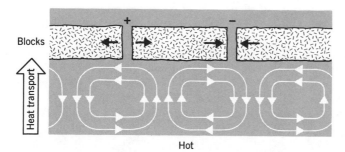

Figure **18-12**   The effect of convection cells on the blocks of the lithosphere.

individual convection cells. Since the geometry of the blocks depends on lines of weakness in the earth's surface, the pattern of the block motion does not necessarily correspond to the pattern of the convection currents.

Summary

The gross dynamic features of the surface of the solid earth thus appear to be the result of the motion of large blocks of the lithosphere. Earthquake activity is largely limited to the margins of these blocks while the interior of the blocks is quiescent. Where the blocks are pulling apart, new oceanic crust is formed, resulting in mid-ocean ridges. Where the blocks converge, areas of mountain building and oceanic trenches occur. The motion is driven by convection currents which result from heat transfer from the interior of the earth. Recent research has outlined the general pattern of this motion during the last ten million years or so. The details of the motions and the dynamic history of the lithosphere in the geologic past have yet to be determined.

Although ocean basins and continents have probably existed during most of geologic time, the geography of the earth is far from static. The Atlantic Ocean is a relatively recent feature, resulting from the splitting apart of a giant supercontinent some 120×10^6 years ago. The continents have drifted about, been torn apart by the formation of new oceans, and reassembled. Before we can understand the history of the oceans, hence the surface environment of the past, we must decifer the dynamic geography of the past.

Study Questions

1. How can latitudes of the past be determined?
2. What is the evidence suggesting that Africa and South America once formed a supercontinent?
3. By tracing over Figure 17-3, indicate the location of the Mid-Atlantic Rift, and show where the rift has been offset by transform faults.
4. What will be the consequences if the present pattern of block motion continues?
5. Why are the areas of convergence mainly continental while the areas of divergence are mostly oceanic?

Supplementary Reading

(Starred items require little or no scientific background.)

* Heirtzler, J. R. (1968). "Sea-Floor Spreading," *Scientific American,* December.
* Hurley, P. M. (1968). "The Confirmation of Continental Drift," *Scientific American,* April.
* Phillips, O. M. (1968). *"The Heart of the Earth."* San Francisco: Freeman, Cooper & Co.
* Takeuchi, H., S. Uyeda, and H. Kanamori (1967). *Debate about the Earth.* San Francisco: Freeman, Cooper & Co. (A historical account of the continental-drift controversy.)
* Wegener, A. (1966). *The Origin of Continents and Oceans.* New York: Dover Publications. (Reprint of a translation of Wegener's book on continental drift.)
Woodford, A. O. (1965). *Historical Geology.* Chapter 13. San Francisco: W. H. Freeman and Co. (Paleontological evidence for continental drift.)

Part IV
The Salt of the Sea

Water is an excellent solvent. By dissolving many substances, water makes them mobile and thus makes possible their utilization by life processes. Rainwater falling on the land partially dissolves rocks and carries the dissolved salts to the sea. The composition of seawater, however, differs significantly from that of river water.

The dissolved salts added to the sea by the rivers are converted back into mineral matter by chemical and biological processes. Carbon dioxide from the atmosphere is absorbed by rainwater, which dissolves mineral matter on land. The carbon dioxide is then added to the sea as river water. Some of the carbon dioxide in the river water is returned to the atmosphere, and the rest is precipitated as calcium carbonate. The carbon dioxide and other geochemical cycles stabilize the composition of seawater.

The earth was once dry and devoid of an atmosphere, and the fluids now on the earth's surface were originally chemically bound within the earth's mantle. The gases, including water vapor, gradually escaped from the mantle to accumulate at the surface, where they reacted with the solid earth to produce sedimentary rocks and seawater. The action of light on the primitive atmosphere gave rise to complex organic molecules from which life originated. The appearance of life altered the chemistry of the ocean and the atmosphere.

The waters of the ocean contain almost 5×10^{22} g of dissolved salts.

Common salt (NaCl) produced by solar evaporation from seawater in San Francisco Bay. Note salt piles in Background (Leslie Salt Co.).

19 Water, the Universal Solvent

The unique thermal properties of water stabilize the climate on earth. Its chemical properties make life as we know it possible. By dissolving matter essential for life, water mobilizes these substances, both outside and within organisms. Before studying the life within the sea, we must therefore consider the chemistry of water and of the saline solution that is seawater.

Solid, Liquid, Gas, and Solution

Pure substances can exist in either solid, liquid, or gaseous state. Thus water occurs as ice, liquid water, or water vapor. Ordinary salt, sodium chloride, is a solid at normal temperatures but melts if it is heated above 801°C. Liquid salt boils at a temperature of 1413°C. If we mix some solid salt with liquid water, the salt dissolves and we obtain a homogeneous solution. Although the original salt and water have different properties, the solution of salt and water has uniform properties that differ, as we shall see, from those of its components before combination.

Unlike a solution, a *mixture,* such as water and oil, does not have uniform properties throughout. The less dense oil will float on top of the water, and we can separate the two readily by pouring the oil off the top. If we subdivide a *solution,* on the other hand, every portion of it has exactly the same properties. To separate salt from water in a solution we must boil off the water to leave the salt behind and recondense the pure water.

A solution consists of two components. The *solvent* is the more abundant constituent, and the *solute* is the less abundant one. Seawater is a solution of various salts in over 96 percent water. Water is therefore the solvent. Although solvents are generally liquid, they may also be solid. An alloy, for example, is a solution of one metal in another.

The solute can be a solid, a liquid, or a gas. Thus we can dissolve liquid alcohol or solid salt in water. The gases of the atmosphere also dissolve in water. If we allow a glass of cold water to warm up, we see gas bubbles forming on the walls of the glass. They form because cold water can dissolve more air than warm water. As the water warms up, the excess air comes out of solution and forms gas bubbles.

The Solvent Power of Water

Of all liquids, water is the best solvent. Most substances are at least slightly soluble in water, and many dissolve to a considerable extent. To understand water's versatility as a solvent, we must look at the water molecule, which consists of one atom of oxygen and two atoms of hydrogen, H_2O.

The hydrogen atom consists of a nucleus having a single positive charge surrounded by one electron. The oxygen atom consists of a nucleus with eight positive charges, surrounded by eight electrons. A survey of the elements demonstrates that atoms with certain numbers of electrons are particularly stable. That is, they hold onto their electrons tightly and interact only weakly with other atoms. These are the noble, or inert gases: helium, neon, argon, krypton, xenon, and radon. Their atoms contain 2, 10, 18, 36, 54, and 86 electrons respectively.

Atoms that have one or two electrons more than a noble gas tend to give up their electrons readily when they interact with other atoms. On the other hand, atoms that have one or two electrons less than a noble gas have a strong attraction for electrons from atoms with which they combine. Oxygen, with its eight electrons, has two less than the noble gas neon. When it combines with two hydrogen atoms to form the water

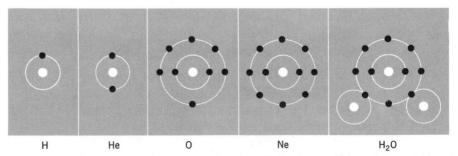

Figure **19-1**   The elements hydrogen (H), helium (He), oxygen (O), and neon (Ne) and the water molecule (H_2O).

Figure **19-2** The water molecule.

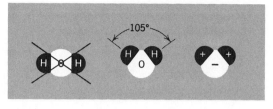

molecule, it holds onto their electrons. Therefore although the molecule as a whole is neutral, its oxygen atom has a net negative charge and the hydrogen atoms have a net positive charge (Fig. 19-1).

One might expect that the water molecule has maximum symmetry, consisting of the three atoms in a line with the oxygen atom in the middle. This is not the case. Rather, the two hydrogen atoms make an angle of 105° with the central oxygen atom (Fig. 19-2). The properties of water result from this shape. Because of the lack of symmetry, one side of the water molecule has a net positive charge while the other side has a net negative charge. Since there is separation of the electrical charges, the water molecule is an *electric dipole,* just as a magnet is a magnetic dipole because the north and south poles are separated from each other. Because of their dipole nature, water molecules interact strongly with each other. Therefore a relatively large amount of heat is required to separate the molecules and convert liquid water to water vapor.

The Conductivity of Seawater

What happens when salt dissolves in water? Sodium chloride consists of a regular array of sodium and chlorine atoms. Sodium contains 11 electrons, just one more than the noble gas neon. Chlorine has 17 electrons, one less than the noble gas argon. In the sodium chloride crystal, the sodium gives up one electron to the chlorine atom. As a result, the crystal consists of a checkerboard-like array of positively charged sodium atoms and negatively charged chlorine atoms. In the crystal, these charged atoms, called *ions,* are held in place by their strong electrical attractions.

When electrical charges move, an electrical current results. Metals contain some electrons that are only loosely held to the metal atoms. If an electric field is applied to the metal, these electrons move, producing a current. Metals are therefore *conductors*. The sodium and chloride ions

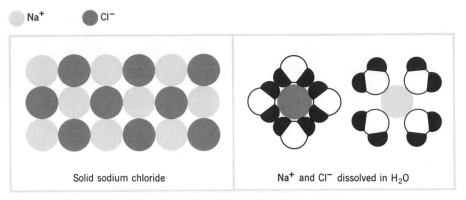

Figure **19-3**   Solid salt (sodium chloride) and dissolved salt.

in solid salt, however, are firmly held in the crystal and so are not able to move. As a result, solid sodium chloride is a good *insulator*.

When sodium chloride dissolves in water, the individual sodium and chloride ions are surrounded by water molecules. The positive sodium ion attracts the negative oxygen of the water molecules. The chloride ion, on the other hand, draws the positively charged hydrogen portion of the water molecules around it (Fig. 19-3). Thus the ions of salt are separated and surrounded by water molecules. If we introduce electrodes into the water, the positive sodium ions will be attracted to the negative electrode and the negative chlorine ions will be drawn to the positive electrode. The ions, surrounded by water molecules, are able to drift through the water to the electrodes, and so an electric current results.

We have said that the salt crystal is an insulator because the ions in the solid are unable to move. In pure water, motion is possible, but there are no ions. The electric field will tend to orient the dipoles of the water molecules, but since their net charge is zero, the molecules do not drift. In a salt solution, however, movement is possible and ions are present. Thus seawater, a solution of ionized salts in water, is a good conductor of electricity although both salt and pure water, taken singly, are insulators. As the temperature of water increases, the ions are able to move more freely through the water; thus the electrical conductivity of seawater increases with temperature.

The conductivity of seawater depends on the concentration of ions as well as on their mobility. The more ions there are in a unit volume, the greater the conductivity. The total salt content of seawater is known as the salinity. Thus the electrical conductivity of seawater varies with its salinity. By measuring the electrical conductivity with a salinometer (Fig. 8-5, p. 98), we can determine the salinity of seawater to better than 0.01‰. Since the conductance also varies with temperature, the salinometer must compensate for the effect of temperature. One type of salinometer is used in the shipboard laboratory to measure the salinity of water samples collected in Nansen or other sampling bottles. Another type of instrument is lowered on an electrical cable to measure the vertical variation of temperature and salinity directly within the water column.

Diffusion

When ionic salts dissolve in water, they dissociate into separate ions which are surrounded by water dipoles. Nonionic solutes are merely dispersed in the water so that the neutral molecules are surrounded by water molecules and are free to move through the water. Thus if we bring pure water, which has been purged of all gases, into contact with air, some of the molecules of the air dissolve in the surface layer of the water. These molecules then move about within the water. This process is known as *diffusion*. As the air molecules diffuse downward, more molecules of air are able to dissolve in the surface layer until the air has diffused through the entire water volume.

While an ion in an electric field moves in a fixed direction, the diffusive motion is random. We can simulate this motion by placing a piece

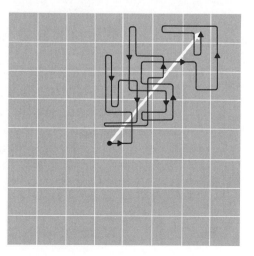

Figure **19-4** Random (diffusive) motion on a chessboard.

in the middle of a chessboard and then tossing two coins to decide whether to move the piece one place to the east, west, north, or south (Fig. 19-4). It can be shown that the average distance the piece has been displaced from its starting point is proportional to the square root of the number of individual moves. On the average, it will require four times the number of moves to move it twice as far. While diffusion over small distances is quite fast, it takes a very long time (i.e., a very large number of "moves") to disperse a solute over large distances.

Solubility

If we pour a lot of salt into a beaker of water, some of the salt will dissolve. Eventually dissolution stops, and the water is said to be saturated with salt. The situation is similar to the saturation of air with water vapor (Chap. 10). Sodium and chloride ions are continuously leaving the surface of the solid salt, while others are being deposited on the solid surface from the solution. When the rate of arrival of ions is equal to the rate of departure, no more dissolution takes place and the solution is saturated. If the rate of arrival on the solid is faster than the rate of departure, the solid grows and the concentration in the solution decreases.

The rate of the ions' departure from the solid depends only on the temperature and pressure. The rate of arrival from the solution depends on the temperature and on the concentration of solute in the solution. Thus, at a given temperature and pressure, the two rates will be equal when the solution has a particular concentration. At this point, the solution is saturated (Fig. 19-5).

If the solution contains dissolved molecules of the solid, the solubility is merely the concentration of dissolved molecules when the solution is saturated. In the case of a salt such as sodium chloride, however, the situation is more complex since the solute consists, not of sodium chloride molecules, but, rather, of sodium and chloride ions. If we produce the solution by dissolving salt in pure water, then the concentration of sodium ions is equal to the concentration of chloride ions, and we can call this the concentration of sodium chloride.

Seawater, however, is not a pure solution of sodium chloride but a complex mixture of several salts, including sodium and magnesium chloride as well as other salts. The concentration of chloride ion is therefore greater than that of sodium ion, and we cannot specify the concen-

Figure **19-5** The exchange of solute between a solid and a solution.

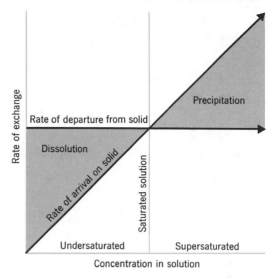

tration of sodium chloride in seawater. One kilogram of seawater contains 0.468 gram-molecular weight, or mole, of sodium ion and 0.5465 mole of chloride ion. What determines the solubility of a salt when the concentrations of its separate ions are not equal?

In order to precipitate sodium chloride, a sodium ion and a chloride ion must leave the solution and be deposited on the crystal. The probability in unit time that a sodium ion will leave the solution for the crystal is proportional to the sodium-ion concentration; the probability that a chloride ion will leave the solution in unit time is proportional to the concentration of chloride ions. The probability that both a sodium and a chloride ion will leave the solution in the same time interval is equal to the product of the individual probabilities.

When the salt crystal is in equilibrium with the solution, the rate of arrival of sodium and chloride equals the rate at which these ions leave the crystal surface. Thus, at equilibrium, the product of the concentrations of sodium and chloride ion must have a specified value, the *solubility product*. One kilogram of saturated sodium chloride solution contains about 265 g or 4.5 moles of sodium chloride. The solubility product of sodium chloride is therefore $(4.5$ moles $kg^{-1})^2$ or about 20 moles^2 kg^{-2}. The concentration product of NaCl in seawater is $0.468 \times 0.547 = 0.256$ mole^2 kg^{-2}. Before sodium chloride will precipitate from seawater, the concentration product must be increased by a factor of 78.

That is, salt will not form until the concentrations of sodium and chloride ions are each increased about ninefold per kilogram of seawater.

The Dissociation of Water

Pure water consists of water molecules. Since these molecules are electrically neutral, pure water should not conduct electricity. Regardless of how thoroughly it is purified, however, water remains slightly conductive, because some water molecules are always dissociating. The result is a positively charged hydrogen ion and a negatively charged hydroxide ion:

$$H_2O \longrightarrow H^+ + OH^-$$

A collision between a hydrogen ion and a hydroxide ion will produce a neutral water molecule. At equilibrium the rates of dissociation and combination are equal; therefore the product of the hydrogen- and hydroxide-ion concentrations in water has a specific value, the *dissociation constant* of water, which is 10^{-14} mole ^{2} kg^{-2}. In pure water, the concentrations of H^+ and OH^- are equal; each is 10^{-7} mole kg^{-1}.

The hydrogen and hydroxide ions are very important, for, in combination with negative and positive ions, respectively, they form acids or bases. Thus hydrochloric acid (HCl) consists of a hydrogen ion plus a chloride ion, while the base sodium hydroxide (NaOH) is a combination of a sodium and a hydroxide ion. When equal amounts of HCl and NaOH are mixed, we obtain a neutral sodium chloride solution, since the hydrogen ion combines with the hydroxide ion to form water.

$$NaOH + HCl \longrightarrow Na^+ + Cl^- + H_2O$$

The resulting NaCl solution will contain 10^{-7} mole kg^{-1} each of H^+ and OH^-.

If a solution contains more hydrogen ions than hydroxide ions it is acidic; an excess of hydroxide ions makes it basic. The product of the two ion concentrations, however, is always 10^{-14}. Thus the concentration of either hydrogen or hydroxide ions is a measure of the acidity of the solution. In order to designate the acidity of a solution, chemists have introduced the pH scale. The pH of a solution is the negative logarithm of the hydrogen-ion concentration. If the solution is neutral, the hydrogen-ion concentration is 10^{-7} and the pH $= -\log 10^{-7} = 7$. If the pH is less than 7, the hydrogen-ion concentration is more than 10^{-7} and the solution is acidic. A pH of 1 indicates that the solution contains 10^{-1} mole of hydrogen ions per kilogram. In such a solution, the concentration of hydroxide ion is 10^{-13} since the product of the two concentrations is always 10^{-14}.

Ions in Seawater

Seawater consists of a neutral mixture of ions in water. In a neutral mixture, the total charge of the negative ions equals that of the positive ions. Although the ions of sodium, chloride, and many other elements are singly charged, other ions, such as those of calcium (Ca^{++}) and magnesium (Mg^{++}), are doubly charged. To keep track of the total charge of the ions, it is convenient to express concentrations in equivalents. An *equivalent* is the amount of an ion required to equal the charge of a gram-molecular weight of a singly charged ion. For a singly charged ion such as Na^+, the number of equivalents is equal to the number of moles; for a doubly charged ion such as Ca^{++}, one mole is equal to two equivalents. In a given amount of seawater, the sum of the equivalents of negative ions must equal the sum of the equivalents of positive ions, so that the solution is electrically neutral.

Most of the ions in seawater do not interact with hydrogen ions. If we add sodium chloride to pure water, then, the concentrations of hydrogen and hydroxide ion are not altered. There are other ions, however, that do interact with the hydrogen ion and in this manner alter the acidity (pH) of the water. The most important of these substances is carbon dioxide, CO_2. This gas, which is a link between life and the inorganic world, plays a dominant role in regulating the acidity of the ocean. To understand the chemistry of the sea, we must pay special attention to carbon dioxide and the products of its interaction with water.

The Carbon Dioxide System

When carbon dioxide dissolves in water, it combines with a water molecule to form a molecule of hydrated carbon dioxide.

$$CO_2 + H_2O \longrightarrow H_2CO_3$$

The carbon atom contains six electrons. In the CO_2 molecule, the carbon atom shares two electrons with each of the oxygen atoms, thus giving

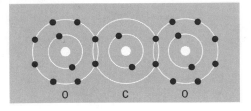

Figure **19-6** The carbon dioxide (CO_2) molecule.

them a neonlike electron configuration (Fig. 19-6). When the CO_2 molecule combines with a molecule of water, the oxygen from the water molecule also shares two electrons with the carbon atom (Fig. 19-7). As a result, the positive nuclei of the hydrogen from the water are only weakly held to the H_2CO_3 molecule. If one H^+ ion is lost, the H_2CO_3 molecule is transformed into the negatively charged *bicarbonate* (HCO_3^-) ion.

$$H_2CO_3 \longrightarrow HCO_3^- + H^+$$

The bicarbonate ion can lose another H^+ ion to become the doubly charged negative carbonate (CO_3^{--}) ion.

$$HCO_3^- \longrightarrow CO_3^{--} + H^+$$

As bicarbonate and carbonate ions form from the dissolved CO_2, they will attract hydrogen ions from the solution to reduce their charge. There is a constant exchange of H^+ ions between the carbonate species and the solution. The relative concentrations of H_2CO_3, HCO_3^-, and CO_3^{--} depend on the H^+ concentration in the solution. At low pH, a high con-

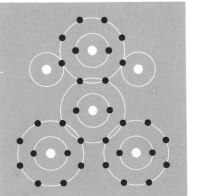
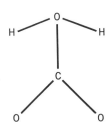

Figure **19-7** The H_2CO_3 molecule.

Figure **19-8** The relative abundance of H_2CO_3, HCO_3^-, and CO_3^{--} in seawater as a function of the pH.

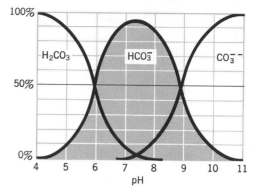

centration of H^+ will result in a high relative concentration of H_2CO_3; at high pH, CO_3^{--} will be the abundant carbonate species. Bicarbonate will dominate at an intermediate hydrogen-ion concentration. The relative concentrations of the three ions in seawater as a function of the pH are shown in Figure 19-8.

Summary

In this chapter we have briefly investigated the chemical properties of water. Because of its strong dipole nature, water is an excellent solvent. Salts become dissociated when they dissolve in water. As a result, salt solutions are good conductors of electricity. Pure water also dissociates slightly, into hydrogen ions and hydroxide ions. The acidity of a solution depends on its hydrogen-ion concentration, and some substances, when they dissolve in water, interact with the hydrogen ion. Carbon dioxide (CO_2) is of particular importance in nature. When it dissolves in water, it gives rise to H_2CO_3, HCO_3^-, and CO_3^{--}. The relative amounts of the three species of dissolved CO_2 depend on the hydrogen-ion concentration.

Study Questions

1. List four examples each of mixtures and solutions.
2. Although sea ice is a poor conductor of electricity, molten salts are good conductors. Explain.
3. In normal seawater there is about ten times as much CO_3^{--} as there is H_2CO_3. What is the approximate pH of seawater and the approximate ratio of HCO_3^- to CO_3^{--}?
4. Carbonated beverages contain high concentrations of H_2CO_3. Are they basic or acidic?

Supplementary Reading

(Starred item requires little or no scientific background.)

GENERAL REFERENCE ON CHEMICAL ASPECTS OF OCEANOGRAPHY

Riley, J. P., and G. Skirrow, eds. (1965). *Chemical Oceanography.* New York: Academic Press.

PROPERTIES OF WATER

*Adams, K. S., and J. A. Day (1961). *Water, the Mirror of Science.* Garden City, N. Y.: Anchor Science Study Series 18, Doubleday and Co.

20 River Water and Seawater

Every second, the rivers of the world add 10^9 kg of fresh water to the ocean. The ancients wondered why this constant flow did not cause the sea to run over or to decrease in saltiness with time. The reason is that the water of the rivers is only one part of the hydrologic cycle: water is evaporated from the sea and then is recondensed in the atmosphere; it falls in part on the land, is gathered in rivers, and returns to the sea. As the rainwater seeps through and over the ground, it dissolves some of the mineral matter and carries it to the sea. Every year, the rivers of the continental United States contribute 3×10^{11} kg of dissolved salts to the ocean (see Table 14-1).

The ocean is not diluted by river water. On the contrary, we would expect its salinity to be increased by the salts leached from the land. Thus we might logically conclude that seawater is concentrated river water. If this were so, the age of the ocean could then be estimated by determining how long it had taken the rivers to deliver the salts now dissolved in the sea. To verify this supposition, let us examine the composition of sea and river water.

The Composition of Sea and River Water

After watching the French Chef on television, Professor K. S. Deffeyes cooked up the following recipe for seawater:

> 20 quarts water
> $20\frac{1}{4}$ cups salt
> 1 teaspoon baking soda
> 10 drops beef broth
> $\frac{1}{2}$ cup epsom salt
> 1 pinch borax
> 1 tablespoon calcium chloride
> 2 teaspoons potassium chloride

Dissolve calcium chloride in 10 quarts of the water. Dissolve the remaining ingredients in the other 10 quarts, then stir in the dissolved calcium chloride. Store in a cool, dark place. Makes 5 gallons. Excellent for boiling crabs.

For river water we require no recipe. We merely draw the water from the nearest stream. To make a quantitative comparison between seawater and river water, however, we must determine the concentrations of the major constituents in the two solutions. Since the various salts in water exist as ions, it is best to list the concentrations of these ions in the water. For the solutions to be electrically neutral, the sum of the concentrations of positive ions (cations) in equivalents must equal the sum of the concentrations of negative ions (anions). In Table 20-1, the concentrations of the major ions in sea and average river water are given in milliequivalents per kilogram. A milliequivalent (meq) is 10^{-3} of an equivalent.

The composition of seawater in Table 20-1 corresponds to a salinity of 35‰. In Chapter 14 (Table 14-1, p. 221) we saw that the concentration of dissolved solids varies widely from river to river. While the Colorado River, which drains the desert regions of the Southwest, contains 760 ppm of dissolved solids, the Columbia River contains only 120 ppm. The average river-water composition listed in Table 20-1 has been taken from a recent summary by Livingston (1963) of data on the rivers of the world.

TABLE **20-1** COMPOSITION OF RIVER AND SEAWATER
Concentration in milliequivalents per kilogram

	Average River Water	35 ‰ Salinity Seawater	Cyclic Salts	River Water − Cyclic Salts
Positive ions (cations)				
Sodium Na$^+$	0.27	468.	0.19	0.08
Potassium K$^+$	0.06	10.	0.00	0.06
Magnesium Mg^{++}	0.34	107.	0.04	0.30
Calcium Ca^{++}	0.75	20.	0.01	0.74
Total cations	1.42	605.	0.24	1.18
Negative ions (anions)				
Chloride Cl$^-$	0.22	546.5	0.22	0.00
Bicarbonate HCO$_3^-$	0.96	2.3	0.00	0.96
Sulphate SO$_4^{--}$	0.24	56.2	0.02	0.22
Total anions	1.42	605.0	0.24	1.18

Figure **20-1** Sea spray produced by breaking wave. (Photographs: B. J. Nixon)

The table shows that seawater is not just concentrated river water. River water is mainly a solution of calcium bicarbonate, while the main salt in seawater is ordinary salt, sodium chloride. As the salts from river water accumulate in the sea, their composition must be modified in order to result in the composition of seawater.

We have assumed that the water that evaporates from the sea and falls on the land is pure H_2O. Actually this is not so. The sea produces much spray (Fig. 20-1). Some of the fine water droplets in the spray dry out, and the tiny crystals of sea salt they contain are then carried into the atmosphere. Because of their small size, these salt crystals are transported by winds and carried over the land where the salt leads to more rapid corrosion of metals near the ocean. The tiny salt crystals cause saturated water vapor to condense around them and so nucleate the formation of cloud droplets. When the droplets coagulate into rain, the sea salt is carried with the rain into the rivers, to be returned once more to the sea. This process results in a recycling of sea salt. The salts so derived are therefore called the *cyclic salts.*

It is difficult to determine the average contribution of cyclic salts to river water. Obviously it is high downwind from the seacoast and decreases inland. Since rocks that are weathered generally contain very little chloride ion, we can obtain a maximum estimate for the cyclic salts by assuming that all the chloride ions in river water come from cyclic salts. The other ions comprising the cyclic salts, then, will have the same ratio to chloride as they have in seawater. Thus we obtain the estimate for the cyclic-salt contribution listed as column three in Table 14-1. Subtraction of the cyclic salt from the river water gives the contribution to the sea of the salts derived from rock weathering. This is shown as the last column in Table 14-1.

Residence Times

The oceans of the world contain 1.4×10^{21} kg of seawater, and the rivers add water at a rate of 10^9 kg sec^{-1}. Thus it takes 1.4×10^{12} sec, or about 4.4×10^4 yr, for the rivers to add to the ocean a mass of water equal to the content of the oceans. This is called the *residence time* of water in the ocean, the time that would be required to double the water of the sea if the rivers continued to flow at their present rate and if there were no loss of water from the ocean.

If we consider only the total content of dissolved salts, we can calculate a residence time for the salt of the sea. The average river water minus the cyclic salts contains about 1.2 meq kg^{-1} of dissolved salts while seawater contains about 600 meq kg^{-1}. Thus the sea has 500 times the concentration of ionic charge in river water. The residence time of total salt is therefore $500 \times 4.4 \times 10^4$ yr, or 2.2×10^7 yr.

Similar calculations for the major constituents of seawater give the results shown in Table 20-2.

Since we assume that all the chloride ion in river water is derived from the cyclic salts, we obtain an infinite residence time for that ion. The residence times of the other ions range from 1.1×10^5 yr for bicarbonate to 2.6×10^8 yr for sodium. Since the geologic evidence indicates that we have had an ocean for more than 10^9 yr, all the residence times, ex-

TABLE **20-2**   RESIDENCE TIMES FOR THE MAJOR CONSTITUENTS
OF SEAWATER

Constituent	Ratio Seawater/River Water — Cyclic Salts	Residence Time 10^6 years
Water	.965	0.044
Total salt	500	22
HCO_3^-	2.4	0.11
Ca^{++}	27	1.2
K^+	170	7.5
SO_4^{--}	250	11
Mg^{++}	360	16
Na^+	5900	260
Cl^-	Infinite	Infinite

cept possibly that for chloride, are short compared to the age of the
ocean. Therefore, the ocean is not just accumulating the salts delivered
by the rivers. Instead, there must be processes by which the ocean can
get rid of the added salts.

Residence Time versus Mixing Time

While the rivers are adding new salts to the ocean, the waters of the
ocean are in constant motion. The surface waters are moving in wind-
driven gyres. In the Antarctic region, dense water is formed and sinks
into the abyss; elsewhere an equal amount of water must be returned
to the surface. By producing circulation, the atmosphere stirs the ocean
and so tends to even out regional differences in the composition of the
sea.

Let us follow a water molecule while it moves through the ocean. At
the surface, where the wind-driven currents are relatively fast, the molecule will circle one of the gyres in a few years. In the deep ocean, the
circulation is much slower. We can get an idea of the rate of motion of
the deep water by studying the concentration of radioactive carbon in it.

Cosmic rays interact with the nitrogen in the atmosphere to produce
radioactive carbon atoms, carbon-14. The radioactive carbon produced
mixes with the carbon in the atmosphere in the form of carbon dioxide.
It decays back to stable nitrogen-14 with a half-life of 5600 yr. That is,
in 5600 yr exactly half of the atoms of carbon-14 decay. To reduce the
radioactivity by half again, to one-quarter of its original value, requires

2×5600, or 10,200 yr. The carbon-14 in the atmosphere has a fixed concentration at which the rate of production of carbon-14 by cosmic rays is equal to the rate of its decay.

In the form of CO_2, the carbon-14 dissolves in the surface water of the ocean and is carried with the water into the deep. Once the water is removed from contact with the atmosphere, no newly produced carbon-14 can be added. With time, the radiocarbon concentration decreases until the water returns to the surface, where it can once more exchange carbon dioxide with the atmosphere. We can therefore use the decrease in the concentration of C^{14} to determine how long ocean water has been out of contact with the atmosphere. By this method we find that the "age" of deep waters ranges from several hundred to a few thousand years. Thus a water molecule spends on the order of 10^3 yr in the deep before it is returned to the surface. Therefore the mixing time of the ocean is of the order of 10^3 yr.

If constituents are added to the ocean at rates that are slow compared to the mixing time, they will become evenly distributed throughout the ocean. If the rate of addition is fast compared to the mixing rate, one would expect to find significant variations from place to place. All the major constituents of seawater listed in Table 20-1 have residence times that are very long compared to the mixing time and therefore should be evenly distributed throughout the ocean. Although the exchange of water vapor between the ocean and the atmosphere produces variations in salinity in the surface water, evaporation does not affect the relative concentrations of the major ions in seawater. Thus the physical properties of seawater depend only on the salinity. The concentration of, for example, Na^+ ion can be accurately determined from the salinity without requiring a chemical analysis.

Example: What is the sodium concentration of seawater with a salinity of 34.0‰?

For $S = 35$‰ $Na^+ = 468$ meq kg^{-1} (see Table 20-1); therefore, for $S = 34$‰ $Na^+ = \dfrac{468 \times 34}{35} = 455$ meq kg^{-1}.

Biological activity in the sea slightly affects the concentration of bicarbonate and calcium ions. As a result, these constituents tend to be somewhat less concentrated in surface water than in deep water. These changes are small, however, and have an insignificant effect on the physical properties of seawater. Unlike the concentration of sodium ion, the concentration of carbon dioxide in its various forms cannot be calculated accurately from a knowledge of the salinity but must be determined by chemical analysis. Fortunately, the variations in carbon dioxide (discussed in more detail in the next chapter) do not significantly affect the physical properties of seawater.

Other Constituents of Seawater

Of the 92 elements found in nature, all but 20 have been identified in seawater and their concentrations measured. Only 14 elements have concentrations larger than one part per million. In addition to hydrogen, oxygen, and the ions listed in Table 20-1 (p. 324), these are bromine, strontium, boron, silicon, and fluorine. The lowest measured concentration, 6×10^{-19} g liter^{-1}, is for the radioactive noble gas radon. One liter of seawater, therefore, contains only 1600 atoms of this gas. To measure the concentrations of the minor constituents in seawater requires large water samples (Fig. 20-2) and very sensitive chemical techniques. The

Figure **20-2** Large water sampler being lowered on hydrographic cable. (Photograph courtesy of Woods Hole Oceanographic Institution)

concentration of radon can be measured only because of its rapid radio-active decay.

Many of the minor constituents have short residence times compared to the mixing time of the ocean. As a result, their concentration in the ocean varies from place to place. They are removed from seawater by being absorbed by organisms in the sea or by being precipitated on sediments that settle out on the ocean floor.

Of particular interest are the so-called nutrients. These are substances that are essential for the growth of plants in the sea. We shall discuss their distribution in Chapter 26, when we consider the ecology of the ocean.

The gases of the atmosphere are also slightly soluble in seawater. The concentration of dissolved nitrogen and oxygen in seawater, saturated with air at 1 atm, is shown in Figure 20-3 as a function of temperature. At $0°C$, 1 liter of seawater holds 0.66 millimole of dissolved nitrogen and 0.36 millimole of dissolved oxygen. By contrast, 1 liter of air at a pressure of 1 atm contains 34.82 and 9.37 millimoles of nitrogen and oxygen, respectively. Note that while the ratio N_2/O_2 in the atmosphere is 3.7, the ratio of the dissolved gases is only 1.8. Thus, mole per mole, oxygen is approximately twice as soluble in seawater as nitrogen.

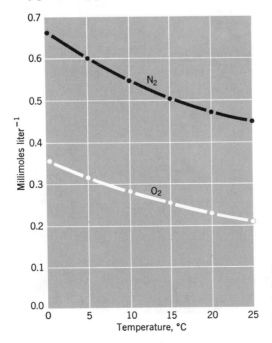

Figure **20-3** The concentration of nitrogen and oxygen dissolved in seawater, having a salinity of 35‰, in equilibrium with air at 1 atm as a function of temperature.

The solubility of the atmospheric gases decreases significantly as the temperature increases. Surface water, which is in contact with the atmosphere, is usually saturated with the atmospheric gases. Once the water sinks below the surface, however, it is no longer able to exchange gases with the atmosphere. Nitrogen gas is largely inert, and therefore its concentration remains constant. Oxygen, on the other hand, is used by animals in the sea for respiration. As a result, the longer the water remains in the deep, the more its dissolved oxygen content becomes depleted. Dissolved oxygen is therefore useful as a tracer for the motion of deep water, for with time in the deep, the oxygen concentration can only decrease.

To get a feeling for how the composition of seawater is stabilized over geologic time, let us now turn to a consideration of how the bicarbonate content of seawater is controlled.

Summary

A comparison of the composition of river water and seawater shows that seawater is not just concentrated river water. The residence time of the major ions in the ocean vary from about 10^5 yr for bicarbonate to over 10^8 yr for sodium. However, these times are all long compared to the mixing time in the ocean. As a result, while the salinity in the ocean varies, the relative concentration of the major salts is effectively constant. Many minor constituents in seawater have very short residence times; therefore their concentration may vary from place to place.

Study Questions

1. How do the following concentration ratios differ for river water and seawater?
   a. Ca^{++}/Mg^{++}
   b. Na^+/K^+
   c. Cl^-/SO_4^{--}
2. Calculate the residence times for Cl^- and Na^+ without making a correction for cyclic salts.
3. Give the concentration of the major constituents in seawater for a salinity of 30‰.
4. A sample of subsurface seawater contains 0.6 millimole liter^{-1} of dissolved nitrogen and 0.1 millimole liter^{-1} of dissolved oxygen. Assuming that no nitrogen has been lost, at what temperature was this water equilibrated at the sea surface, and how many moles of oxygen have been lost per liter by respiration?

21 The Carbonate Cycle

Of the major ions in seawater, bicarbonate has the shortest residence time. Thus the stabilization of the bicarbonate content of the sea poses the most immediate problem. Carbon dioxide plays an important role in the heat balance of the atmosphere and in the weathering of rocks. It is the carbon source for plants and is produced by the oxidation of organic matter. A study of how this material is transferred from land to sea to the atmosphere and through living organisms also sheds light on other geochemical cycles. Most components of seawater undergo similar cycles. Rather than examine a number of cycles superficially, let us concentrate on the geochemical cycle of carbon dioxide.

Inventory of Carbon Dioxide

Carbon dioxide exists in different forms in different parts of the surface environment. The atmosphere contains 0.03 percent CO_2. Since carbon dioxide strongly absorbs infrared radiation, it affects the heat balance of the earth. Carbonate ion, CO_3^{--}, when combined with ions of the alkaline earths, primarily calcium, Ca^{++}, and magnesium, Mg^{++}, forms large deposits of carbonate rock, the limestones and dolomites. These exist as both loose sediments and consolidated rocks. When they have been deeply buried, they are transformed into marble. Carbon dioxide is the carbon source for all plants, and when the plants are consumed by animals, most of the organic carbon is respired as carbon dioxide. Some of the organic carbon becomes buried in sediments. If the organic content is sufficiently high, the fossilized organic matter can be extracted to yield the fossil fuels: coal, petroleum, and natural gas. When carbon dioxide dissolves in water, it is hydrated to H_2CO_3 and by dissociation gives rise to the bicarbonate ion, HCO_3^-, and the carbonate ion, CO_3^{--}. An estimate of the quantities of the various forms of carbon that exist near the earth's surface is given in Table 21-1.

TABLE **21-1**   INVENTORY OF CARBON NEAR THE EARTH'S SURFACE*

Living organic matter: marine	8×10^{14} g moles
nonmarine	7×10^{14}
Atmosphere	5.4×10^{16}
Ocean: 0–100 m	7.4×10^{16}
below 100 m	3.3×10^{18}
Carbonate rocks	1.5×10^{21}
Organic matter in sediments	6×10^{20}
Fossil fuels	6×10^{17}

* After Rubey, (1951).

Our task here is to study the carbonate system in the ocean and to see how the CO_2 content of the ocean is altered by interaction with the atmosphere, river water, the lithosphere, and the biological processes within the sea. The problem is complicated by the fact that CO_2 exists in seawater as three species, H_2CO_3, HCO_3^-, and CO_3^{--}. There is a ready exchange among these three forms, and their relative amounts depend only on the acidity of the water (Fig. 19-8, p. 321). The problem can be greatly simplified if we employ a diagram proposed by K. S. Deffeyes (1965), which enables us to discuss the system in terms of two new variables, the concentration of total CO_2 and the alkalinity.

The Deffeyes Diagram

The concentration of total CO_2 is the sum of the concentrations of H_2CO_3, HCO_3^-, and CO_3^{--} expressed as millimoles per kilogram. This concentration can be altered only if we add or subtract carbonate in any of its forms to the water.

In Chapter 19, we saw that there are two kinds of ions in water: those like Cl^- that do not interact with H^+ ion, and those like HCO_3^- that do interact with H^+ ion. The *alkalinity* is a measure of the concentration of the ions that interact with H^+. It is the net negative charge of all the ions that interact with hydrogen ion or are H^+. The alkalinity is expressed in milliequivalents per kilogram. We must sum the milliequivalents (meq) of all the ions like HCO_3^-, CO_3^{--}, and OH^- and subtract from them the concentration in meq of H^+ per kilogram of solution. Thus pure water has zero alkalinity, since the concentration of negative OH^- is equal to the concentration of positive H^+; both are 10^{-4} meq kg^{-1}.

In seawater, the alkalinity is due primarily to the presence of HCO_3^-.

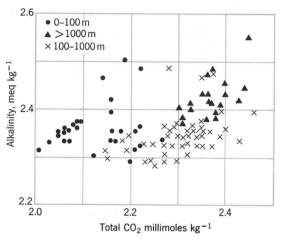

Figure **21-1** Alkalinity-total CO_2 (Deffeyes) diagram for seawater off the west coast of South America (R. V. *Eltanin* cruise III).

In addition, there are small amounts of CO_3^{--} and of various forms of boric acid (H_3BO_3, $H_2BO_3^-$, HBO_3^{--}, BO_3^{---}). Note that while the total CO_2 content is given in millimoles, the alkalinity is expressed in milliequivalents. The Deffeyes diagram consists of a plot of the concentration of total CO_2 versus the alkalinity. A particular seawater will be represented by a point on this diagram. The location of different waters on the Deffeyes diagram is shown in Figure 21-1. The data are for water from various depths in the Pacific Ocean off the west coast of South America.

We see from the figure that the total CO_2 content of the water increases with depth. The alkalinity is also somewhat higher below 1 km. To understand these changes, we must investigate how various processes alter the composition of seawater.

Consider seawater whose alkalinity and total CO_2 content are represented by point A on the Deffeyes diagram (Fig. 21-2). Now we add carbon dioxide to the water. As a result, the total CO_2 will increase by the amount of the addition. Since the CO_2 does not carry any charge, the alkalinity will not change. CO_2 addition thus moves the point representing the water to the right. Similarly, removal of CO_2 will move the point to the left.

Now consider the effect of bicarbonate addition (Fig. 21-3) such as would result from adding river water to seawater and then evaporating the excess water. In this case, if we add 1 millimole of HCO_3^- to a kilo-

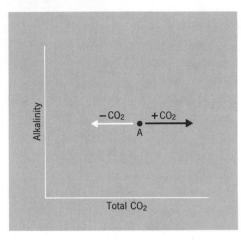

Figure **21-2**   Removal or addition of CO_2.

gram of seawater, we increase the total CO_2 content by 1 millimole, and the alkalinity is increased by 1 meq. The point A thus moves at an angle of 45° up and to the right. Removal of 1 millimole of HCO_3^- would move the point the same distance in the opposite direction.

Now let us consider the effect of adding CO_3^{--}. Such an addition could be the result of dissolution of limestone.

$$CaCO_3 \longrightarrow Ca^{++} + CO_3^{--}$$

If we add 1 millimole of CO_3^{--} to a kilogram of seawater, we increase the total CO_2 content by 1 millimole. Since the carbonate ion carries two negative charges, the change in the alkalinity will be 2 meq. Therefore, our point is displaced at an angle of 60° up and to the right (Fig.

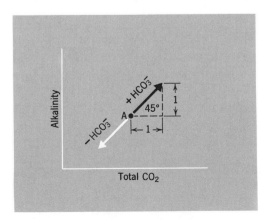

Figure **21-3**   Removal or addition of HCO_3^-.

Figure **21-4**   Removal or addition of CO_3^{--}.

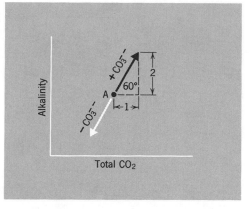

21-4). Precipitation of calcium carbonate, hence removal of CO_3^{--} from the solution, will produce motion in the opposite direction.

Finally, we can add acid or base to the seawater. If we add a strong acid such as HCl, the added H^+ ion will combine with the bicarbonate to form H_2CO_3. As a result, the alkalinity of the water will be reduced, while the total CO_2 content will remain constant. Thus the point will move downward. If we add too much acid, we will end up with an excess of H^+ ion over the sum of the negative ions, HCO_3^- and OH^-, and the alkalinity will become negative (Fig. 21-5).

Addition of a base, such as NaOH, increases the alkalinity of the water without altering the total CO_2 content. Thus it causes the point to move vertically upward (Fig. 21-5). When seawater interacts with some minerals, the effect is equivalent to adding acid or base to the water. Thus

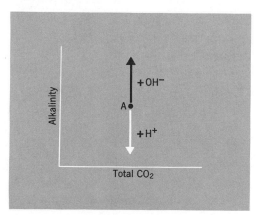

Figure **21-5**   Addition of acid or base.

some clay minerals may incorporate Mg^{++} and OH^- ions from the water into their crystal lattice. This is equivalent to removing negative OH^- or adding positive H^+.

The Carbonate Cycle on the Surface of the Earth

Let us consider the effect of the long-time addition of river water to the ocean. Since the water will be reevaporated, the net effect is to increase the HCO_3^- content of seawater; hence the composition on the Deffeyes diagram will move up and to the right at a 45° angle. How can this change be counteracted? The main process in the sea is the precipitation of calcium carbonate. This is accomplished mainly by plants and animals, which build their skeletons out of $CaCO_3$.

Carbonate precipitation, however, will remove alkalinity and total CO_2 in the ratio 2:1. Thus we cannot return to our starting point (Fig. 21-6). Rather, we will end up with the same alkalinity and an excess of total CO_2. To complete the cycle, the excess CO_2 is transferred to the atmosphere, where it is dissolved in rainwater. Pure water has zero alkalinity and zero CO_2. When CO_2 dissolves in pure water, it adds total CO_2 but no alkalinity.

The CO_2-charged rainwater falls on the land, where the carbonated water reacts with limestones to dissolve them and produce Ca^{++} ion and HCO_3^- ion:

$$CO_2 + H_2O \longrightarrow H_2CO_3$$
$$H_2CO_3 + CaCO_3 \longrightarrow Ca^{++} + 2HCO_3^-$$

The river water is then returned to the sea to repeat the cycle. The net result of the cycle is to dissolve carbonate rocks on the land and precipitate them in the sea. The concentrations in the atmosphere and the ocean remain the same.

The actual cycle is more complex than is indicated in Figure 21-6. Weathering by the CO_2-charged rainwater dissolves not only carbonate rock but also some of the silicate rocks. In the sea, the silicates interact with the water, changing its alkalinity. The details of the process vary from place to place depending on the rocks exposed to weathering by the rainwater. On isolated carbonate islands in the ocean, such as low coral atolls, the process will be essentially as shown in Figure 21-6. The rainwater is continually dissolving the exposed limestone on the island,

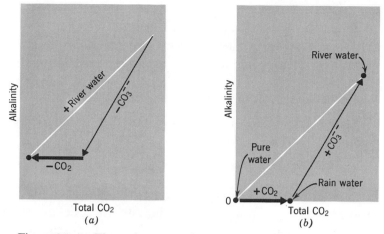

Figure **21-6** The carbonate cycle: (*a*) in the ocean; (*b*) on the land.

while plants and animals are reprecipitating calcium carbonate in the shallow water around the island.

The Carbonate Cycle within the Sea

In Figure 21-1, we saw that the alkalinity and total CO_2 concentrations in the ocean vary with depth. What causes these variations, and how are the variations limited? Below the sea surface and away from shore, there is no addition of HCO_3^- from rivers and no exchange of CO_2 with the atmosphere. The organisms within the sea, however, interact with the carbonate system in two ways. Plants remove carbon dioxide to synthesize organic matter, while animals burn organic matter and so give off CO_2. The removal of CO_2 by plants requires light and therefore is restricted to the upper regions of the sea; however, animals can add CO_2 to the seawater at all depths.

The fact that respiration by animals in the deep, dark water increases the content of total CO_2 while the alkalinity remains unchanged is apparent in Figure 21-1, in which the points from deeper levels shift progressively to the right. In addition, however, the figure shows a slight increase in alkalinity with depth. This results from the interaction of seawater with calcium carbonate. This substance is precipitated within the cells of many plants and animals to form an internal or external skeleton. Many of these animals float about within the sea, and when they die their skeletons sink to the sea floor. Other plants and animals live on the floors of the shelves, where the calcium carbonate they deposit accumulates on the shallow sea floor. In this manner, as volcanoes sink below sea level in tropical waters, the calcium carbonate deposited by plants and animals maintains the top of the mound near sea level.

The Solubility of Calcium Carbonate in Seawater

When calcium carbonate dissolves in water, it dissociates into a calcium and a carbonate ion:

$$CaCO_3 \longrightarrow Ca^{++} + CO_3^{--}$$

For a solution to be in equilibrium with solid calcium carbonate, the product of the calcium- and the carbonate-ion concentrations must have a specified value, the solubility product of calcium carbonate. Seawater contains 20 meq kg^{-1} of Ca^{++} and 2.3 meq kg^{-1} of HCO_3^-. While most of the carbon dioxide in seawater is in the form of HCO_3^-, a fraction of it is in the form of CO_3^{--} and H_2CO_3. The apportionment of the total CO_2 among the three species depends on the pH in the water (Fig. 19-8, p. 321). At a pH of 8, about 10 percent of the total CO_2 is in the form of CO_3^{--}, while only 1 percent is in the form of H_2CO_3.

The concentration of Ca^{++} is almost constant in seawater, varying only in response to changes in salinity. The concentration of total CO_2 is also relatively constant, although it is somewhat more variable than that of Ca^{++} (between 2.0 and 2.5 meq kg^{-1} for the data shown in Fig. 21-1). The concentration of carbonate ion, however, can change markedly in response to changes in the pH of the water. Thus, when the pH changes from 7.5 to 8.0 to 8.5, the fraction of the total CO_2 that is in the form of CO_3^{--} increases from 4 to 10 to 28 percent. The concentration of carbonate ion in seawater depends much more strongly on the pH than on the total CO_2 content of the water.

This strange behavior of the carbonate system is well illustrated when we add CO_2 to seawater, increasing the total CO_2 content of the water without altering the alkalinity. The CO_2 addition causes the pH to drop since most of the H_2CO_3 is converted to HCO_3^- and thus adds H^+ ion to the water. Some of the added hydrogen ion combines with CO_3^{--}, reducing the concentration of carbonate ion. In this manner the addition of CO_2 reduces the carbonate-ion concentration of water.

The addition and removal of CO_2 are responsible for the formation in

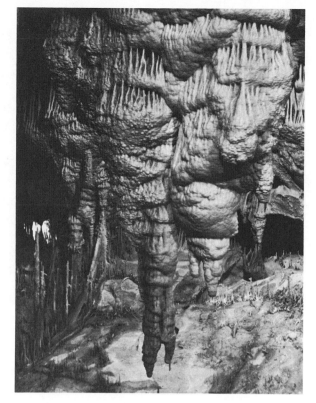

Figure **21-7** Limestone cave with stalactites (Magnum)

limestone caves of stalactites and stalagmites (Fig. 21-7). As water percolates through the soil above the cave, it picks up CO_2 produced by the decay of organic matter. As a result, it is able to dissolve more calcium carbonate. When the CO_2-charged solution, saturated in calcium carbonate, enters the air space of the cave, it loses CO_2 since the CO_2 content of the soil was much greater than that of the atmosphere. As a result, the carbonate concentration increases, the solution becomes supersaturated in $CaCO_3$, and limestone is precipitated by the dripping water in the cave.

What about normal seawater? Is it undersaturated with calcium carbonate, supersaturated, or in equilibrium? To find out, one must measure the solubility product of calcium carbonate and compare it with the concentration product found in the ocean.

A number of chemists have measured the solubility product of calcium carbonate in seawater. Careful experiments using different techniques have obtained values that differ by a factor of 2. We can also look

Figure **21-8** Oolite bars in the Bahamas.

at carbonate beaches to discover whether seawater is undersaturated or supersaturated with respect to calcium carbonate. If seawater is undersaturated, carbonate sediments should dissolve; if it is supersaturated, the precipitation of calcium carbonate on the loose sediments should cement them together into hard rock.

The evidence from carbonate sediments is confusing. In tropical surface water we see relatively little indication of either dissolution or precipitation. Tidal bars in tropical areas, devoid of land-derived sediments, contain sand-size spherical grains of calcium carbonate, called oolites (Fig. 21-8). These appear to be the result of very slow precipitation of calcium carbonate from seawater. On some beaches, lime sediments are cemented together by calcium carbonate to form solid beach rock.

However, the evidence of precipitation from tropical surface seawater is not universal. Many carbonate shell fragments have existed in seawater for thousands of years without apparent overgrowth or cementation. Elsewhere we find nips carved into the limestone in the intertidal zone (Fig. 21-9).

This latter phenomenon has been "explained" by the contention that the seawater near the surface becomes charged with CO_2 from the atmosphere and that, as a result, this water is more acidic and thus is able

Figure **21-9** A tidal nip in carbonate rock on the west coast of Bonaire, N. A. (Photograph: K. S. Deffeyes)

to dissolve the limestone. Careful measurements, however, have shown that seawater near the nips does not differ significantly in pH from the water below. Close examination of such a nip reveals that it is inhabited by an extensive fauna. Emery (1962), in his study of the Island of Guam, noted that the nip was densely populated by limpets, and that each occupied a slight depression in the rock surface which conformed to the shape of the animal. When he measured the soles of freshly collected limpets, he found them to be considerably more acid than seawater. Other animals live in tiny caves that they have dissolved or worn out of the rock in the intertidal zone. Thus the nip is of biologic origin and is not evidence that the seawater itself dissolves limestone.

Marine plants and animals are able to etch limestone and to precipitate it as skeletal material. However, seawater by itself does not seem to interact significantly with limestone, for the fine skeletal material on a limestone beach remains virtually unaltered for periods of the order of a thousand years. Fresh water behaves very differently, for dissolution and precipitation take place rapidly in it, leading to the decoration of limestone caves shown in Figure 21-7.

Searching for an answer to this puzzle, the author (Weyl, 1967) found that this difference in behavior appears to be due to the high magnesium concentration of seawater. Magnesium ion is chemically similar to calcium ion except that it has a slightly smaller radius. With calcium, it forms the double-carbonate mineral dolomite, $CaMg(CO_3)_2$. Magnesium also occurs up to 20 percent in calcium carbonates precipitated by some plants and causes their skeletons to be considerably more soluble than pure calcium carbonate. Thus magnesium can be coprecipitated with calcium carbonate, resulting in a disordered crystal of higher solubility.

If a crystal of calcium carbonate is placed in seawater which is supersaturated with pure calcium carbonate, precipitation takes place; however, some magnesium is coprecipitated. As a result, the solubility of the solid is increased until it matches the concentrations in the seawater. Thus the growing crystal comes to equilibrium with the seawater by growing a more soluble layer over its surface. After very little growth, precipitation ceases, since the altered crystal surface now contains magnesium and has come to solution equilibrium with the seawater. In fresh water, on the other hand, precipitation continues until the concentration of calcium times carbonate in the water has come to equilibrium with the pure crystal. Therefore, it appears that while surface seawater is generally supersaturated with pure calcium carbonate, the coprecipitation of magnesium produces a very thin surface layer that is in equilibrium with the seawater. Over a range of a factor of 2 in the carbonate concentration, limestone grains can reside in seawater without measurable dissolution or precipitation. This explains why different investigators obtained large differences in the measured solubility product of calcium carbonate in seawater, while the results in fresh water show no such discrepancies.

The Snow Line

When we leave the shallow tropical waters for the deep ocean, the situation becomes simpler. In the upper layers of the ocean, there are tiny animals with carbonate shells. After the animals die, the shells sink slowly to the ocean floor. As long as the water through which they sink is saturated or supersaturated, the particles end up on the bottom and form a large fraction of the deep sea sediment. If the deep water is undersaturated, however, the tiny shells dissolve before they reach the ocean floor.

When we examine deep-sea sediments, we find that there is a boundary below which calcium carbonate is absent. Because the small, slowly settling shells are white, the line that separates calcareous from noncalcareous deep-sea sediments is known as the snow line. Submarine hills above the snow line are covered by white lime sediments, but only a few etched particles of calcium carbonate are able to survive the trip through the undersaturated water below the line.

On the average, the snow line is at a depth of 4.2 km in the Pacific and 4.7 km in the Atlantic. The undersaturation of the deep water is due to two factors. The animals increase the CO_2 content of the water and so reduce the pH, leading to a smaller CO_3^{--} concentration. In addition, the increase in pressure and the lower temperature in the abyss increases the solubility of calcium carbonate.

The deeper water has a larger concentration of total carbonate, and the dissolution of calcium carbonate increases the alkalinity. If the change resulted only from the dissolution of calcium carbonate, the increase in alkalinity should be twice the increase in total CO_2 (Fig. 21-4). Actually, the increase in total CO_2 is larger than the change in alkalinity (Fig. 21-1). Thus the addition of CO_2 by animals must exceed the addition of CO_3^{--} contributed by the dissolution of calcium carbonate.

The Effect of Man on the Carbonate Cycle

The bicarbonate content of seawater is stabilized by interaction of the ocean, the land, the atmosphere, and biological processes. Since the beginning of the industrial revolution, man has interfered significantly with the natural carbonate cycle by burning fossil fuels such as coal and petroleum at an ever-increasing rate. Revelle and Suess (1957) have analyzed the statistics of fuel consumption. Their data on the rate of fuel burning for the world and their projections into the future are shown in Figure 21-10.

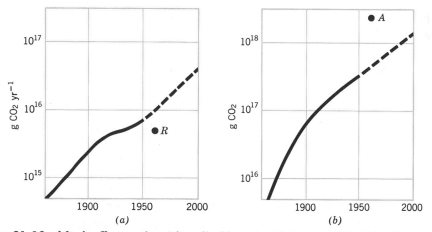

Figure **21-10**   Man's effect on the carbon dioxide cycle: (*a*) the rate of fossil fuel burning; (*b*) the cumulative addition of CO_2 to the atmosphere since the beginning of the industrial revolution. (Revelle and Suess, 1957) (*A*) CO_2 in the atmosphere, (*R*) measured rate of CO_2 increase in the atmosphere. (Pales and Keeling, 1965)

At the present time we are adding carbon dioxide to the atmosphere at a rate of 10^{16} g yr^{-1}. Since the atmosphere contains 2.4×10^{18} g of CO_2, the present rate of fuel burning would double the CO_2 in the atmosphere in 240 years. Meanwhile the rate of fuel consumption is continually increasing. If we extrapolate the total CO_2 produced by man's industrial activity, we find that it would equal the present content of the atmosphere by the year 2020.

What is the fate of the CO_2 we are adding to the atmosphere? Does it stay there or is it partially absorbed by the ocean (which contains 65 times as much CO_2 as the atmosphere)? In order to answer this question, it is necessary to monitor the CO_2 content of the atmosphere. If all the added CO_2 remained in the atmosphere, its concentration would increase by 0.5 percent per year. It is therefore necessary to make very accurate measurements over a long period of time. One must be careful that the data are not influenced by local sources of CO_2 but represent average conditions for the earth.

To get as far away from industrial areas as possible, investigators have monitored the CO_2 content of the atmosphere near the summit of Mauna Loa, on the island of Hawaii (Pales and Keeling, 1965), and at the South Pole (Brown and Keeling, 1965). The data from Hawaii for 1958–1963 are shown in Figure 21-11. The monthly average values shown in the upper curve indicate an annual cycle with an amplitude of 6 ppm having a maximum in May and a minimum in September.

The annual cycle is the result of biological activity. Between May and September in the northern hemisphere, the plants are removing CO_2 from the atmosphere to produce organic matter at a rate faster than the rate at which the organic matter is reoxidized. Between September and May, the rate of oxidation of organic matter exceeds the rate of new plant growth. Over Antarctica, the amplitude of the cycle is about one-fourth that at Hawaii. Because the seasons in the southern hemisphere are opposite to those in the northern hemisphere, the Antarctic maximum occurs in November while the minimum occurs in March.

To estimate the long-time trend, we can eliminate the annual cycle by averaging the data over a 12-month period. The running yearly average (Fig. 21-11b) clearly shows a gradual increase of CO_2 at a rate of 0.68 ppm yr^{-1}. The data from the South Pole give a rate of increase of 0.72 ppm yr^{-1}, in good agreement with the data from Hawaii. On the average,

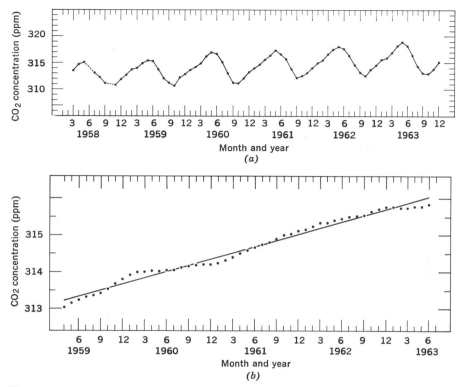

Figure **21-11**   Concentration of atmospheric CO_2 at Mauna Loa Observatory, Hawaii: (*a*) monthly average; (*b*) 12-month running average. (Pales and Keeling, 1965)

the CO_2 content of the atmosphere is increasing by about 5×10^{15} g yr^{-1}, which is about half the amount of CO_2 added per year by fossil fuel burning. The remaining half of the CO_2 added by man must therefore be absorbed by the ocean.

The CO_2 in the atmosphere is a strong absorber of infrared radiation and thus contributes to the greenhouse effect which increases the surface temperature on earth (see Chapter 7). We might therefore expect that the addition of CO_2 by man's industrial activity will lead to a gradual warming of the climate. Just such an increase was observed between 1900 and 1940. Since 1950, however, in spite of the increased rate of fossil fuel burning, the average surface temperature of the northern hemisphere has been declining.

The factors that control the climate on the surface of the earth are very complex, and CO_2 is only one of them. Until we have a better understanding of the interplay of the climate-determining factors, we cannot predict the future trend of climate and how it will be affected by man's industrial activity. We shall examine this problem in more detail in Chapter 30.

The rate at which we are adding CO_2 to the atmosphere is about ten times the rate at which the rivers are carrying bicarbonate to the sea. The fossil fuels also contain about 1 percent by weight of sulphur. This sulphur is oxidized and contributes sulphate ion to the river water. The sulphate so contributed amounts to at least half the total sulphate in river water. If we plot the distribution of the sulphate concentration of rivers, we obtain a map of industrial activity. The rivers of Europe and North America have average concentrations of 0.50 and 0.42 meq kg^{-1} of sulphate, while the rivers of South America and Australia have average concentrations of only 0.01 and 0.05 meq kg^{-1}. Having examined the various processes by which the bicarbonate content in the ocean is stabilized, let us now investigate how seawater attained its present composition.

Summary

Carbon dioxide in various forms permeates the surface environment of our planet. The way in which CO_2 interacts with seawater can best be understood by considering the alkalinity–total CO_2 diagram. Different biological and chemical processes shift the composition of seawater on this diagram in different ways. The long-term stability of seawater results from cyclic processes that convert the salts leached from the land to marine sediments. While calcium carbonate in fresh water dissolves and precipitates readily, its behavior in seawater is more complex. As a result, biological processes tend to dominate the deposition and removal of calcium carbonate in seawater. Man, since the industrial revolution, has been drastically altering the CO_2 balance in nature. The consequences of the addition of CO_2 by fossil fuel burning cannot yet be predicted reliably.

Study Questions

1. If the organic matter in sediments has been accumulating at a uniform rate for 10^9 years, what fraction of the current living organic matter must be preserved per year, and what fraction per year gives rise to fossil fuels? (See Table 22-1, p. 334.)

2. A 1 kg sample of seawater has an alkalinity of 2.4 meq kg^{-1} and a total CO$_2$ content of 2.2 millimoles kg^{-1}. What will be the alkalinity and total CO$_2$ content of this water if:
   a. We dissolve 0.1 millimole kg^{-1} of CaCO$_3$?
   b. We remove 0.2 millemole kg^{-1} of CO$_2$?
   c. We add 0.2 meq kg^{-1} of CaHCO$_3$?
   d. We add 0.2 meq kg^{-1} of HCl?
   e. We simultaneously do a, b, c, and d?
3. Suppose that average river water — cyclic salts (Chap. 20, Table 20-1, p. 324) were added to seawater, and that the only processes were the precipitation of all the added Ca^{++} as CaCO$_3$ and the evaporation of the added water. What would be the composition of seawater at the end of 10^6 years?
4. How would the appearance of tropical carbonate islands differ if the behavior of seawater relative to calcium carbonate were the same as that of fresh water?
5. Compare the seasonal variation in the CO$_2$ content of the atmosphere over Hawaii with the estimate of carbon in the form of living organic matter, assuming that the average variation for each hemisphere is reflected by the measurements.

Supplementary Reading

Weyl, P. K. (1964). "The Solution Alteration of Carbonate Sediments and Skeletons." In J. Imbrie and N. Newell, eds., *Approaches to Paleoecology.* New York: John Wiley and Sons. Pp. 345–356.

——— (1966). "Environmental Stability of the Earth's Surface—Chemical Considerations," *Geochimica et Cosmochimica Acta,* Vol. 30, pp. 663–679.

Revelle, Roger, and Fairbridge Rhodes (1957). "Carbonates and Carbon Dioxide," *Geological Society of America Memoir #67,* 1, pp. 239–296.

22 The Geological History of Seawater

The residence times of most of the constituents of seawater are short compared to the geologic record of life in the sea. Therefore the concentrations of most constituents in seawater must be stabilized by geochemical cycles. Although these cycles help to explain the composition of the present ocean, they do not enlighten us about the ultimate origin of seawater. Has the earth always had an ocean, or did the seas accumulate gradually after the solid earth was formed? Were the early ocean and atmosphere more or less as they are today, or did they undergo a chemical evolution? How did the origin and early evolution of life interact with the chemistry of the surface of our planet? In this chapter, we will explore some answers to these questions.

Was There a Primordial Ocean?

The earth accumulated as a planet some 4.5×10^9 years ago. Did this primordial earth already have an ocean and an atmosphere, or was the earth originally without a fluid envelope? The earth's retention of volatile components such as air and water depends on its temperature. Because of their thermal motions, gases are continuously diffusing outward, but this tendency is counteracted by the gravitational attraction of the earth. In order to escape from the earth, a molecule, like a spaceship, must have a velocity that is greater than the escape velocity from earth, 11.2 km sec^{-1}.

The average kinetic energy of a gas molecule is proportional to the absolute temperature:

$$\text{KE} = \tfrac{1}{2}mv^2 \text{ proportional to } T$$

Therefore, the average velocity, $\bar{v}$, is proportional to the square root of the absolute temperature divided by the mass:

$$\bar{v} \text{ proportional to } \sqrt{Tm^{-1}}$$

The lighter the molecule and the higher the temperature, the more likely it is that the gas can escape from the earth's surface. Under present conditions, hydrogen and helium gas are rapidly lost from the atmosphere, while the heavier gases such as oxygen and nitrogen are retained. The moon, because of its weaker gravitational attraction and greater maximum surface temperature, has not been able to retain either water or an atmosphere.

If the earth has never been hotter than it is at present, it could have preserved any primordial ocean and atmosphere. If, on the other hand, the earth had been much hotter during its early history, it would have lost any volatiles originally accumulated. A clue to the early thermal history of the earth is offered by the relative abundance of the noble gases on earth and in stars. These gases, unlike water, do not combine chemically and so always have been in a gaseous state. We must compare the abundance of these permanent volatiles with that of an element that is chemically bound in the solid matter of the earth, such as silicon, the major metallic element in rocks.

Since the earth originally accumulated from stellar material, the original ratio of the noble gases to silicon was probably similar to the ratio that is currently observed in stars. If the material of the primitive earth is then heated, there will be a loss of the volatile components, resulting in a decrease in the ratio of noble gas to silicon. The depletion of noble gases will be greatest for those elements of low atomic weight, helium and neon, and less for those of increasing atomic weight. The relative abundance ratios of the noble gases are shown in Figure 22-1 as a function of the atomic weight. We note that neon is depleted on earth by a factor of 10^{10} and that the depletion factor decreases with increasing mass, to 10^6 for xenon.

The depletion in noble gases suggests that the surface of the earth must once have been very much hotter than it is today. If the earth lost most of its argon, with a molecular weight of 40, it must also have lost its water vapor, nitrogen, and oxygen, with molecular weights of 18, 28, and 32, respectively. Thus the atmosphere and the ocean cannot be a remnant of the primordial earth; these gases must originally have been chemically combined within the solid earth; they can have accumulated on the surface only since the earth cooled to near its present temperature.

The data on the abundance of noble gases suggest that all the water

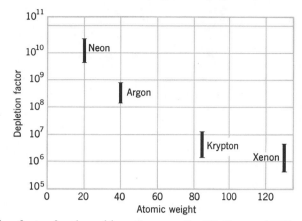

Figure **22-1** Depletion factor for the noble gases on earth. (H. Brown, 1952)

in the ocean and the gases of the atmosphere must once have been held within the solid earth. Is this a reasonable assumption? The mantle of the earth has a volume of 10^{27} cm³. Assuming a density of 4, this amounts to a mass of 4×10^{27} g, while the water of the ocean has a mass of 1.4×10^{24} g. Thus the mantle must have lost 0.035 percent water, on the average. We must compare this figure with the average water content of the mantle.

Although the mantle is not accessible for sampling, meteorites offer us samples of mantle-like material. Meteorites are fragments of a planet-like object in the solar system that broke up. Pieces of this material are frequently captured by the earth's gravitational field. Some meteorites consist mainly of iron with a high nickel content and are believed to resemble the material of the core of the earth. Others, the stony meteorites, contain silicates and are believed to resemble the mantle. By examining the water content of the stony meteorites, we can therefore obtain an estimate of the water content of the mantle. Although individual analyses differ, the average water content is about 0.5 percent or 10 times as much as the loss from the mantle that is required to account for the present ocean. Thus the mantle could be an adequate source for the water in the ocean.

The Geochemical Balance Sheet

The abundance of the noble gases indicates that the surface of the earth was once dry and devoid of an atmosphere. Prior to the evolution of the ocean and the atmosphere, there could have been no chemical weathering, hence no sediments. Thus at one time the earth was covered by *primary crystalline* rocks. As the ocean and the atmosphere accumulated, the denudation of the solid earth began (see Chap. 14). Sediments were formed, and the dissolved salts began to accumulate in the sea.

The weathering process consists of converting primary crystalline rocks into sediments and the contents of the sea. Once sedimentary rocks are formed, they also may be weathered and converted back to sediments again. In this manner there is a constant recycling of old sediments as well as a weathering of fresh crystalline rock. Since the beginning of weathering, primary crystalline rock has been converted to sediments and salts which have remained in solution in the water of the ocean and the water that fills the interstices of the sediments. In addition, some of the volatile components have given rise to the atmosphere.

Since mass must be conserved, the total mass of weathered crystalline rock must equal the sum of the masses of the products of weathering. Not only must the total mass be conserved, but the total mass of each element must also be conserved, since, except for the radioactive elements, atoms are not converted from one element to another. Thus the total number of atoms of each element, hence their mass, is conserved (Fig. 22-2).

The mass of an element in crystalline rock is equal to the total mass of crystalline rock multiplied by the average concentration of that element in crystalline rock. Thus the requirement for conservation of mass leads to the following equation: (the total mass of weathered crystalline

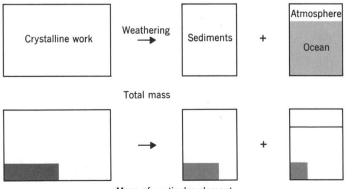

Figure **22-2** The method of geochemical balances.

rock) × (the average concentration of an element in it) = (the mass of sediments) × (the average concentration of the element in sediments) + (the mass of the ocean) × (the average concentration of the element in the ocean) + (the mass of the atmosphere) × (the average concentration of the element in the atmosphere).

A conservation-of-total-mass equation can be written for each of the stable elements. The accuracy with which the budget can be balanced depends on how well one can estimate the total masses and average concentrations. Since the volume of the ocean is fairly well known and since it is well mixed, the ocean is the most accurate entry in the balance sheet. Crystalline rocks are more heterogeneous, and so the estimate of their average composition is more uncertain.

Sedimentary rocks are of various types and they may have undergone a variable amount of later alteration. The primary sediment types are carbonate rocks, sandstones, and shales. To obtain an average composition, it is necessary to estimate the total amount of each type of sediment as well as its average composition. A number of geochemists have drawn up balance sheets. The surprising thing is not that the results differ from one another but, rather, that in spite of the fragmentary data, they agree fairly well. For the major rock-forming constituents, the geochemical equation balances within the accuracy of the estimates.

For some components, however, regardless of how hard one tries, the equation does not balance. These elements are much more abundant in the ocean, atmosphere and sediments than can be accounted for by weathering of crystalline rocks. Because these components are mostly volatile at the surface, they are called the *excess volatiles*. Estimates of these "volatile spirits" are shown in Table 22-1. In each case, the

TABLE **22-1**   GEOCHEMICAL BALANCE SHEET*

	Element or Compound in 10^{20} g				
	H_2O	C as CO_2	Cl	N	S
In atmosphere and ocean	14,600	1.5	276	39	13
In sedimentary rock	2,100	920	30	4	15
Total	16,700	921	306	43	28
From weathering of crystalline rock	130	11	5	0.6	6
Excess	16,600	910	300	42	22

* Rubey, (1951).

amount in the ocean, atmosphere and sediments exceeds the amount derived from crystalline rock by a large factor.

The Excess Volatiles

When we look at the elements that exist in excess at the surface of the earth, we note that they are all volatile. Either they exist as gases at normal temperatures or they become gases at moderately elevated temperatures. They cannot be the remainder of some primitive atmosphere, nor can they have been derived from the weathering of crystalline rock. Therefore we must look for another source for these gases.

At the present time, volcanoes and hot springs vent gases from the earth's interior into the atmosphere. These gases consist mainly of water with lesser amounts of CO_2, sulphur, and nitrogen. Some of the vented gas may be nothing more than ground water and recycled volatiles derived from sedimentary rocks such as carbonates. As a result, the composition of the gases is variable. Some of the gases, however, are derived from molten rock. When molten rock cools, solid minerals crystallize out, and some of the volatiles escape. Some of the gases thus represent new material brought up from the mantle.

We can analyze small bubbles of volatiles that have become occluded in igneous rocks. That is, instead of escaping when the rock crystallized, these small bubbles became frozen into the rock. When we analyze the gases in these bubbles, we find that their relative composition is similar to the excess volatiles. Since these rocks generally crystallized at considerable depth, there is little likelihood that the gases contained in them were derived from ground water or sediments.

The chemical composition of the gases derived from molten rock matches the composition of the excess volatiles rather well. Is the present rate of escape of these gases adequate to supply the mass of the excess volatiles over a period of 3×10^9 years? We can look at the water being discharged by hot springs in the United States (Fig. 22-3). If we assume that this is typical for the continents and that the ocean floor contributes hot-spring water at half the rate per unit area, we find that the total flow is adequate to fill the ocean 100 times in 3×10^9 years.

In order to be an adequate source, 1 percent of the water in the hot springs must be new water added from within the earth. We can get a rough estimate of the amount of new water if we consider the tempera-

Figure **22-3**   Grotto Geyser, Yellowstone National Park. (Photograph: John Clawson)

ture of the water. The average temperature of such water in the western United States is 29°C while the yearly average temperature of the ground is 9°C. The water is therefore heated by 20°. If we assume that the water of the hot springs is a mixture of steam added from the interior at 600°C and ground water at 9°C, we obtain the following:

> 1 g new steam will add 1100 calories in condensing and cooling to 29°C. This is adequate to heat 55 g of ground water from 9° to 29°C. This simple calculation indicates that about 2 percent of the water is added from below. Thus it appears that the outgassing of molten rock derived from the interior of the earth is an adequate source for the excess volatiles.

The Volume of the Oceans through Geologic Time

Geochemical data on the noble gases and on the excess volatiles suggest that the water of the oceans has accumulated gradually over geologic time. About 4.5×10^9 years ago it was zero; at present it comprises 1.4×10^{24} g. We have no way of estimating the volume of the ocean in the past; however, the present rate of outgassing is not inconsistent with a uniform rate of increase of ocean volume through geologic time. Does this imply that the ocean of the past covered a smaller fraction of the earth's surface?

The surface area of the ocean depends on the volume of seawater and the distribution of elevation of the solid earth. The depth of the ocean results from the difference in thickness between continental and oceanic crust. If, as is probable, there was originally no differentiation between oceanic and continental crust, then the continental crust has also been accumulating over geologic time. It is probably created by the convergence of the blocks of the lithosphere (see Chap. 18). At the same time, this convergence gives rise to volcanism, which vents volatiles from below. Thus the addition of volatiles and the formation of continental crust may be two aspects of the same phenomenon.

It appears, then, that as water is added to the ocean, more continental crust is formed, and the relief of the surface of the solid earth gradually increases. Thus the increase in ocean volume is accompanied by an increased average depth, and the aerial extent of the oceans may not have changed drastically. The geologic evidence suggests that there have been times in the past when large portions of the present continents were covered by shallow seas. Because of the nature of the hypsometric curve (Fig. 13-7, p. 209), however, large areas of the continental crust can become inundated without large changes in the ratio of continental crust to ocean volume.

Another important problem is the salinity history of the ocean. Has the ocean always had the salinity it has today, or have there been large changes in salinity over geologic time? The dominant negative ion in seawater is Cl^-, and most of the chlorine outgassed ends up in seawater (Table 22-1). The problem of the salinity of the past, therefore, depends mainly on the relative outgassing of chlorine and water vapor. If these two volatiles have escaped in constant ratio throughout geologic time, then the salinity must have been relatively constant. There is no evidence to suggest that the chlorine-to-water ratio has varied significantly in the past. We are therefore led to the conclusion that seawater has always had a salinity not drastically different from that which it has today.

The general picture that emerges is one of uniformity. The salinity of the ocean and the relative areas of oceanic and continental crust have probably not changed much over geologic time. The total volume of the ocean and the volume of continental crust, on the other hand, have increased gradually during the last 4.5×10^9 years. There is one aspect of the history of the surface environment, however, that has undergone drastic changes. These changes are the result of the origin and later evolution of life on earth. We have seen that life processes at present have a profound effect on the carbon cycle in nature. Before there was life, the surface environment was very different, and this difference was due mainly to a lack of oxygen in the primitive atmosphere.

The Primitive Atmosphere and the Origin of Life

In rock, iron occurs in two forms—as *ferrous* iron, FeO, and as *ferric* iron, Fe_2O_3. When ferrous iron is exposed to the atmosphere, it combines with oxygen to form ferric iron:

$$4FeO + O_2 \longrightarrow 2Fe_2O_3$$

If we examine the state of iron in volcanic rock, we find that it is largely in the ferrous state. Thus any oxygen that might occur in volcanic gases would be removed to oxidize some ferrous iron to ferric (Holland, 1965). We must therefore conclude that the excess volatiles could not contain free oxygen and that the primitive atmosphere could not, then, have contained this vital gas.

The absence of oxygen in the primitive atmosphere has important consequences. Oxygen gives rise to ozone in the upper atmosphere, and this ozone layer, as we noted in Chapter 7, shields the surface of the earth from ultraviolet radiation. Without oxygen, the ultraviolet sunlight is not absorbed by the atmosphere and so reaches the surface of the ocean (Berkner and Marshall, 1965). In the absence of oxygen, the hydrogen, H_2, and nitrogen, N_2, released by outgassing give rise to small amounts of the gases methane, CH_4, and ammonia, NH_3.

Ultraviolet radiation is inimical to life, for it breaks down complex organic molecules within living cells. However, if we irradiate water in an oxygen-free atmosphere containing some methane and ammonia with ultraviolet radiation, we find that a diverse assortment of complex organic molecules is formed (Miller, 1959). Thus the primitive oxygen-free atmosphere-ocean system produces the organic building blocks required for the first self-replicating biological organism.

In some manner, more than 3×10^9 years ago, a self-replicating organism arose from the organic building blocks. It had to be protected from the ultraviolet radiation near the sea surface, while the continued production of organic matter there provided a constant supply of food (Weyl, 1968a). The first organisms for which we have a record were bacteria-like (Chap. 3), and these were followed by blue-green algae. These most primitive of plants are able to utilize light to convert carbon dioxide and water to organic matter with the emission of free oxygen. Thus the evolution of plants leads to the biochemical generation of free oxygen. Most of the oxygen liberated will first be used to oxidize the ferrous iron of weathered rocks to ferric. Other oxygen is utilized to oxidize components of the volcanic gases. Once these oxygen require-ments have been met, free oxygen can accumulate in the atmosphere.

Oxygen is utilized by animals to burn organic plant matter back to carbon dioxide and water. Thus the net production of oxygen consists of the excess production of organic matter over and above the amount con-sumed. In Table 21-1, we estimated the total organic matter now buried in sediments as 6×10^{20} moles of carbon. For every carbon atom laid down as organic matter, the plants emitted one molecule of free oxygen gas. The atmosphere at present contains 0.38×10^{20} moles of oxygen, or about 6 percent of the total net oxygen produced by the flora of the earth through time. The other 94 percent of the oxygen liberated was utilized to oxidize the weathered rocks and the volcanic gases emitted from within the earth.

Summary

At one time, the surface of the earth was devoid of an ocean and an atmosphere. The emission of volatiles from within the earth led to the gradual accumulation of the ocean and the atmosphere. Once water was present at the surface, weathering of crystalline rocks could commence. As a result, crystalline rocks were transformed to sediments and the salts of seawater. The dynamics of the lithosphere then led to the formation of continental crust so that, as the volume of water at the surface in-creased, the difference in elevation between the floor of the ocean and the surface of the continents also increased. Originally the atmosphere was devoid of oxygen, and the ultraviolet irradiation of the sea surface led to the synthesis of complex organic molecules. Life evolved from, and was originally nourished by, the organic matter produced by solar radiation near the surface of the sea. The evolution of the first marine plants led to the biologic emission of free oxygen and the gradual oxidation of the surface environment. Eventually this led to the present oxygen-containing atmosphere. Thus life processes have transformed the surface environment on our planet.

Study Questions

1. The crust of the earth weighs about 4×10^{25} g. Could the water of the ocean have been derived from outgassing of the crust?
2. The method of geochemical balances is based on certain assumptions. Identify these assumptions and discuss some processes that violate them.
3. Most of the carbon from the excess volatiles is now in the form of sediments while most of the chlorine is found in seawater (Table 22-1). Explain.
4. If all the fossil fuels were burned to CO_2 (Table 21-1), what would be the percentage of reduction of oxygen in the atmosphere?

Supplementary Reading

(Starred item requires little or no scientific background.)

Brancazio, P. J., and A. G. W. Cameron, eds, (1964). *The Origin and Evolution of Atmospheres and Oceans.* New York: John Wiley and Sons.
* Oparin, A. I. (1964). *Life: Its Nature, Origin, and Development.* New York: Academic Paperbacks, Academic Press.

Part V
Life in the Sea

Life originated in the sea more than three billion years ago and has been evolving ever since. Living organisms in the ocean range in size from bacteria and microscopic plants to the largest living animal, the whale. Marine plants utilize sunlight to convert water, carbon dioxide, and inorganic nutrients to organic matter. The plants provide food for herbivores, which, in turn, are eaten by carnivores. Respiration by animals converts the organic matter back into carbon dioxide. The regeneration of the nutrients is facilitated by bottom scavengers and bacteria which decompose most of the organic debris that settles on the sea floor. Thus the complex web of life in the sea continuously recycles the chemical constituents essential for life. When man interferes with the naturally evolved cycles, he may unwittingly overtax the capacity of the system and transform parts of the living sea into a stinking mess.

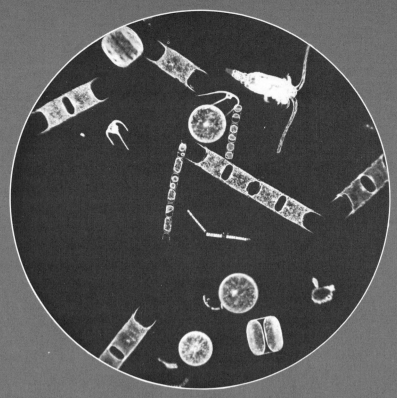

Life in a drop of sea water as seen under the microscope.

Whales, marine mammals. Note the air bubbles being exhaled by these air breathing animals. (Morris Huberland, National Audubon Society.)

23 The Basis of Life

A living organism is a more representative sample of the average composition of the universe than is a piece of rock. Although the earth is deficient in volatile components, organisms consist mainly of oxygen, carbon, and hydrogen, with lesser amounts of nitrogen, sulfur, phosphorous, and potassium. On the oxygen-free surface of the primitive earth, these elements were in the form of simple compounds, such as water, ammonia, methane, carbon dioxide, and carbon monoxide. Ultraviolet radiation from the sun converted some of these simple molecules into organic compounds, for example, amino acids, purines, pyrimidines, and carbohydrates. These simple organic molecules then combined into larger molecules such as proteins and nucleic acids.

More than 3×10^9 years ago, the complex organic building blocks gave rise to the first self-replicating living system. Then the synthesis of organic matter became regulated by the genetic code. Through natural selection, the modification and diversification of the genetic code has given rise to the variety of life we now find on the surface of our planet. The record of life of the past, preserved as fossils in sedimentary rocks, appears to represent a single chain of evolving forms of life. Thus every plant, animal, and bacterium traces its ancestry back continuously to some of the earliest forms of life. Although part of the geologic record is marked by widespread extinctions of types of organisms, the broad biologic record is one of continuity. Organisms have generally been well adapted to the environment, and the life processes, in turn, have interacted with the environment to permit the continuity of life.

Time

Every living organism is the current manifestation of a continuous chain that reaches back to the beginnings of life. During the nineteenth century, some scientists believed that life could originate spontaneously

365

from nonliving matter. This belief was based on the sudden appearance of microorganisms in originally sterile cultures which were not purposely inoculated. Louis Pasteur (1822–1895) proved that the organisms had resulted from airborne bacteria unintentionally introduced into the cultures. By showing that no growth took place when contamination of the cultures by germs in the air was prevented, he demonstrated that the medium itself could not give rise to a bacterial culture without inoculation with the bacterial cell.

The chain of life is maintained by reproduction. To ensure the survival of the species, the population of each type of organism must give rise to a sufficient progeny. Thus each species has a natural tendency to multiply. The resulting population is limited by the finite resources of the environment. Therefore, there is a constant conflict between the tendency of a population to grow in a geometric progression and the availability of fixed resources to support this growth.

To illustrate the problem of population growth, consider a single microscopic organism having a volume of one cubic micron (10^{-12} cm^3). Assume that this organism doubles its number once a day. After only 10 days there will be 10^3 organisms, and after 40 days the total volume occupied by the population will be 1 cm^3. Only 61 days later, the volume of the ocean will no longer be adequate to contain the progeny of the single minute cell.

Populations of organisms are thus in a constant state of flux. An increase in resources leads to an expansion of the population until the population has saturated the resources. Each organism, in turn, is a resource for other organisms that feed on it; thus populations are often held down by predation rather than by a limitation in resources. Under favorable conditions, a population increases until resources are fully utilized or until an increase in predation leads to a population decrease. Some populations are fairly stable; others vary dramatically with time.

Matter and Energy

In order to remain part of the chain of life, organisms must reproduce. The genetic code contains the information to construct a new organism. In addition to a blueprint, construction requires matter and energy. The various chemical constitutents of the organism must be synthesized from material derived from the environment. Furthermore, the process of synthesis requires energy as well as the elemental constitutents.

An organism is either autotrophic or heterotrophic. *Autotrophic* organisms are independent of outside sources of organic matter; they can manufacture the necessary organic matter from inorganic constituents, deriving the energy required for organic synthesis either from sunlight (photosynthetic autotrophs) or from inorganic chemical reactions.

Heterotrophic organisms require an environmental source of organic matter. Thus animals require food, which serves as a source of both matter and energy. They are not able to synthesize organic matter from inorganic sources. The distinction between autotrophs and heterotrophs is often not obvious. Many marine plants that are able to synthesize most substances from inorganic sources nevertheless require trace amounts of vitamins and other organic substances. Such plants are not able to grow in purely inorganic cultures. When organic growth factors, or bacteria that produce these factors, are added, however, these plants grow readily and synthesize all other organic compounds from inorganic sources. Whether a plant is truly autotrophic or requires certain organic compounds can be determined only by careful experimentation.

Space and Motion

The living space in the ocean extends from the intertidal zone along the shore to the bottom of the deepest trenches and comprises the sea surface, the water of the ocean, and the sea floor. These diverse environments are classified by depth and by habitat (Fig. 23-1). The *benthic* organisms are those that live on the bottom or within the sediment; organisms that live within the water column away from the bottom are called *pelagic.* The latter, depending on their capacity for movement, are subdivided into *plankton,* the drifters, and *nekton,* the active swimmers.

Continued life is not possible without relative motion. If the organism is stationary relative to its medium, it will eventually deplete the environment of the necessary nutrients and food and will become surrounded by its own excreta. Land plants are stationary; however, the air moves over their leaves, carrying in fresh carbon dioxide and removing oxygen and water vapor. Meanwhile, groundwater is percolating through the soil, bringing water and dissolved nutrients to the roots. Similarly, the constant motions within the sea continually bathe the sedentary marine organisms, the benthos, in fresh seawater.

We have noted that the nekton have active swimming mechanisms which permit them to pursue their prey while the smaller plankton drift

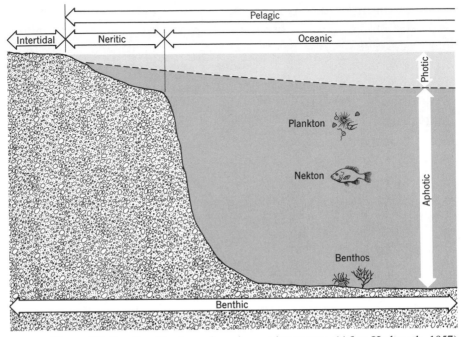

Figure **23-1**   Classification of marine environments. (After Hedgpeth, 1957)

about with the moving seawater. If their movements were completely passive, however, they would soon strip their immediate environment of food or nutrients. To survive, the organisms of the plankton, while they are not able to swim actively, must be able to move relative to the water. Their motion is adequate to contact fresh seawater; however, it is not sufficiently rapid for the plankton to avoid being swept along by the currents of the ocean.

The life zones of the ocean are classified by depth as well as habitat. The shallowest region is the intertidal zone, which ranges between the high- and the low-water line. The shallow ocean over the continental shelves to a depth of about 200 m is known as the *neritic* zone, to distinguish it from the *oceanic* region, which extends from the edge of the continental shelf to the deep trenches. Finally, it is convenient to subdivide the ocean into the region illuminated by sunlight, the *photic* zone, and the *aphotic* zone, which is in continuous darkness. The depth of the photic zone depends on the clarity of the water and can range from over 100 m to only a few meters in some coastal areas.

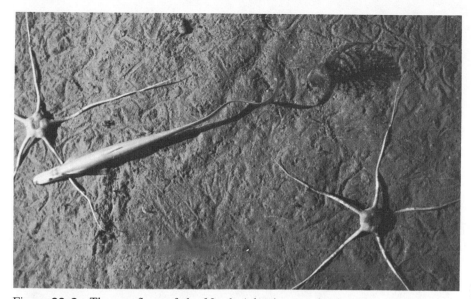

Figure **23-2** The sea floor of the North Atlantic at a depth of 1.5 km showing a Halosaur of the genus Aldovandia and large brittle stars. (Woods Hole Oceanographic Institution)

Figure 23-2 is a photograph of the sea floor of the North Atlantic at a depth of 1.5 km. It therefore shows the oceanic, benthic environment. A fish is swimming just above the bottom, occupied by a large, brittle star. The muddy bottom is marked by trails and depressions, indicating biological activity. To take this picture, a flashtube and the camera were triggered when a weight suspended below the camera touched the sea floor. Figure 23-3, a photograph taken in midwater at a depth of 1 km in the Gulf of Mexico, shows an example of the pelagic oceanic environment. The automatic flash-camera used to photograph this deep-sea animal (called a *medusa*) was triggered when light from the luminous organs of the animal actuated a special photocell in the camera.

Because the pelagic organisms can roam within the three dimensions of the sea, they do not have a space problem. By contrast, benthic organisms, which live on or in the bottom, compete for the limited space available. Particularly in the intertidal zone, on a rocky coast, all available space is densely occupied (Fig. 23-4). Where there is competition for space, all available surfaces are soon covered by a community of organisms (Fig. 23-5). Only by coating the hulls of ships with special antifouling paints and by scraping them frequently can one keep them free of attached organisms.

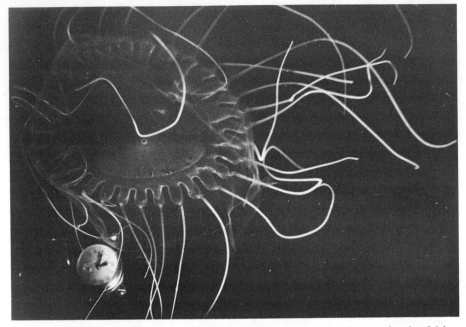

Figure **23-3** Self-portrait of rare jellyfish (Solmissus *incisia*) taken at depth of 1 km off George's Bank. The luminescence of the medusa triggered the electronic flash of the special deep sea camera designed by Breslau and Edgerton. (George L. Clarke, Harvard University and Woods Hole Oceanographic Institute)

Figure **23-4** Typical fauna from the rocky intertidal zone, showing mussels, acorn barnacles, and limpets. (Photograph: Douglas P. Wilson, F.R.P.S.)

Figure **23-5** Barnacles cover stakes that had been in seawater for one summer. (Photograph: American Museum of Natural History)

Diversity

Pliny the Elder (A.D. 23–79) took an inventory of the diversity of life in the sea. After listing 176 species, he wrote:

Surely, then, everyone must allow that it is quite impossible to comprise every species of (land) animal in one general view for the information of mankind. And yet, by Hercules! in the sea and in the ocean, vast as it is, there exists nothing that is unknown to us, and, a truly marvelous fact, it is with those things which Nature has concealed in the deep that we are best acquainted.

Today, no one would claim that all the varieties of living organisms in the sea are known. A rough estimate suggests that Pliny's inventory of the number of species in the ocean is low by at least a factor of 10^3. The task of listing and arranging the great diversity of living organisms was undertaken by the Swedish botanist Linnaeus (1701–1778), in his *Systema Naturae* (10th ed., 1758). He grouped related species of organisms into genera and introduced the binomial system, whereby an organism is named by giving the Latin version of the generic name followed by the name of the species. Thus the scientific name of the Maine lobster (Fig. 23-6) is Homarus *americanus*. Genera are then further classified into larger groupings. Thus the Maine lobster belongs to the suborder Reptantia, the order Decapoda, the subclass Malacostraca, the class Crustacea, and the phylum Arthropoda (Table 25-1, p. 401). To

Figure **23-6** The Lobster, Homarus *americanus*. (Photograph: John H. Gerard, National Audubon Society)

give the reader a feeling for the diversity of marine life, the major marine phyla are listed in Table 23-1. For many phyla, the important marine classes are also listed. The higher classifications are in a state of flux, since research results often require revision of previous classification schemes. Interpretation of the relative affinity of organisms is to some extent subjective, and so there is frequent controversy in questions of classification. The classification presented in Table 23-1 is not the ultimate summary but merely one of several schemes currently used by biologists.

Each marine phylum tends to be primarily benthic, planktonic, or nektonic. Thus the dinoflagellates (Dinophyta) are mainly planktonic plants while the echinoderms (Echinodermata) primarily inhabit the sea floor. The dominant life habit of each group in the adult stage is also indicated in Table 23-1. While many adult organisms live on the bottom of the sea or are active swimmers, most juvenile forms are planktonic.

Many plants and animals have mineral skeletons which, after the death of the organism, collect on the sea floor. Although a number of other mineral constituents have been identified, three compounds—calcium carbonate, calcium phosphate, and silica—make up the bulk of all skeletal material. Table 23-1 indicates the dominant composition of the skeletal material for each class or phylum. Not all representatives of a particular phylum have a mineral skeleton. Thus many marine green algae (Chlorophyta) do not have a skeletal structure. Those green algae

TABLE **23-1**   CLASSIFICATION OF PRINCIPAL MARINE ORGANISMS

Phylum	Class	English Equivalent	Figure Reference	Habitat			Skeletal Material		
				Plankton	Nekton	Benthos	Calcium Carbonate	Calcium Phosphate	Silica
Schizophyta		bacteria	3-1	X		X			
Cyanophyta		blue-green algae	3-2	X		X			
Chlorophyta		green algae	27-10			X	X		
Phaeophyta		brown algae	24-12			X			
Rhodophyta		red algae	24-13			X	X		
Chrysophyta		diatoms	24-6	X		X			X
		coccolithophorids	24-9	X			X		
Dinophyta		dinoflagellates	24-11	X					
Rhizopoda	Foraminifera	foraminifera	25-1	X		X	X		
	Radiolaria	radiolarians	25-2	X					X
Cnidaria	Hydrozoa	hydras	25-4	X		X			
	Scyphozoa	jellyfish	25-5	X					
	Anthozoa	corals, etc.	25-6			X	X		
Ctenophora		comb jellies	25-7	X					
Plathelminthes		flatworms	25-8a	X		X			
Chaetognatha		arrow worms	25-8b	X					
Annelida		segmented worms	25-8c	X		X			
Arthropoda	See Table 25-1	arthropods		X		X	X	X	
Porifera		sponges	25-13			X	X		X
Bryozoa		moss animals				X	X		
Brachiopoda		brachiopods				X	X		
Mollusca	Monoplacophora					X	X		
	Amphineura	chitons	25-14a			X	X		
	Scaphopoda	tusk shells	25-14c			X	X		
	Bivalvia	bivalves	25-14d	X		X	X		
	Gastropoda	snails	25-14b	X		X	X		
	Cephalopoda	squids	25-14e,f	X	X	X	X		
Echinodermata	Crinoidea	crinoids	25-15a			X	X		
	Asteroidea	starfish				X	X		
	Ophiuroidea	brittle stars	23-2			X	X		
	Echinoidea	sea urchins				X	X		
	Holothuroidea	sea cucumbers	25-15b			X	X		
Protochordata		tunicates	25-12	X					
Chordata	Agnatha	jawless fish	25-16		X			X	
	Elasmobranchii	cartilagenous fish	25-17		X	X		X	
	Osteichthyes	bony fish	23-2		X	X		X	
	Reptilia	reptiles			X			X	
	Aves	birds			X			X	
	Mammalia	mammals	25-20		X			X	

that are mineralized, however, invariably contain calcium carbonate, and the mineral remains of these plants contribute significantly to the calcium carbonate sedimentation in the tropical neritic zone.

Summary

The organisms that currently inhabit the earth are the diversified direct descendants of the earliest forms of life. There is a constant tendency of populations to expand in geometric progression, which is counteracted by limited resources and by the tendency of predator populations to respond to changes in the availability of prey. To reproduce, organisms require matter and energy. Autotrophic organisms derive these resources from inorganic matter and sunlight, while heterotrophic organisms make use of organic foodstuff. If the organism is to derive the necessary nutrients and avoid being surrounded by its own excretion products, there must be relative motion between the organism and its environment. Depending on their life habit, marine organisms are divided into the plankton, the nekton, and the benthos. These organisms range the waters from the intertidal zone, to the neritic zone over the continental shelves, to the oceanic environment.

The variety of life in the sea is enormous. Organisms are classified by species, genera, order, class, and phylum. Many organisms have mineralized skeletons composed primarily of either calcium carbonate, silica, or calcium phosphate.

Study Questions

1. Compare the impact on nature of a population explosion by a primitive human population and a technically advanced human population.
2. Give examples of terrestial autotrophic and heterotrophic organisms.
3. What prevents a forest from exhausting its resources and drowning in its decay products?
4. What kind of environment is provided by a fishing float adrift on the surface of the ocean?
5. Make a list of various seafoods arranged by class and phylum.

Supplementary Reading

(Starred items require little or no scientific background.)

* Simpson, G., and W. S. Beck (1965). *Life: An Introduction to Biology,* 2nd ed. New York: Harcourt, Brace and World. (Introductory biology text.)
* Malthus, Thomas Robert (1798). *An Essay on the Principle of Population as It Affects the Future Improvement of Society.* London: Johnson.

24 Plant Life in the Sea

All life in the sea depends ultimately on the photosynthetic auto-trophs, the plants of the sea. Using sunlight for photosynthesis, the plants convert the inorganic constituents of seawater into organic matter.

Photosynthesis

Light is necessary for photosynthesis. Plant cells contain special pig-ments, the chlorophylls, that absorb part of the incident sunlight, pri-marily in the violet ($.45\mu$) and red ($.68\mu$) portions of the visible spec-trum. Absorption of these components of the incident white light causes the plants to look green. Chlorophylls are complex organic molecules that contain magnesium: for example, the chemical formula for chloro-phyll-a is $C_{55}H_{72}O_5N_4Mg$. The pigments in the plant cell are contained in special organelles, the chloroplasts. Chlorophyll is transformed into a higher energy state by absorption of light. The excitation energy is then transferred to other molecules in a series of steps that involve the excitation of two pigments. While the detailed process is complex and not yet completely understood, we know that the net effect of the ab-sorption of light is that water and carbon dioxide are converted into sugar and oxygen.

$$6CO_2 + 6H_2O \xrightarrow[\text{chlorophyll}]{\text{light}} C_6H_{12}O_6 + 6O_2$$

The sugar so synthesized acts as an energy source for the production of all the other organic compounds the plant requires.

To determine how plant growth is affected by light intensity, it is necessary to work with an artificial culture in which the necessary nutri-ents are supplied in abundance and all factors except light intensity are held constant. The results of such an experiment for a planktonic green algae are shown in Figure 24-1. Because of the large range in light in-

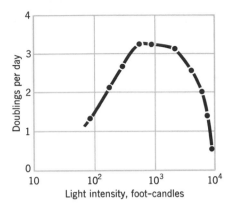

Figure **24-1** Growth rate of a culture of the green algae Chlorella *pyrenoidosa* at 25°C as a function of the light intensity. (After Sorokin and Krauss, 1958)

tensity used, it is plotted on a logarithmic scale. The rate of plant growth is expressed by noting how many times a day the cells of the culture double their number. At low light intensities, the rate of multiplication is slow. It increases with light intensity up to about 550 foot candles, when a rate of about 3.2 doublings per day is reached. Further intensification of the light does not increase the rate of multiplication. When the illumination exceeds 3000 foot candles, cell division begins to decline, and in very intense light, growth is strongly inhibited.

This and similar experiments demonstrate that the rate of photosynthesis increases with increasing light intensity, as long as the light level is low. In bright light, the rate of photosynthesis becomes independent of the light intensity over a significant range of illumination. Finally, very bright light has an adverse effect on the rate of growth. In the ocean, we can therefore expect that the rate of photosynthesis will decrease with depth as the light intensity drops off. In very bright light, near the surface, the rate of photosynthesis may be inhibited. With depth the rate reaches a broad plateau and finally declines toward the bottom of the photic zone.

Except for the shallow portions of the neritic zone, sunlight does not penetrate to the bottom of the ocean. Thus benthic plants can live only on the continental shelves. In the oceanic realm, the plants must be planktonic rather than benthic, for they are limited to the photic zone. Once they sink below the region of light, all photosynthesis will cease.

In addition to sunlight, photosynthesis requires CO_2 and water. Obtaining enough water can be a problem for land plants; however, there is no shortage of H_2O in the ocean. Although dissolved CO_2, in the form of H_2CO_3, is scarce in seawater, there is an ample supply of bicarbonate; when CO_2 is withdrawn by the plant, therefore, more H_2CO_3 is formed from the bicarbonate, with a resultant increase in the pH of the seawater.

$$HCO_3^- + H^+ \longrightarrow H_2CO_3$$

Determining the Rate of Photosynthesis

To see how the rate of photosynthesis is measured, we must examine the chemical equation for photosynthesis and its inverse, the equation for respiration:

$$\text{light} + CO_2 + H_2O \xrightarrow{\text{photosynthesis}} CHOH + O_2$$

$$CHOH + O_2 \xrightarrow{\text{respiration}} CO_2 + H_2O + \text{energy}$$

Thus the rate of photosynthesis can be determined by measuring any one of the following: the rate at which carbon dioxide is used up, the rate at which oxygen is liberated, or the rate at which carbon is converted to organic matter. For example, a sample of seawater containing algae is placed in a transparent bottle and suspended in the sea for a day. If we measure the oxygen content at the beginning and end of the experiment, the change will represent the sum of the oxygen changes of the plankton due to photosynthesis and respiration. The gross photosynthetic rate is obtained by subtracting the amount of respiration due to the animals and plants of the plankton. To do this, we fill an opaque bottle with the same water and suspend it next to the clear bottle. Since no light can enter the opaque bottle, there will be no photosynthesis, and any change in oxygen content will reflect only the respiration during the experiment. The sum of the two, then, is the gross rate of photosynthesis. Such an experiment was carried out in June 1935 in the North Sea by R. S. Wimpenny (1966). The results were as follows:

	O_2 Millimoles liter^{-1}	
	Transparent Bottle	*Opaque Bottle*
Initial 6 A.M.	0.288	0.288
Final 9 P.M.	0.292	0.282
Difference	+0.004	−0.006
Gross productivity	0.010 millimoles O_2 liter^{-1}	
Net productivity	0.004 millimoles O_2 liter^{-1}	

Per cubic meter ($1 \text{ m}^3 = 10^3$ liters), the gross productivity amounts to 0.01 mole of O_2. Thus 0.01 mole of carbon $= 0.12$ g of carbon are converted to organic matter per day per cubic meter. On the particular day of the experiment, there was adequate light to about 15 m. Assuming a uniform rate of gross productivity over the top 15 m, a total of 1.8 g m^{-2} day^{-1} of carbon were converted to organic matter by the plants.

In coastal water, where the primary productivity of the phytoplankton is high and restricted to a relatively thin layer, this method of measuring photosynthesis works fairly well. Even so, we have to make very accurate measurements in order to detect a small difference in the total oxygen concentration. In the open ocean, the situation is much more difficult. The amount of photosynthesis per unit area of the sea surface is generally much less and the depth over which active photosynthesis can take place is greater. This is a direct result of the greater light penetration in ocean water relative to the more turbid coastal water.

The oxygen changes we can expect in our bottles in the open ocean are much less than we can hope to detect by the most careful analytical techniques. The artificial production of a radioactive isotope of carbon, carbon-14, has made it possible to measure the primary productivity in the open ocean. Since all the life in the sea ultimately depends on the production of organic matter by plants, the *primary productivity*, it is important to know this rate. It directly affects the total amount of fish we can hope to harvest.

A method using carbon-14 was first worked out by Stemann Nielsen in 1956 and used by him on the Galathea Deep Sea Expedition, which circled the globe between 1950 and 1952. The method depends on the fact that carbon-14 behaves chemically just like the stable isotope of carbon, carbon-12. The atoms of carbon-14, however, are unstable; in a period of about 5600 years, half of them decay to nitrogen, with the emission of an energetic electron. It is possible to detect these energetic electrons with a Geiger counter and thus to determine the rate at which electrons are given off. If we have only 10^{-6} g of carbon-14, this will give off 2×10^7 electrons per minute. Since we can count these electrons with high efficiency, we can detect extremely small quantities of carbon-14.

A small amount of carbon-14, in the form of bicarbonate, is added to a seawater sample. The phytoplankton, the plants of the plankton, cannot tell the difference between the carbon-14 and ordinary carbon and so will convert some carbon-14 into organic matter by photosynthesis. After the plankton in a bottle are exposed to light for a known length of time, the phytoplankton are filtered out of the water and placed in a radiation detector to determine the amount of carbon-14 assimilated. Since the ratio of carbon-14 added to the amount of total bicarbonate in the bottle is known, one can calculate the amount of carbon that has been converted from bicarbonate into organic matter by photosynthesis.

The Matter Requirements of the Plants

The phytoplankton consist mostly of water. If we dry 100 g of wet phytoplankton, we end up with 8 to 20 g of dry plant matter. This dry matter is a mixture of organic matter and the mineral skeletons, siliceous and calcareous, of the organisms. The amount of mineral matter will depend on the types of plant present and their stage of development. The dry matter may contain 25 to 75 percent mineral matter.

The organic matter of the phytoplankton consists of 25 to 65 percent protein, 2 to 10 percent fat, and 0 to 35 percent carbohydrate. These materials contain mostly the elements carbon, oxygen, and hydrogen. In addition, the organic matter of the phytoplankton also contains significant amounts of the elements nitrogen and phosphorus. About 8 percent of the dry organic matter is nitrogen and about 1 percent is phosphorus. For every atom of phosphorus there are approximately 15 atoms of nitrogen and 100 atoms of carbon.

About 5×10^{-4} mole of nitrogen gas are dissolved in a liter of seawater (Fig. 20-3, p. 330). The plants of the sea, however, are not able to utilize nitrogen gas directly. To form organic nitrogen compounds, they require nitrogen combined as ammonia (NH_3), nitrite (NO_2), or nitrate (NO_3). The available nitrogen in seawater exists primarily as nitrate, while phosphorus is in the form of phosphate (PO_4). The concentration of these nutrients is generally low in surface seawater. Thus the rate of growth of the phytoplankton is often limited by the availability of these essential elements.

If it were not for the relative movement between the plant and the water, the continued absorption of the necessary nutrients from the surrounding seawater would soon deplete it of nutrients. Since the ocean is always in motion relative to the solid earth, plants attached to the sea bottom are continually bathed by fresh seawater. Planktonic plants, however, are carried with the moving water. To avoid depleting the surrounding water, therefore, they had to evolve special mechanisms for independent movement.

There are two methods by which planktonic plants can move. The flagellates have special threadlike projections from their cells, the flagella, whose lashing motion propels the cells through the water. Other plants, such as the diatoms, are slightly denser than the seawater and therefore sink slowly through the water column.

The flagellates can propel themselves horizontally or vertically and thus can remain within the illuminated surface layer. Diatoms, on the other hand, move by adjusting their buoyancy; hence they can move only in the vertical direction. As they slowly sink through the water, they are carried out of the photic zone. In still, deep water, a sinking population, even if it is reproducing, will inevitably be swept out of the photic zone and become extinct. The turbulent motion of the surface layer of the sea, however, makes survival possible for the diatoms.

The problem of survival by the sinking diatoms was analyzed by Riley, Stommel, and Bumpus (1949). Consider two diatoms at the same depth within the sea (Fig. 24-2). Let the mean time for cell division be

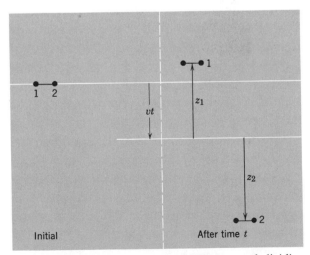

Figure **24-2** Diatoms sinking, diffusing, and dividing.

such that after t sec each of the two diatoms has divided, to give four diatoms in all. In the intervening time, at a sinking rate of v cm sec^{-1}, they will have sunk a distance vt cm. If there were no turbulence, all four diatoms would find themselves vt cm deeper in the ocean. The turbulent motion in the seawater, however, causes the diatoms to wander randomly about their mean position. The law of such random movements states that the square of the mean displacement increases proportionally to a constant, characterizing the turbulence (the diffusion constant D) and the time. Thus the mean square displacement $\bar{z}^2$ is proportional to Dt.

Now, it is just as likely for the diatoms to be carried upward as downward, so that, on the average, one of the two original diatoms will be carried toward the surface while the other will sink. The turbulent motion of the water is superposed over the general sinking motion of the diatoms. After time t, the two diatoms divide, giving rise, on the average, to two diatoms above and two below the mean position. If the mean distance the diatoms have diffused vertically from their initial position ($\bar{z}_1$, $\bar{z}_2$) is greater than the distance the diatoms have sunk (vt), there will be two diatoms at equal or shallower depth than originally, and thus the population in the photic zone can be maintained. However, if the random displacements are less, on the average, than the distance they have sunk, the population will ultimately be carried out of the light and be eliminated.

The condition for preservation is, therefore, that vt be smaller than the average of $\bar{z}$, which, in turn, is proportional to the square root of Dt. The condition for preservation turns out to be

$$v^2 \leqq 4D/t.$$

Diatoms are also removed by the grazing of marine animals so that t should not be the actual time for cell division but, rather, the effective doubling time in the presence of grazers. If the two terms of the equation are equal, the population will remain constant while a slower sinking rate, or a more rapid rate of cell division, will lead to a population increase in the photic zone.

In order to give rise to two diatoms, the individual cell must double its nutrient content by absorption from the surrounding seawater. Because of depletion of the surrounding water, the rate of nutrient uptake

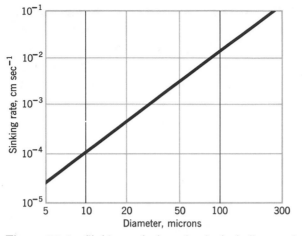

Figure **24-3** Sinking velocity of spherical diatoms having an excess density of 2 percent as a function of their diameter. (After Munk and Riley, 1952)

can increase only if the sinking rate of the diatom is increased. Thus the requirements for light and nutrients are contradictory. To stay in the light, the diatom should sink slowly and multiply rapidly. To permit sufficient nutrient uptake for rapid cell division, however, the diatom should sink rapidly. These conflicting constraints have been analyzed by Munk and Riley (1952).

To illustrate the problem, let us consider the case of spherical diatoms with a density 2 percent greater than that of seawater. Figure 24-3 shows the sinking rate of such spheres in seawater as a function of size. Note that the rate of sinking increases as the square of the diameter, (as it does in the case of fine sediment particles, Fig. 14-3, p. 227). The rate of nutrient uptake will be determined by the nutrient having the smallest relative concentration. If we take a typical value for nutrient-poor surface water (10^{-7} mole phosphate per kilogram), we obtain the curve for the minimum times to double the nutrient content (Fig. 24-4). The larger the diatom, the longer is the time required to double the nutrient content.

Next, we must determine the doubling time required to maintain the population in the light. As the size of the cell, hence the sinking rate, increases, the population must increase more rapidly. The doubling time also depends on the turbulence of the water. For a vertical coefficient of turbulent diffusion of 10 cm^2 sec^{-1}, we find that the doubling time must be at least as short as is indicated by the line t_T in Figure 24-4.

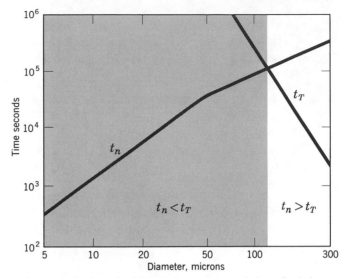

Figure **24-4** Minimum time required to double nutrient content (t_n) and minimum duplication time to maintain spherical diatoms of diameter d in the photic zone (t_T). Assume: $D = 10$ cm² sec⁻¹, excess density 2 percent, phosphate concentration 10^{-7} mole kg⁻¹. (After Munk and Riley, 1952)

With these assumptions, for sizes less than 130 microns, the time required to replace the nutrients is shorter than the required doubling time to maintain the population in the light. However, if the diameter of the spherical diatom is larger than 130 microns, nutrients cannot be supplied at a sufficient rate to permit an adequate rate of cell division. Thus the physical and chemical conditions in the sea limit the maximum cell size of diatoms and tend to favor smaller cells.

Diatoms

Having discussed some of the conditions imposed on planktonic plants that are denser than seawater, let us now examine the diatoms which belong to the phylum Chrysophyta. In order to obtain a plankton sample, we must filter seawater through a very fine net. The plankton net is a long, slightly tapered cone with a bottle attached to its narrow end. As the net is slowly pulled through the sea, the plankton is swept into the bottle (Fig. 24-5).

When we examine the contents of the bottle under a microscope, we find a great variety of microscopic plants. Among the most abundant

Figure **24-5** Plankton net. (Photograph: courtesy of Woods Hole Oceanographic Institution)

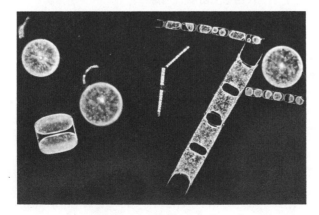

Figure **24-6** Various diatoms from the marine plankton. Approximately ×40. (Photograph courtesy of Douglas P. Wilson, F.R.P.S.)

Figure **24-7**   Cell division of a diatom.

will be diatoms, unicellular algae enclosed in a tiny box of glasslike silica (Fig. 24-6). The glass box has an intricate pattern of perforations that permits the cell to absorb substances from the sea. Within the box are the chloroplasts, the pigments that absorb sunlight to convert carbon dioxide and water to organic matter.

The glass box of the diatom is filled with the cell fluid, the protoplasm. The nucleus of the cell contains the genetic code that controls the development of the cell. When the cell divides, the nucleus first reproduces itself. The glass box of the diatom is constructed like a pillbox with a bottom and a lid. These two now slide apart and form the covers of the two new cells. Meanwhile new silica bottoms are formed by the dividing plant (Fig. 24-7). As the diatoms continue to divide, they become diverse in size. One cell will always have the size of the original cell, but the others will become smaller and smaller (Fig. 24-8). The process of

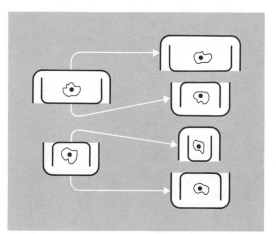

Figure **24-8**   Variation in the size of diatoms as a result of cell division.

division and diminution in cell size obviously cannot go on indefinitely. When the cell gets too small, it discards its silica box and temporarily becomes a growing bare blob of protoplasm. When it has grown to mature size, it secretes a new silica box.

Diatoms vary in shape as well as size. In many, the dividing pillboxes remain attached to one another, forming chains of two, four, eight, or more cells (Fig. 24-6). Attachment of the cells is often facilitated by fine gelatinous threads which protrude from the cells.

The siliceous boxes of the diatoms gradually sink to the sea floor. If the silica does not dissolve during the descent, the tiny silica shells accumulate on the sea floor to form a *diatom ooze.* Such oozes are transformed in time into a sedimentary rock; thus we obtain diatomaceous earth. Because it consists of countless tiny, porous silica vessels, diatomaceous earth can be used as a filtering material.

Between 1866 and 1867, the Swedish chemist Alfred Nobel was looking for ways to produce a safe high explosive. Nitroglycerine was an effective explosive; however, it had a tendency to explode spontaneously when jarred. By using diatomaceous earth to absorb the liquid nitroglycerine and thus prevent a premature explosion, Nobel produced a safe explosive—dynamite. Both the Nobel fortune and the famous prize established by the will of Alfred Nobel are based on the lowly diatom.

Coccolithophores

When Thomas Henry Huxley (1825–1895) examined mud brought up from the floor of the deep sea, he found very fine grains of calcium carbonate, about 1 micron in diameter. At first he thought that these particles were merely the result of the disintegration of larger shells. He noticed, however, that the particles had a definite form and size. The electron microscope reveals that these particles, which Huxley called coccoliths, have a very intricate structure (Fig. 24-9).

These little plates of calcium carbonate are enclosed within a tiny spherical plant, called a *coccolithophore.* The plant is equipped with a tiny hair, a *flagellum,* used for locomotion. Occasionally these plants are so numerous that they give the sea a milky appearance.

Other calcareous plants are the Rhabdoliths. Their architecture consists of a geometric elaboration of a simple sphere, with long, spiny calcareous plates (Fig. 24-10).

Figure **24-9** Electron micrograph of a coccolithophore, Coccolithus *huxleyi.* Magnification ×10,250 (Photograph: Andrew McIntyre, Lamont Geological Observatory)

Dinoflagellates

Next to the diatoms, the most numerous organisms in the plankton are the dinoflagellates. These have two whips, or flagella. One beats in a groove that runs across the surface of the cell while the other projects behind from a longitudinal groove. The cell wall consists of a number of cellulose plates fitted together. Dinoflagellates come in various shapes; some are spherical and others have long, hornlike protuberances

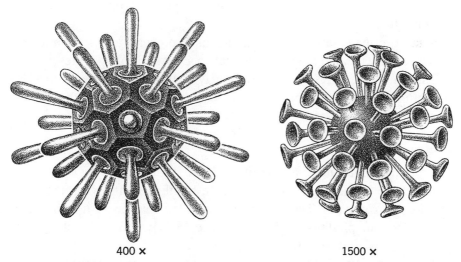

400 ×   1500 ×

Figure **24-10** Rhabdospheres. (Plate from volume by E. H. Haeckel from the *Challenger* reports)

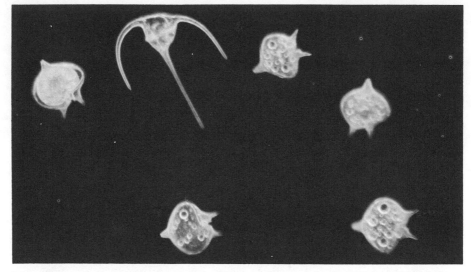

Figure **24-11**   Some dinoflagellates. (Photograph: Douglas P. Wilson, F.R.P.S.)

(Fig. 24-11). Dinoflagellates can be either photosynthetic autotrophs or heterotrophs. The autotropic forms contain chlorophyll while the heterotrophic dinoflagellates lack pigments and derive their energy from dissolved organic matter in the seawater which they absorb through the cell wall. Still others can ingest particulate organic material. Other members of this versatile phylum can support themselves either by photosynthesis or heterotrophically, depending on the availability of sunlight.

Some dinoflagellates have a light-sensitive organ and so are able to move actively toward the light. Several species of dinoflagellates produce a brilliant bioluminescence, particularly if the water is agitated. Thus breaking waves on the ocean are often lit up at night by the glow of countless dinoflagellates.

Benthic Plants

From the shoreline to a maximum depth of about 100 m, light is able to penetrate to the sea floor. As a result, benthic plants are able to exist in this environment. These are the so-called seaweeds, marine algae. There are four phyla of algae (see Table 23-1, p. 373)—the blue-green, green, brown, and red algae.

Blue-green algae (phylum Cyanophyta) are very primitive plants. Filamentous blue-green algae often form an algal mat in the intertidal zone. Fine-grained sediments tend to be collected and bound within this mat. Blue-green algae from such mats of sediment are among the oldest fossils from the Precambrian (see Fig. 3-3, p. 18).

Green algae (phylum Chlorophyta), as the name implies, are green in color and contain two types of chlorophyll, a and b. Algae of the genus Halimeda (see Fig. 27-10, p. 461), which is abundant in tropical seas, contain calcareous plates. When the plant dies, these plates are a dominant component of the sand-size sediments. Other tropical green algae contain small needle-shaped crystals of calcium carbonate. These algae are a major source of mud-size carbonate sediments.

Brown algae (phylum Phaeophyta) are generally brown in appearance because the green color of the chlorophyll is masked by yellow and brown pigments. Algae of this phylum are often very large, and they form offshore *kelp* beds (Fig. 24-12). Some brown algae attain a length of over 35 m. They are anchored to the bottom by a *holdfast*. This is not a true root but, rather, a branched structure which holds the algae in place. From this rises a cylindrical hollow stipe which extends from the sea floor to near the surface. The stipe ends in a gas-filled bulb which acts like a float from which ribbon-like fronds issue. These are the primary sites for photosynthesis. The great length of the stipe permits the fronds to receive ample sunlight even though the holdfast is anchored at considerable depth.

Figure **24-12** The brown algae Laminaria *digitata*. (Photograph: Douglas P. Wilson, F.R.P.S.)

Figure **24-13** The red algae Chondrus *crispus* (Irish Moss). (Photograph: Hugh Spencer from National Audubon Society)

The *red algae* (phylum Rhodophyta) have various colors, such as red, purple, brown, and green. In addition to chlorophyll, they contain several pigments, including red pigments. These algae are generally small in size and large in number and variety (Fig. 24-13). Many red algae are calcareous including encrusting forms that tend to cement sediments together. Other red algae contribute to the sand-size sediment. The calcium carbonate in these plants often contains a considerable amount of magnesium.

Human Use of Marine Plants

Because of their small size (10 to 100μ), the plants of the plankton cannot be harvested economically for human use. The small size of the plants is compensated for by their numbers; there are between 10^8 and 10^{10} phytoplankton cells per square meter of sea surface. Since planktonic plants multiply rapidly, efficient harvesting would require filtering the seawater once a day to extract the plant matter. Therefore, we depend on marine animals to graze the plants and concentrate the food resources into a form that can be more readily extracted from the sea.

Benthic algae, on the other hand, grow to significant size and can be harvested readily. In the Orient, particularly in Japan, seaweed is used extensively as a human food. Seaweeds are also used as animal feed—for example, by sheep on the coast of Ireland—and as fertilizer, particularly on rocky islands, where soil is scarce.

The products that can be extracted from seaweeds range from iodine and potassium chloride to various organic colloids. The colloids, such as agar and algin, are used as thickeners in such foods as ice cream and soups. In California, kelp is harvested by giant barges that can collect 300 tons of algae per day. As the surface mat is cut and harvested, the kelp grows new fronds in a few weeks.

Summary

Marine plants produce organic matter from inorganic sources and so are the primary source of all food in the sea. Because they are restricted to the photic zone, benthic plants can exist only in shallow water. Most of the plants of the sea are therefore planktonic, comprising the phytoplankton.

The rate of photosynthesis can be determined by measuring the increase in dissolved oxygen or by determining the rate at which radioactive carbon-14 is fixed in plant matter.

The rate of plant growth is limited by the availability of essential plant nutrients, such as nitrate and phosphate. Since the nutrients in the water around the plant cells are rapidly depleted, plants must move relative to the water, either with the aid of flagella or by sinking. The phytoplankton includes diatoms, coccolithophores, and dinoflagellates; the benthic plants consist of algae of various phyla.

Study Questions

1. Contrast grasses on land and diatoms in their role as a primary food source for animals.

2. What is the surface area per gram for a cubical plant with a density of 1 g cm^{-3} and with edge dimensions of:
   a. 1 micron (10^{-6} cm)
   b. 10 cm
3. Why is there an upper limit on the size of diatoms?

Supplementary Reading

(Starred items require little or no scientific background.)

* Hardy, Sir Allister (1965). *The Open Sea: Its Natural History.* Boston: Houghton Mifflin Co., Part I, Chap. 3.
* Huxley, Thomas Henry (1967). *On a Piece of Chalk.* New York: Charles Scribner's Sons. (Well-illustrated edition of a work originally published in 1868.)
* Nielsen, Stemann E. (1956). "Measuring Productivity in the Sea." In Anton F. Bruun et al., eds., *The Galathea Deep Sea Expedition.* London: George Allen and Unwin Ltd. Pp. 53–64.
Raymont, John E. G. (1963). *Plankton and Productivity in the Oceans.* London: Pergamon Press.
* U. S. Dept. of Interior (1962). *Seaweeds Are Not Weeds.* Conservation Note 7, Fish and Wildlife Service, January.
* Wald, George (1959). "Life and Light," *Scientific American,* October.

25 The Animals of the Sea

Animal life in the sea includes more than 15 phyla and ranges in size from tiny unicellular organisms to the largest of all animals, the whale. Equipped with an aqualung, man himself becomes a member of the marine fauna, at least temporarily.

An adequate description of the bewildering variety of known animals in the sea would require many volumes. In this chapter it is possible to present only a minute sample. For a more detailed and extensive survey, the interested student should examine one or more of the references listed at the end of the chapter.

The Distribution of Animal Life

The need for sunlight limits the plants of the ocean to the photic zone. The animals, on the other hand, are limited only by the availability of food and oxygen, hence range over the entire depth of the ocean, from the surface layer to the bottom of the deepest oceanic trenches. The animals that live within the water are called the pelagic animals; those that inhabit the bottom are benthic. The pelagic animals consist of the zooplankton, which drift with the currents of the sea, and the nekton, the actively swimming forms. Benthic animals live either on the bottom or within the bottom sediment, close to the water-sediment interface.

The food for animals in the sea is concentrated primarily in the surface layer and on the bottom. In the surface layer, plants provide a constant supply of organic matter by photosynthesis. Some of the dead organic matter sinks to the floor of the ocean and provides food for the benthic scavengers. The organic matter not assimilated by animals is decomposed by bacteria. The bacteria, in turn, become food for animals.

In shallow waters, the photic and benthic zones merge, and plants are able to live on the bottom. Vertical mixing is also enhanced in shallow water, so that the continental shelves are often very productive and support large commercial fisheries. In contrast, the open ocean, particularly its intermediate depths, is a relative desert, since most planktonic animals live near the surface where they can graze on phytoplankton, and the benthos feed by ingesting the detritus on the sea floor or by filtering organic matter from the water adjacent to the bottom.

Planktonic Animals

Many marine organisms are planktonic during part of their life cycle; however, a number of phyla consist mainly of permanent inhabitants of the plankton (Table 23-1, p. 373). The most primitive of the phyla whose members are dominantly planktonic is the phylum Rhizopoda, represented by Foraminifera and Radiolaria.

Foraminifera. The White Cliffs of Dover are one of the outstanding features of the southern coast of England. The cliffs are composed primarily of very soft calcium carbonate, known as chalk. Examination of the chalk under the microscope reveals that it is not just a collection of crystals of calcium carbonate. Rather, it is an aggregate of numerous tiny, delicate shells, about 1 mm in diameter. Since the chalk beds are about 500 m thick, they represent about 5×10^5 layers of these tiny shells. The shells are the remains of unicellular animals called Globigerinae, of the class Foraminifera. We often find the same tiny shells forming a calcareous *Globigerina ooze* when we examine sediments from the ocean bottom. The English chalk deposits consist of a Globigerina ooze that was laid down on the floor of the ocean during Cretaceous time, about 10^8 years ago.

When we examine the tiny shells carefully, we see that they contain many small holes; hence the name Foraminifera, which means holebearers. The protoplasm of the cell is exuded through these holes to form tiny arms that capture and digest food. At first, the Foraminifera consist of a blob of protoplasm surrounded by a calcareous shell. As the animals grow, the protoplasm protrudes from the shell and covers itself with more calcium carbonate. Thus the Foraminifera gradually add one chamber after the other to its growing shell. The chambers are connected and the protoplasm occupies them all (Fig. 25-1).

Radiolaria. While Foraminifera build tiny, snail-shaped, calcareous shells, other Rhizopoda, the Radiolaria, construct a silica shell. Although the housing of the Foraminifera is a perforated shell, the

Figure **25-1** Glass model
of a Foraminifera. (Photograph: American Museum
of Natural History)

enclosure of the Radiolaria is a space outlined by a network of silica
rods decorated with spines. Some of these tiny animals, 50 to 400
microns in diameter, appear suitable for illustrating a text in solid
geometry. Others look like the work of a modern sculptor. The
Radiolaria collected by the *Challenger* expedition were studied by the
German biologist Ernst Heinrich Haeckel (1834–1919). He was able to
enumerate 4000 new species and illustrated his report with 140 plates
of these beautiful animals (Fig. 25-2). Like the remains of the diatoms
and the Foraminifera, the remains of the Radiolaria form an ooze on
the floor of the deep sea.

Cnidaria (Coelenterata). The organisms of the phylum Cnidaria consist of a tubelike, two-layered wall surrounding a digestive cavity. Tentacles are used to move food toward the single opening, the mouth.
Cnidaria may be found attached, as polyps (Fig. 25-3*a*), or free-floating,
as medusae (Fig. 25-3*b*). Many species are transformed from a polyp to
a medusa during different times of their life cycle.

The class Hydrozoa includes many polyps that grow attached to rocks
and seaweeds in shallow water and then bud off small planktonic
medusae. Other Hydrozoans are adapted to floating on the sea surface.
These include colonial forms, such as the Portuguese-man-of-war

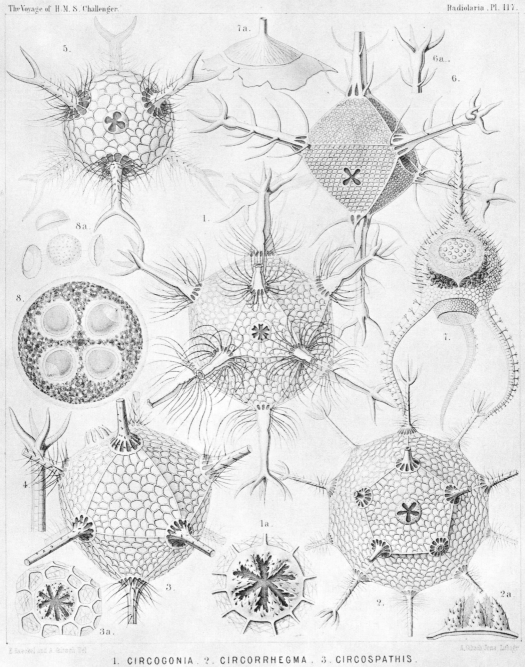

1. CIRCOGONIA . 2. CIRCORRHEGMA . 3. CIRCOSPATHIS .
4 · 6. CIRCOPORUS . 7. CORTINETTA . 8. CATINULUS .

Figure **25-2**　Various Radiolaria. (Plate from volume by E. H. Haeckel from the *Challenger* reports)

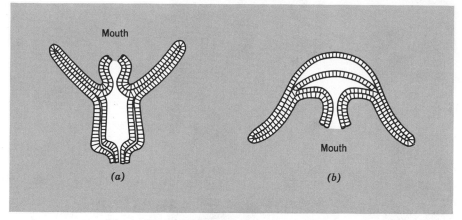

Figure **25-3**   The two forms of Cnidaria: (*a*) polyp; (*b*) medusa.

(genus Physalia) and the Jack-sail-by-the-wind (Velella) (Fig. 25-4). In these forms, different polyps of the colony have different functions. Flotation is provided by special gas-filled polyps, while others form a sail that enables the animal to drift relative to the water. The tentacles which hang below the float capture food and carry it to the opening in the digestive cavity. The Velella has a delicate blue color, and the little floats, with their sails, measure about 5 cm in diameter. The Portuguese-man-of-war is larger (about 30 cm long) and has very long tentacles with powerful stinging cells.

The large, floating jellyfish belong to the class Scyphozoa. They propel themselves through the water by pulsations of their large bell. Some of the larger jellyfish range up to 2 m in diameter (Fig. 25-5).

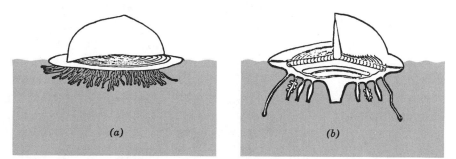

Figure **25-4**   The colonia Hydrozoa Velella: (*a*) adult sailing at the surface; (*b*) section.

Figure **25-5** A jellyfish. (Photograph: William Amos)

   Whereas the Scyphozoa are mainly planktonic medusae, the third class of Cnidaria, the Anthozoa, exist mainly as benthic, colonial polyps. To this class belong the sea anemones, the corals, and the alcyonarians. Sea anemones are frequently found in tidal pools in colder waters (Fig. 25-6) while corals and alcyonarians abound in tropical waters. We shall consider these forms in more detail in Chapter 27, where we discuss the coral-reef environment. Solitary corals are found in the aphotic benthic environment; but the reef-building corals are limited to the photic zone.

   Ctenophora. The small, jelly-like spherical planktonic forms, the sea gooseberries, belong to the phylum Ctenophora. They are often very abundant and feed actively on the smaller constituents of the plankton. Pleurobrachia, an example of the Ctenophora also known as comb jellies, is shown in Figure 25-7.

   Marine Worms. Wormlike animals of the plankton belong to three phyla—the flatworms, Plathelminthes; the arrow worms, Chaetognatha;

Figure **25-6** Sea anemones Metridium *senile* attached to a large shell. (Photograph: Douglas P. Wilson, F.R.P.S.)

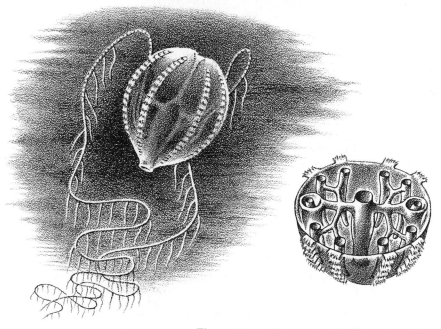

Figure **25-7** The comb jelly Pleurobranchia.

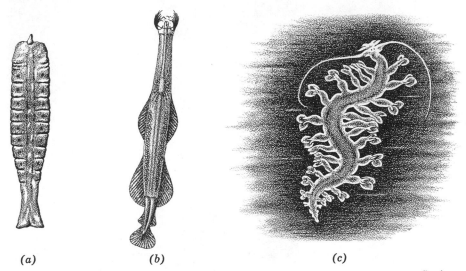

(a)   (b)   (c)

Figure **25-8**   Planktonic worms: (*a*) flatworm Nectonmertes; (*b*) arrow worm Sagitta; (*c*) segmented worm Tomopteris.

and the segmented worms, Annelida. Although the flat and segmented worms (Figs. 25-8*a* and *c*) are primarily benthic, the arrow worms (Fig. 25-8*b*) are planktonic. The arrow worms have fins for motion through the water, movable bristles that resemble jaws, and teeth for holding their prey. They range widely in depth and latitude but are restricted to the marine environment.

Arthropoda. The Arthropoda have greater diversity than any other phylum, and contain more than 80 percent of all the known living species. Most of these, the insects, are terrestrial; however, the class Crustacea is marine and is an important constituent of the plankton, both in variety and in number (Table 25-1).

Two subclasses of Crustacea, the Cephalocarida and the Mystacocardia, were discovered only recently in Long Island Sound. Such new discoveries in a relatively well-explored area suggest that we can expect other important additions to the list of known organisms from the ocean.

To the subclass Branchiopoda belongs the tiny brine shrimp, Artemia, the only arthropod found in highly saline waters (Fig. 25-9*a*). The subclasses Ostracoda and Cirripedia are highly modified in form. The Ostracoda (Fig. 25-9*b*) have a small, bivalved carapace and bear more superficial resemblance to bivalve mollusks than to arthropods. They

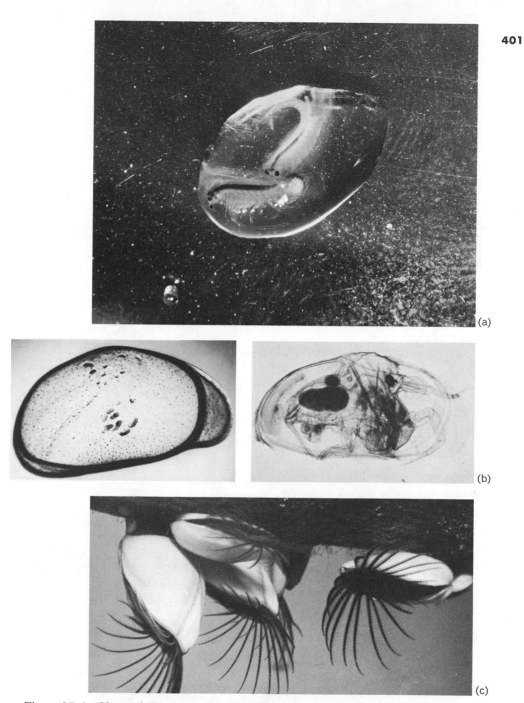

Figure **25-9** Unusual Crustacea: (*a*) The brine shrimp, Artemia, two individuals in a drop of water. (*b*) The ostracod Chlamydotheca *arcuata;* left) interior of right valve, right) median section x 50 (R. Smith SUNY Stony Brook); (*c*) Goose barnacles attached to floating driftwood.

TABLE **25-1**   MARINE ARTHROPODA

		Figure
Trilobita	Extinct marine arthropods	3-5a
Arachnomorpha	Horseshoe crab, *Limulus,* benthic	
Crustacea		
Cephalocarida ⎫	Recently discovered subclasses	
Mystacocarida ⎭		
Branchiopoda	Brine shrimp, *Artemia,* planktonic	25-9a
Ostracoda	Bivalved carapace, planktonic, benthic	25-9b
Copepoda	Planktonic, benthic	25-10
Cirripedia	Barnacles, benthic	25-9c
Malacostraca		
Euphausiacea	Krill, planktonic	25-11
Decapoda	All larger Crustacea	
Nantantia	Swimming forms, shrimp	
Reptantia	Crawling forms, crabs, lobsters	23-6

live both within the water and on the bottom. The Cirripedia are the barnacles whose calcareous shells are attached to rocks and other surfaces (Fig. 25-9c). Many attach themselves to drifting objects, including other animals.

The Copepoda are often the most numerous of the small Crustacea in the plankton (Fig. 25-10). They range from 0.3 to 10 mm in size, and

Figure **25-10**   The copepod Calanus.

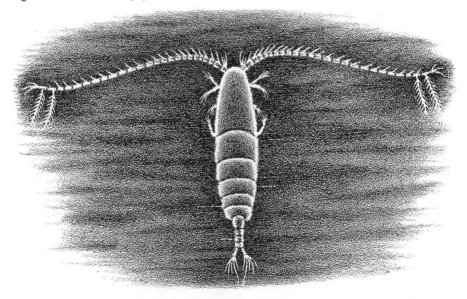

in many areas they make up over 70 percent of the volume of the zooplankton. Some copepods are benthic and a few are parasitic.

The larger Crustacea belong to the subclass Malacostraca. Among these are the Euphausiacea, the krill, which are extremely abundant. In the Antarctic region, a single species, Euphausia *superba* (Fig. 25-11), is the main food for the whale. In Antarctic waters the krill occur in tightly crowded patches that give a red appearance to the water. These patches are 1 to 2 m thick with lateral dimensions ranging from several meters to several hundred meters.

Figure **25-11**   The krill, Euphausia *superba,* the principal food of the whale: (*a*) adult; (*b*) various growth stages. (After Hardy, 1967)

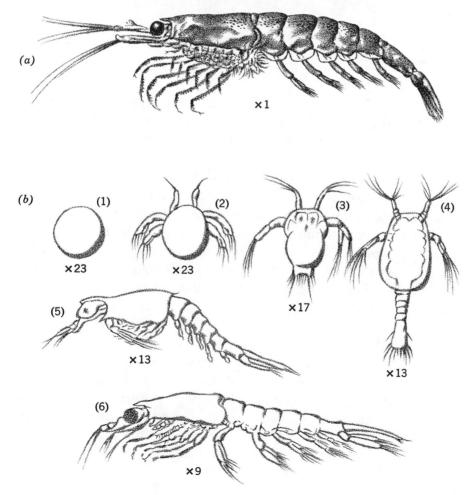

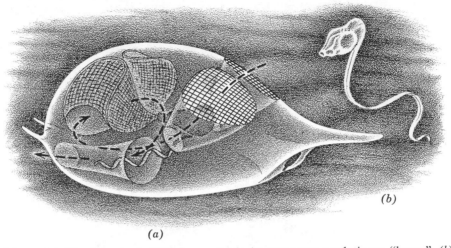

(b)

(a)

Figure **25-12**   A tunicate Oikopleura: (*a*) in its transparent gelatinous "house"; (*b*) swimming freely.

The largest Crustacea belong to the order Decapoda, which in turn is subdivided into the Natantia, the swimming forms such as shrimp, and the Reptantia, the crawling forms. To the latter belong the crabs and lobsters (Fig. 23-6).

Protochordata. The Protochordata include many planktonic and benthic marine forms. Of particular interest is the tunicate Oikopleura (Fig. 25-12). This small animal, about 3 mm in length, builds a delicate gelatinous house containing an elaborate mechanism through which water is pumped to filter out the very fine phytoplankton. When the filters become clogged, the animal abandons its house and constructs a new one. The filtering membrane of the house is extremely fine and is able to filter out particles as small as 0.1 micron in diameter.

Benthic Animals

Porifera. Having discussed the major phyla that are primarily planktonic, we now turn to the animals of the benthos. The most primitive sessile organisms are the sponges, of the phylum Porifera (Fig. 25-13). Some sponges are vase-shaped animals that live attached to the sea floor. They feed by causing currents of seawater to flow in through pores on their outer surface and out through a large opening, the osculum.

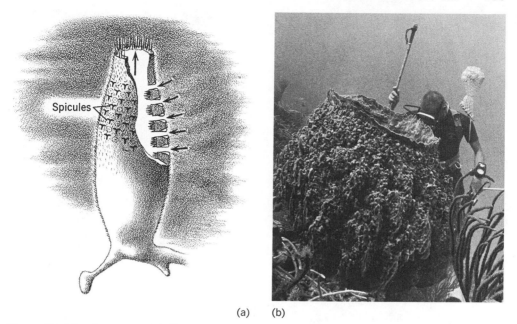

(a)   (b)

Figure **25-13**   A sponge: (*a*) diagram showing flow of water through a sponge; (*b*) a large sponge, Xestospongia, at a depth of 25 m in reef on the north shore of Jamaica. The diver is carrying a bag of flash bulbs for underwater photography. (Photograph: T. F. Goreau)

Sponges can grow to very large size; some are large enough for a diver to enter the central cavity. The cell framework is stiffened by numerous spicules which usually consist of silica but may also be calcareous.

Bryozoa. The Bryozoa, or moss animals, are small encrusting forms. They live as colonies within protective cases of horny or calcareous material out of which they extend their tentacles for feeding. Bryozoa are much more complex in internal structure than the corals, which they resemble in life habit.

Brachiopoda. The phylum Brachiopoda consists of shelled organisms which resemble the more common bivalves. However, unlike clams, which have a right and left shell, the Brachiopoda have a top and a bottom shell. They are attached to rocks by means of a stalk. Since the shape of their lower shell resembles that of an ancient oil lamp, Brachiopods are also known as lamp shells.

Figure **25-14** Variations of Molluscan architecture: (*a*) chiton; Amphineura; (*b*) snail, Gastropoda; (*c*) tusk shell, Scaphopoda; (*d*) clam, Bivalvia; (*e*) nautilus, Cephalopoda; (*f*) squid, Cephalopoda.

Mollusca. The members of this phylum present a highly varied appearance, as is evident in Figure 25-14. Although most mollusks are benthic animals, some are pelagic. Mollusca are characterized by a large muscular region below the viscera. The most primitive class, Monoplacophora, was known only as Paleozoic fossils until a living specimen, Neopilina, was dredged up from deep water in 1957.

The Amphineura, the chitons, are flat animals with a shell composed of eight plates. They creep with a broad foot and are often found in the intertidal zone.

The class Gastropoda includes the slugs, snails, whelks, and limpets. Most forms have a spiral shell. Although most gastropods are benthic and creep on the sea floor, some others, the pteropods, are part of the plankton, and their foot has become adapted for swimming. In some areas the pteropods are so abundant that their tiny shells form a pteropod ooze on the sea floor. Sea slugs, or Nudibranchiae, are gastropods without a shell. They are often brilliantly colored.

The Scaphopoda have a long, slightly curved, cone-shaped shell, open at both ends. Hence they are known as tusk shells. The Scaphopoda were used by West Coast Indians as wampum. They lived buried in soft sediment with only their posterior end projecting into the water.

The Bivalvia, also known as the Pelecypoda, include clams, oysters, and mussels. They are characterized by two symmetrical shells and a hatchet-shaped foot. The two valves are hinged and held together by muscles. Many bivalves live buried in the bottom mud and feed by means of long siphons that carry seawater to the mouth. Others bore into rock or wood.

The Cephalopoda constitute a highly varied class that includes squid, octopi, and the nautilus. Squid have an internal shell and can move very rapidly by expelling jets of water from a cavity in their mantle. Some squid attain very large size; they can measure up to 16 meters in length when their tentacles are spread out. The octopus has large suckers on its eight tentacles and lives close to the sea floor. The nautilus has an external chambered cell. As the animal grows, it adds a new chamber to its spiral shell and seals off the previous one.

Echinodermata. The Echinodermata are a benthic phylum with calcareous spines or spicules imbedded in the exterior surface of the organisms. The class Crinoidea, the sea lilies, live attached to the sea floor (Fig. 25-15a). They are found over a wide range in latitude and depth. In the geologic past, the crinoids were a dominant fauna in shallow seas, and the remains of their calcareous stalks form massive beds of crinoidal limestone.

The Holothuroidea, or sea cucumbers, are tubelike and live on the sediment surface (Fig. 25-15b). Some holothurians gather in living and dead organic matter from the sea floor with their tentacles. Others simply ingest the surface sediment, digest the usable organic matter, and expel the remainder through their anus.

The Asteroidea, the sea stars or starfish, are abundant in shallow water but also occur at depth. They have five arms that radiate out from the central plate, which contains the mouth opening. The Ophiuroidea differ from the Asteroidea in that they have very elongated, brittle arms that are much longer than the central disk (Fig. 23-2). Also known as brittle stars, they are particularly abundant in deeper water.

The final class of Echinodermata are the Echinoidea, the sea urchins and sand dollars. These consist of a spherical or disk-shaped body composed of closely fitted plates armed with numerous radial, movable spines. Some live in hollows they excavate within reefs. The spines of some sea urchins have light-sensitive organs that direct the spine toward a shadowing object overhead, thus protecting the animal from predation.

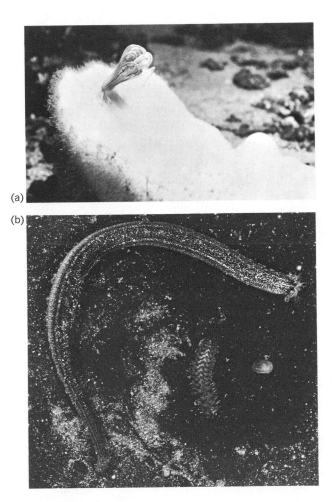

(a)

(b)

Figure **25-15** Two classes of Echinodermata. (*a*) Sea lily resting on a sponge (Ron Church) (*b*) The sea cucumber, Holothuroidea Labidoflax *digitata*. (Photograph: Douglas P. Wilson, F.R.P.S.)

The Nekton

Of the invertebrates, only some mollusks, such as the squid and the octopus, are capable of rapid, extended motion. Most nekton belong to the phylum Chordata, whose members range from the lamprey to the whales and porpoises. Since the development of the aqualung, man himself has become an active predacious member of the nekton. In addition to fish and aquatic mammals, some reptiles and birds derive their food from the sea, and many swim actively near the sea surface.

The most primitive fish are the Agnatha, the jawless fish. The sea lampreys (Fig. 25-16) and hagfish have a circular sucking mouth without jaws with which they attach themselves to their prey. The hagfish frequently enter the mouth or gill cavity of a dead or dying fish and eat their way through the fish, leaving nothing but skin and bones.

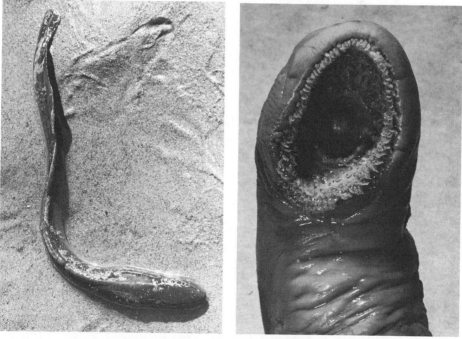

(a)   (b)

Figure **25-16** The lamprey, a jawless fish. (*a*) Top view. (Photograph courtesy of Michigan Conservation Department) (*b*) Close-up of sucking mouth. (Photograph courtesy of United States Department of the Interior, Fish and Wildlife Service)

Figure **25-17** A ray, Raja *clavata*, seen from below. (Photograph: Douglas P. Wilson F.R.P.S.)

Next come the Elasmobranchii, the sharks, rays (Fig. 25-17), and skates. They have a cartilagenous skeleton, and their skin is covered with denticles rather than scales. Their teeth are nothing more than modified denticles growing from a part of the skin tucked into their mouths. As the sharp teeth become worn, they are replaced by a fresh growth from the folded skin.

The bony fish, or Osteichthyes, have teeth anchored in the jaw. They have a bony skeleton and are covered by scales. They are streamlined for speed and occur in a great variety of shapes. Most Osteichthyes are part of the nekton, but some, such as the flounder, live on the bottom, partially buried in the sediment.

The Reptilia are represented in the ocean by snakes and turtles. They breathe air and therefore can stay below the surface for only a limited time. The sea turtles return to the shore to lay their eggs, and their feet are especially adapted for swimming.

The class Aves, the birds, contains many groups that depend exclusively on the sea for their food. Some return to the land only for nesting and to rear their young. Some birds, such as the penguins, cannot fly but are excellent swimmers. In many fertile near-shore areas, such as off the coast of Peru, the birds play an important role as predators, and their excrement, the guano, accumulates in minable quantities near their nesting sites.

The Mammalia contain various groups that have become adapted to life in the sea. The Arctic Ocean is the habitat of the polar bear, a good swimmer who depends on marine life for its food. Other mammals, such as seals, porpoises, walruses, manatees, and whales, are exclusively marine. The blue whales grow to a maximum size of 34 m, and some weigh over 10^5 kg. They grow to this enormous size on a diet of small crustaceans, the krill (Fig. 25-11). All mammals breathe air and must therefore return to the surface to breathe.

Respiration

All animals require oxygen in order to obtain energy by the oxidation of organic matter to carbon dioxide. Except for the air-breathing mammals, all marine animals utilize the oxygen dissolved in seawater. The simple animals absorb this dissolved oxygen through their cell walls and similarly return the carbon dioxide to the sea. Animals with a circulatory system have gills in which oxygen and carbon dioxide are exchanged between the internal fluid and seawater. The gills are delicate structures with a very large surface area which permits the rapid exchange of gases. A constant flow of seawater must be maintained over the outside surface of the gills while the body fluid is pumped through their interior.

Fish pump water over the gills by first opening the mouth and then expelling the water with the mouth closed. Gills provide for an efficient flow of water over the gill filaments within which capillaries carry the blood for gas exchange. In the Elasmobranchii, the sharks and rays, the water leaves through separate gill slits, usually five on each side (Fig. 25-18a). The gills of the bony fish, the Osteichthyes, on the other hand, are covered by a flap with a single opening on either side (Fig. 25-18b).

While the gills permit the exchange of CO_2 and O_2 between seawater and the body fluids, the lungs of mammals, on the other hand, exchange these gases between air and blood. Since air has a very low viscosity, it

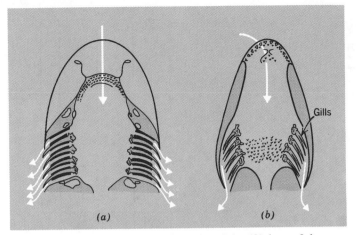

Figure **25-18**   Gills; (*a*) cartilagenous fish; (*b*) bony fish.

can readily be drawn into a complex network of small spherical cavities and exhaled again. The small cavities are surrounded by capillaries through which the blood is circulated.

Man in the Sea

Let us now consider some of the problems man faces when he desires to move about freely within the sea. In order to breathe, he must have a continuous supply of oxygen, and carbon dioxide must be removed from his lungs. For ordinary breathing when submerged near the surface, man can use a simple tube, a snorkel, that reaches above the sea. Near the surface the air in the diver's lungs is at a pressure of 1 atm. But within the sea the water pressure on the diver's body increases by 1 atm for every 10 m of depth. The resulting pressure difference presses in on the abdomen and rib cage, making breathing impossible. The strength of the chest muscles used for inspiration is such that we can develop at most a pressure difference of about 50 cm of water. The snorkel works well as long as one swims just below the water surface but becomes inadequate at slightly greater depth.

If a man wishes to dive to greater depth, he must close his mouth. As the external pressure increases with depth, the chest is squeezed in and the air within the lungs is compressed so that the pressure within the lungs is equal to the external pressure. As a result, the pressure of oxygen within the lungs is increased and oxygen continues to diffuse from the air in the lungs to the blood within the capillaries. However, the compression also increases the pressure of carbon dioxide within the lungs. As a result, CO_2 can no longer diffuse from the blood to the air but diffuses into the blood. The primary factor restricting free diving is, therefore, not the lack of oxygen but the inability of the compressed lung to remove CO_2 from the blood (Fig. 25-19). With practice, a diver can hold his breath for eight to ten minutes when he is resting and for about five minutes if he is actively swimming.

If a diver is to breathe underwater, he must be supplied with air at a pressure equal to that of the water around him. This is made possible by

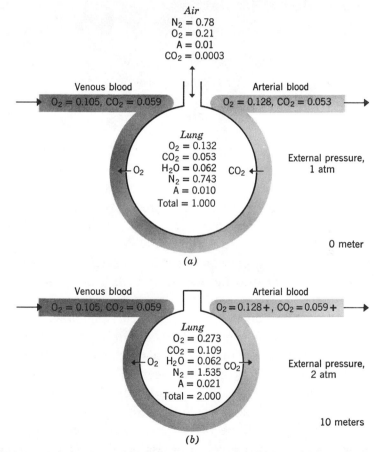

Figure **25-19** The exchange of O_2 and CO_2 between blood and air in the lungs: (*a*) while breathing at the surface; (*b*) during a rapid free dive at 10 m. Numbers are the partial pressures of the gases in atmospheres.

Figure **25-20** Diver with self-contained underwater breathing apparatus SCUBA. (Photograph: ESSA)

the aqualung, developed by Cousteau and Gagnan. The aqualung consists of one or more tanks of compressed air connected to a mouthpiece through a special regulator (Fig. 25-20). The tanks are worn strapped to the diver's back. The regulator delivers air to the diver on demand at the environmental pressure regardless of depth. The spent air from the lungs is exhaled into the water, and fresh air is drawn into the lungs from the air tanks.

The aqualung diver can operate well at shallow depths; however, as he attempts to go deeper, he faces a number of problems related to the increased pressure of the gases in his lungs with depth. At elevated pressures, the atmospheric gases nitrogen and oxygen have deleterious physiological effects. Oxygen becomes toxic when its pressure exceeds three atm, and nitrogen has a narcotic effect at pressures in excess of three to four atm. To overcome these limitations, special gas mixtures are required for extended dives in excess of 30 m. These mixtures contain a lower than usual percentage of oxygen, and the nitrogen is replaced by a gas, such as helium, that does not have a narcotic effect.

The lungs of the aqualung diver are filled with gases whose total pressure must equal the environmental pressure. These gases dissolve in the blood, and their solubility in blood is directly proportional to the gas pressure. Thus, as the depth of the dive increases, the amount of gas that dissolves in the blood increases. If the diver stays at a particular depth for a sufficient time, his blood will become saturated with the gases at the ambient pressure. If the diver returns to a shallower depth, the pressure drops and the blood becomes supersaturated with gas. As a result, the blood is no longer able to keep the gases in solution, and small gas bubbles form in the blood vessels. These bubbles interfere with the circulation of the blood and lead to a very painful condition known as the bends.

To avoid the formation of gas bubbles as a result of reduction in pressure, the diver's ascent must be very gradual. Slow decompression permits release of the excess gas in the blood through the lungs to the environment and prevents the formation of gas bubbles in the bloodstream. Thus the diver returns to the surface in short steps, remaining at each depth long enough to permit his blood to get rid of the excess gas. If the diver had remained at depth for only a short time, his blood would not have become saturated, hence the time needed for decompression would be substantially reduced. In returning to the surface, the diver must follow a safe decompression schedule that depends on the depth and duration of his dive.

The fact that a large fraction of the diver's time must be spent in decompression very seriously limits the usefulness of diving. To increase the efficiency of divers, various undersea habitats have been employed in which divers can live at a fixed depth for an extended period of time Because the underwater habitat is at the ambient pressure, decompression is not required so long as the diver remains at a fixed depth. Thus the diver can work at depth and must undergo the slow process of decompression only as he returns to the surface.

Summary

Animal life in the sea is extremely diverse in life habit and in form. Many planktonic animals have calcareous or siliceous skeletons that accumulate on the sea floor to form a soft sediment or ooze. The most diverse planktonic phylum is the Arthropoda; the phyla Mollusca and

Enchinodermata are mainly benthic. Most of the active swimmers in the sea belong to the phylum Chordata, which includes the jawless fishes, the cartilagenous fishes, the bony fishes, and the marine mammals.

Most animals in the sea utilize the oxygen dissolved in seawater for respiration. The higher organisms have gills that permit the exchange of oxygen and carbon dioxide between the circulating body fluid and the seawater. Mammals, because they are air breathers with lungs, must return to the surface to eliminate CO_2 and obtain oxygen. The aqualung permits man to explore the shallow portions of the sea; however, these endeavors are severely limited by his physiology.

Study Questions

1. What are the possible adaptive advantages of a shell to a benthic organism?
2. Sketch some of the diverse shapes of bony fishes.
3. In Figure 25-19, why is the pressure of water vapor the same at the surface as at 10 m? What determines this pressure?
4. Explain how divers get the bends and how this condition may be avoided.

Supplementary Reading

(Starred items require little or no scientific background.)

* Buchsbaum, Ralph (1948). *Animals Without Backbones, An Introduction to the Invertebrates,* rev. ed. Chicago: University of Chicago Press.
* Hardy, Sir Allister (1967). *Great Waters.* New York: Harper and Row. (A voyage of natural history to study whales, plankton and the waters of the Southern Ocean.)
* ———— (1965). *The Open Sea, Its Natural History.* Part I, The World of Plankton; Part II, Fish and Fisheries. Boston: Houghton Mifflin.
* Kylstra, J. A. (1968). "Experiments in Water-Breathing," *Scientific American,* August, pp. 66–74.
* Parker, E. K. (1965). *Complete Handbook of Skin Diving.* New York: Avon Books.
 U. S. Navy Diving Manual, NAVSHIPS 250–538. Washington, D. C.: U. S. Government Printing Office.
* Yonge, C. M. (1963). *The Sea Shore.* New York: Atheneum, paperback.
 Wilmoth, James H. (1967). *Biology of Invertebrata.* Englewood Cliffs, N. J.: Prentice-Hall.

26 Marine Ecology

To study marine life, we can concentrate on the biology of individual organisms, or we can investigate the way in which these organisms interact with one another and with their environment. The latter approach is the focus of the science of ecology. Ecologists attempt to integrate their understanding of the physics and chemistry of the environment with their knowledge of the biology of individual species in order to comprehend the dynamic interaction between species and their environments. Ecology is concerned with the distribution of organisms in time and space and with the flux of mass and energy through the environment.

The Distribution of Organisms in Time and Space

The ocean is populated by a wide variety of plants and animals that live within the sea and near the sea-sediment interface. Let us compare the distribution of organisms within the ocean with the distribution of a mixture of gases within a container. The gases exist both as free gas molecules within the container, the "pelagic population," and as absorbed molecules on the walls of the container, the "benthic population." If we sample a unit volume within the container and analyze it for the numbers of each type of gas molecule, we obtain the same result regardless of when and where we take the sample. Similarly, if we sample a unit area of the surface of the container and analyze it, we find that the concentration of adsorbed molecules is independent of space and time although the adsorbed population may differ from the population within the gas.

If we sample the pelagic population within the ocean by catching the organisms in a given volume with a net, the results are quite different. The numbers of each type of organism caught and the variety of organisms present vary widely from place to place. Furthermore, if we sample at the same place at various times, we discover differences between day and night, from season to season, and from year to year. The same is true if we photograph the sea bottom. The probability that a specific benthic organism will appear in the picture will vary from place to place. A particular organism will be present only in a limited depth range and will be associated with a particular type of sediment.

The organisms of the sea are not randomly distributed in space and time. Rather, their distribution depends on the physical and chemical conditions, the presence of other organisms, and the past changes in these parameters. One must distinguish between distributions of ways of life and specific organisms. For example, planktonic plants are limited to the photic zone. Specific plants, however, may be restricted to small geographic areas of the photic zone. Animals that burrow in soft sediments are limited to those parts of the ocean floor where the sediments are soft. Particular species of these types of organisms, however, are limited to only a small fraction of the available area. Now let us examine some aspects of the distribution of marine organisms in space and time.

Vertical Migration of Marine Animals

The zoologists of the *Challenger* expedition made extensive collections in all portions of the world ocean. Their descriptions of this variety of life fill most of the 50 volumes of the *Challenger* reports. In Chapter 4 (p. 45), we read that Sir John Murray had been able to catch many animals in the surface layer of the sea at night; however many of the animals disappeared during the day. By lowering his net to 200 m, he discovered that the animals migrated to that depth during daylight hours.

When the echo sounder came into use, "false bottom" reflections were often observed, in addition to the reflection from the true sea floor. The false bottom is a so-called *deep scattering layer* within the sea that strongly reflects sound pulses. This layer is observed only during the daytime and mysteriously disappears at night. Careful observations showed that the deep scattering layer rises toward the surface at sunset and descends from the surface at dawn (Fig. 26-1). Attempts were made to observe the scattering layer from a research submersible. The surface ship would detect the layer, measure its depth, and direct the submarine to the proper location. To observe the layer, the submarine required

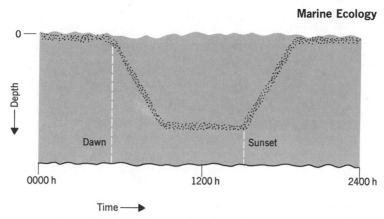

Figure **26-1** Diurnal migration of the deep scattering layer.

lights, but the bright lights frightened away the animals in the scattering layer. When the submarine turned off its lights and then flashed them on again, it was seen that the layer consisted of numerous small fish and crustaceans.

Many planktonic animals in the sea undergo diurnal vertical migrations; many others do not. For example, in the Antarctic Ocean, the copepod Clausocalanus *laticeps* is found at about 100 m during the day and rises close to the surface during the night, while the copepod Rhincalanus *gigas* always remains between about 80 and 150 m (Hardy, 1967, p. 344). By rising to the surface at night, Clausocalanus *laticeps* is able to graze on the phytoplankton of the photic zone without being exposed to light.

Avoidance of light is a common behavior in marine animals and can have a number of adaptive advantages. The animal cannot be seen by predators in the dark. Also, just as the speed of ships through the water is impeded by the many plants of the photic zone which attach themselves to the ship bottoms, so fouling by plants would be detrimental to many small animals. It is therefore advantageous for the animal to remain in the dark, where these plants cannot grow. Another reason for light avoidance is that some phytoplankton, during photosynthesis, leak substances into the water that are inimical to some animals.

Some vertically migrating animals remain at a level of constant illumination within the sea. Thus they synchronize their vertical motions with the surface illumination. Hardy (1967, p. 359) has pointed out that the vertical migration of zooplankton modifies their rate of horizontal

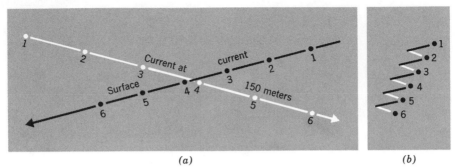

Figure **26-2** The effect of diurnal migration on horizontal drift: (*a*) daily progress of an animal drifting with the current at the surface and at a depth of 150 m; (*b*) progress of an animal that migrates diurnally between the surface and 150 m. (After Hardy, 1967)

drift. Surface currents differ in speed and often in direction from the currents at depth. As a result, the net drift of a vertically migrating species will differ from that of an animal which remains at constant depth (Fig. 26-2). The former animals may remain in a favorable location for a longer time, and they will always encounter fresh volumes of surface water during their nightly ascent.

The Horizontal Distribution of Marine Animals

When we attempt to assess the populations within the ocean, it soon becomes apparent that the horizontal distribution is not random. For example, the Antarctic krill occurs in dense patches, with few animals in the region between patches. Many species of fish live in large schools, in which millions of like fish swim parallel to one another in close formation. Once a school has been located, fisherman can fill their nets rapidly with a huge quantity of fish. If the fish were randomly distributed in space, most fisheries would not be economical. Obviously fish do not school for the benefit of fishermen; clearly there must be an adaptive advantage to schooling. One advantage is that schooling makes fertilization of the spawned eggs more likely. The social proximity in a school also seems to aid the development of the fish. Experiments by Shaw (1962) have shown that normally schooling fish have a very high mortality rate when reared in isolation. In some manner, the feeding habit is stimulated by the presence of like fish. Precisely how schooling is beneficial to particular species is not yet well understood.

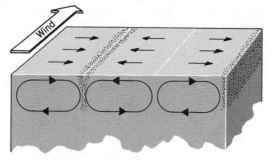

Figure **26-3** Langmuir convection induced by the wind.

The distribution of zooplankton can also vary widely in the horizontal direction as a result of physical forces. On a vacation cruise through the Sargasso Sea, Langmuir (1938) observed seaweed in neat rows aligned in the direction of the wind. He reasoned that this alignment must result from a convection pattern induced by the wind. Each pair of counter-rotating cells forms a line of flotsam in the water where the surface flow converges (Fig. 26-3). Such convection cells would cause the Sargassum weed to be swept to the lines of convergence. To complete the circulation, the deep water would have to flow in a direction opposite to that at the surface. When he returned from his cruise, Langmuir made observations on Lake George, near his home, and confirmed that the postulated convection cells were in fact produced by the wind.

The plant life in the sea not only gives rise to plant cells but also produces a large amount of dissolved organic matter in the sea. Much of this organic matter is concentrated at the air-sea interface. Langmuir received the Nobel Prize in Chemistry in 1932 for the study of such organic surface films.

Sutcliffe, Baylor, and Menzel (1963) demonstrated that the Langmuir circulation sweeps organic matter in these films at the sea surface together along the lines of convergence. As a result of lateral compression, the organic matter is converted into small organic aggregates which are carried downward by the sinking water.

The particulate organic matter so formed is a food source for many members of the zooplankton. Since the particles are concentrated under the lines of convergence, one would expect the zooplankton also to gather there. By counting the zooplankton with a sensitive sonic detector that is towed through the water, Baylor demonstrated that this is indeed the case. The density of the zooplankton below the zones of convergence can be greater by a factor of 100 than its density in adjacent areas.

This nonuniform distribution of the zooplankton presents a special problem when one tries to estimate the amount of animal life in the sea. When pulling nets through the water, we generally like to sail in the direction of the wind in order to reduce the roll of the boat. If we sail along a line of convergence, we will get a much greater plankton concentration—up to 100 times more—than we would obtain 10 m on either side. Obviously, if we are to obtain reliable data, the sampling plan must be carefully designed.

In Chapter 22 we discussed how life may have originated from organic matter produced by ultraviolet irradiation of the sea surface. It may be that the Langmuir circulation has also played a role in aggregating this organic matter (Weyl, 1968a). The aggregates so formed would be swept downward from the sea surface and so would be protected from further ultraviolet radiation. Depending on their density, the aggregates would come to rest either on the sea floor or along density gradients within the sea. Thus the organic matter would be concentrated. Once a self-replicating organism arose, it could feed heterotrophically from the fresh supply of aggregates.

Benthic Communities

To sample the populations within the sea, one must use a net to sweep out a certain volume of water. To sample the bottom, one must use a special bottom sampler. Since many of the animals live buried in the sediments with only syphons protruding from the mud, the sampler must remove an adequate thickness of sediment. When we compare such samples taken from various areas, we find that certain kinds of organisms usually occur together in particular zones. These are known as benthic communities.

An example is the Macoma community, which is found in shallow water near shore. The community consists of two genera of clams, Macoma and Mya, which feed on the organic matter deposited in the sediment. In addition, one genus of cockles, Cardium, is present which feeds by filtering out organic matter suspended in the seawater. The clams will dominate in muddy bottoms, while an increase in silt and sand will favor the cockles. Also, there is usually present a genus of the

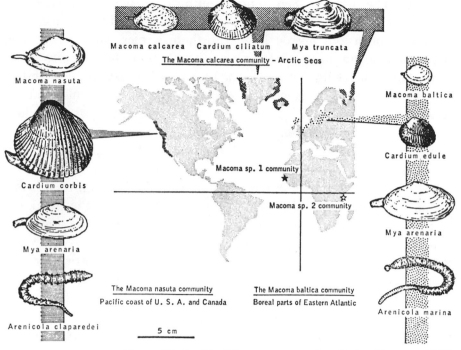

Figure **26-4** The Macona community. (After Thorsen, 1957)

polychaete worm, Arenicola, which lives in a U-shaped tunnel in the sediment. The same community, differing only at the species level, is found on the west coast of North America, in the Arctic region, and in the North Sea-Baltic area (Fig. 26-4).

Other assemblages of organisms make up different communities. A community is bounded by a range in depth, a particular sediment type, and a limited range in water temperature and salinity. In Chapter 27, we shall examine some of the communities of the coral-reef environment.

Biogeography

The geographic distribution of marine organisms depends on many factors. To survive, an organism must be able to grow and to reproduce. These two requirements for survival are not necessarily met by the same environment. Thus the whales of the southern hemisphere feed in summer near the Antarctic continent on the abundant krill. They migrate north in winter to breed off the coasts of Africa, Australia, and South

America. The whales then return south to their summer feeding grounds. Since the gestation period of the whale is about 11 months, the calf is born and nursed in the northern winter quarters. At five months of age, the young whale goes south with its mother and is weaned in the Antarctic feeding grounds. The blue whale measures 7 m at birth and 16 m by the time it is weaned. After two years it is sexually mature and able to breed. During the two Antarctic feeding seasons it has grown to a length of 23 m. Because the whale migrates between the breeding and feeding grounds, its life cycle has to be adapted to the seasonal changes in environment.

In contrast to the nekton, the plankton is not able to move actively in the surface current. The geographic distribution of the plankton therefore depends on the currents and the tolerances of the organism as the water circulates in gyre systems.

In general, the conditions required for repopulation may be more restrictive than the conditions required for survival. Thus organisms may be carried by the currents into areas where they can survive but where they are not able to breed. For example, a marked decline of the rock lobster has been observed in Bermuda. After various conservation measures proved to be useless, it was discovered that the lobsters were not able to breed in Bermuda. The Bermuda lobster population depended on a supply of larval stages that drifted in from the Bahamas. Thus, to preserve the lobster in Bermuda, the breeding grounds in the Bahamas had to be protected.

The factors that control the distribution of species in the ocean are very complex. The geographical distribution of organisms depends not only on the present oceanic environment but also on its history. For example, at the present time, the tropical Atlantic is separated from the tropical parts of other oceans by continents. Tropical marine organisms therefore cannot migrate between the Atlantic and the tropical Pacific and Indian oceans without crossing cool waters. This isolation, however, is relatively recent. The Isthmus of Panama became a land bridge only a few million years ago. Before that time a seaway linked the Caribbean with the tropical Pacific. The present distribution of species in the oceans cannot be understood without considering the physical evolution of the oceans. At the same time, marine biogeography provides important data that must be utilized if we wish to reconstruct the geographic history of the oceans.

The Balanced Aquarium

To understand how matter and energy pass through the marine environment, let us first consider a simple system: a sealed aquarium, filled with seawater and populated by plants and animals. Illumination provides a primary energy source permitting photosynthesis by the plants. What conditions are necessary in order that this system be able to support a continuous chain of life?

Photosynthesis by plants produces organic matter and oxygen. Respiration by the animals then recombusts the organic matter to carbon dioxide, utilizing the oxygen generated in photosynthesis. As long as there is no leak in the cycle, it can continue to operate indefinitely. It is essential, however, that no organic matter accumulates on the bottom of the aquarium. Also, all the nutrients contained in dead organic matter, such as phosphate and nitrate, must be converted into such a form that they can once more be utilized by the plants. If any essential material is deposited in the aquarium in a form that cannot be recycled, this material will gradually be stripped from the seawater, and life in the aquarium will come to a stop.

The regeneration of organic matter into a form that can once more be utilized by plants is accomplished in part by microorganisms. In a forest, the micro-organisms on the forest floor decompose the fallen leaves and trees and return their nutrients to the soil, where they can once more be absorbed by the roots of the growing trees. In the sea, bacteria in the water, and particularly on the sea floor, assist in the decomposition of the organic matter. The bacteria, in turn, are a source of food for bottom-living animals.

If our aquarium contains phytoplankton, zooplankton, bottom scavengers, and bacteria, it can operate indefinitely, provided that there is 100 percent efficient regeneration of nutrients. Stable operation of the aquarium does not necessarily mean that the standing crops of all populations remain constant. Rather, the various feedback relationships and the differences in rates of reproduction may lead to cycles. Such cycles are possible even if the physical parameters, such as temperature and light intensity, remain constant.

If the system starts with a high concentration of nutrients in the water, the plant population will multiply rapidly. Because of the increased food supply, the herbivores will also multiply. Increased grazing combined with a depletion of nutrients will then lead to a reduction in the plant population. The lesser amount of food available will in turn affect the animal population. As the organic matter is decomposed, the concentration of nutrients in the water increases once more to start a new cycle. Now let us investigate the population dynamics in a natural system.

Long Island Sound

The seasonal variations of the central region of Long Island Sound have been studied over a number of years by G. A. Riley and associates (1956) of the Bingham Oceanographic Laboratory of Yale University. Their findings for the period October 1952–July 1953 are summarized in Figure 26-5.

The input of solar radiation decreases to a minimum in December, when the amount of daylight is at a minimum. It increases to about six times the minimum value during the middle of summer. The water temperature lags behind the radiation curve by about two months, reaching a minimum in February. The concentration of the nutrients, nitrate and phosphate, is high during the winter and begins to decrease markedly in February. Between January and June, the water loses 14.5×10^{-6} mole liter^{-1} of nitrate and 1.5 of phosphate. This rapid decline in nutrients can only be due to the fact that nitrogen and phosphate have been incorporated into plant material by photosynthesis. Let us, therefore, examine next the time variation of the population densities of plants and animals.

The animal population, shown on a logarithmic scale at the right in Figure 26-5c, ranges from 10 to 100 individuals per liter, while the much smaller phytoplankton ranges from 10^5 to 2×10^7 plant cells per liter. The plants multiply at an accelerating rate from October on reaching a peak in March, after which their concentration declines. The animal population behaves quite differently. The minimum animal density occurs in January, and the population increases as the plants become more abundant; however, the decline in plants that begins in March is not accompanied by a similar decline in the animal population. How can these seasonal cycles in the populations be explained?

In October and November, the nutrients are abundant and there is adequate light; however, the plant population is small, probably because of the relatively large population of animals, which limits the net

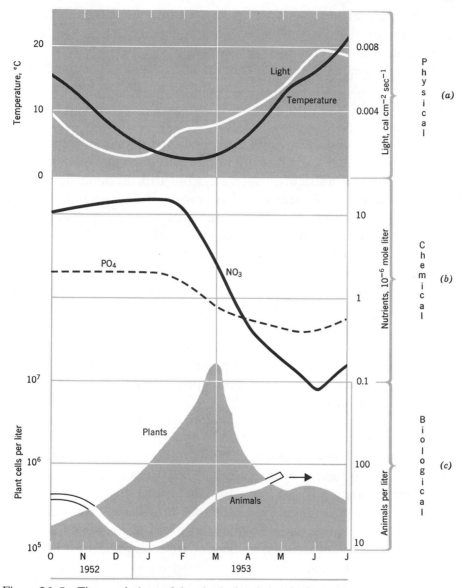

Figure **26-5** Time variations of the physical and chemical parameters and of the populations of phyto- and zooplankton in the central region of Long Island Sound. (After Riley et al., 1956)

multiplication of the plants by grazing. During the winter the animal population declines, and thus the plant population is able to multiply rapidly once the light intensity is adequate.

The rapid decline of the plants at the end of March is more difficult to assign to a clear cause. It could be due to the decrease in the nutrients, or it could result from excessive grazing by the increased animal population.

The doubling time of phytoplankton is shorter than that of zooplankton. Therefore, it is logical to think that the plant population, given enough light and nutrients, can always outstrip the food demand of the animals. This argument would be correct if the number of plants consumed per day by an animal were equal to its fixed physiological requirements. Observations of herbivores, however, have shown that in many species, grazing is independent of need.

Many grazers tend to clear approximately a fixed volume of seawater. In one experiment, the concentration of phytoplankton in a given body of water was increased by a factor of 10, but the volume that was cleared of plants by an animal decreased by only 21 percent (Ryther, 1954a). Thus the animals in the richer water were eating eight times as much as they had when there were fewer plants. Many grazers devour all the plant food they encounter. If the food supply is larger than the organism requires, the food is only partially digested, and the excess matter is excreted. The still nutritious fecal matter then slowly sinks and serves as food for other planktonic animals and for the benthos.

Cushing (1959) has developed a theoretical model for the population dynamics of a system consisting of plants and grazers. He assumes that the plant population is not limited by the availability of nutrients but only by grazing. Using observed values for the rates of reproduction of plants and grazers and a reasonable model for the grazing efficiency as function of the plant density, he obtained the results shown in Figure 26-6. Although his calculations were based on a different area of the ocean, his model reproduces the main features of the population changes in Long Island Sound quite well. Note that the herbivore population is not significantly affected by the sharp decline in phytoplankton. According to the assumptions made in the model, the steep decline in the plant population shown here is due entirely to increased grazing. The fact that the plant population can be reduced by grazing alone does

Figure **26-6**   Time variation of the concentration of phytoplankton and herbivores of a theoretical model in which plants are only limited by grazing. (After Cushing, 1959)

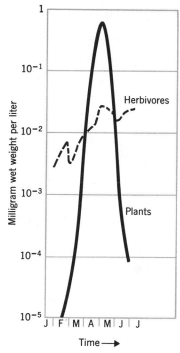

not prove that in Long Island Sound the decline in nutrients did not have a significant effect on the phytoplankton population.

The English Channel

The data for Long Island Sound show that the relationship between the populations of plants and animals and the physical and chemical parameters is complex, and that there is a strong seasonal cycle of life in the sea. To study the economy of nature in more detail, one would like to know how much of the organic matter synthesized by plants is consumed by the animals and where. It is necessary to distinguish between the amounts of different kinds of organic matter present and their fluxes, just as in economics one must distinguish between accumulated wealth and income.

Only a few studies exist which have attempted to draw up a complete balance sheet for the life in the sea. Harvey (1950), studying the English Channel off Plymouth, estimated the amount of dry organic matter present per unit area of the sea, in the form of phytoplankton, zooplank-

TABLE **26-1**   STANDING CROP AND PRODUCTIVITY IN THE
ENGLISH CHANNEL*

| | *Dry Weight of Organic Matter* $(g\ m^{-2})$ | | |
	Standing crop	*Production per Day*	*Loss per day through Respiration*
Phytoplankton	0.4	0.4–0.5†	
Zooplankton	1.5	0.15	0.06
Fish	3	0.0026	0.04
Benthos	17	0.03	0.2–0.3

* Harvey, 1950.
† Net production = photosynthesis − respiration.

ton, fish, and benthos. In addition, he determined the rate of net production of phytoplankton per unit area. Finally, he estimated the rate at which food was consumed by the zooplankton, fish, and the benthos. Part of the food serves as a source of energy for the animals; another portion is used to produce organic matter by growth of the organisms. Harvey's budget sheet is given in Table 26-1.

When we graph the standing crop against the rate of production (Fig. 26-7), we find almost an inverse relationship. The phytoplankton has the largest rate of production but the smallest standing crop. The crop of fish weighs almost 10 times that of phytoplankton, but new plant matter is produced 100 times faster than fish. Since the fish lead a more active life than the bottom fauna, a larger fraction of their food input is used for energy. The zooplankton, on the other hand, is more efficient in converting plant food into animal matter.

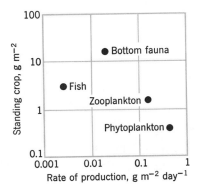

Figure **26-7** The relationship between standing crop and rate of production in the English Channel. (After Harvey, 1950)

In the English Channel, about half of the organic matter produced by the phytoplankton is consumed by zooplankton and half by the bottom fauna. About one-third of the zooplankton is consumed by fish and the rest by the benthos. In addition to the organisms listed in Table 26-1, there were 0.04 g of bacteria per square meter floating within the sea and 0.1 g m^{-2} on the bottom. On the average, the marine bacteria lose about one-third of their body weight per day by respiration. Thus bacterial respiration amounts to about 0.04 g m^{-2} day^{-1}, or about 10 percent of the primary productivity. Although bacterial respiration represents a small fraction of the total, it is essential in keeping the nutrient elements in circulation.

The Oceanic Environment

Thus far we have considered only the neritic zone, which has relatively good vertical mixing, and where a significant fraction of the water is in the photic zone. The ocean, on the other hand, is much deeper than the illuminated region, and its water is not well mixed vertically. Although photosynthesis is limited to the surface layer, respiration and regeneration of nutrients occur at all depths. The life processes in the ocean, therefore, result in a chemical differentiation between the surface water and the deep water.

The phytoplankton is continuously fixing the nutrients in the surface layer of the ocean and is grazed by zooplankton. Because of the diurnal migration of the animal populations, much of the plant matter is reoxidized below the surface zone. In addition, some of the plant matter sinks below the surface, to be consumed at greater depth and on the bottom. Thus, on the average, the nutrients are regenerated at greater depth than that at which they are incorporated into plant matter. The biologic cycle thus acts like a pump, continuously removing nutrients from the surface layer and depositing them lower down in the ocean.

If the "nutrient pump" were not counteracted by other processes, the surface of the ocean would soon be stripped of all nutrients, and plant life, hence all life, in the sea would cease. Counteracting the nutrient pump, vertical circulation and mixing return the nutrients from deeper water to the surface layer, where they can once more be utilized for photosynthesis. The vertical distribution of nutrients in the ocean is the result of the interaction between the biological and physical processes.

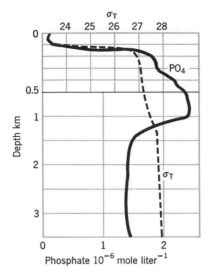

Figure **26-8** Vertical variation of the phosphate concentration and the density in the western tropical Atlantic, 6°N 45°W. (Equalant III, R. V. *Mikhail Lomonosov*). Note change of scale at 500 m.

A typical vertical distribution of the phosphate concentration is shown in Figure 26-8. The vertical change in density (σ_T) is also indicated. In the well-mixed low-density surface layer, the phosphate concentration is only 10^{-7} mole liter^{-1}. Between 100 and 200 m, the density and the phosphate concentration increase rapidly with depth. The phosphate concentration reaches a maximum of 2.4×10^{-6} mole liter^{-1} at 1 km and then declines to 1.4×10^{-6} mole liter^{-1} between 2 km and the bottom. There is a slight increase in concentration near the bottom.

The distribution of phosphate indicates that the photic zone has a very low concentration of nutrients. The maximum at intermediate depth suggests that dissolved phosphate is being produced in this region by the decomposition of organic matter. The very sharp phosphate gradient in the region of density increase implies little vertical mixing; otherwise the steep phosphate gradient would have been dissipated. The slight increase in phosphate near the bottom results from the decomposition of organic matter by the benthos. To verify these interpretations, let us examine the vertical distribution of dissolved oxygen at the same location.

Oxygen is produced by photosynthesis in the photic zone and is removed by respiration to form carbon dioxide. In addition, oxygen is exchanged between the sea and the atmosphere at the surface. The vertical variation of the concentration of dissolved oxygen in the tropical

Figure **26-9** Vertical variation of the oxygen and phosphate concentrations in the western tropical Atlantic (same location as Figure 26-8).

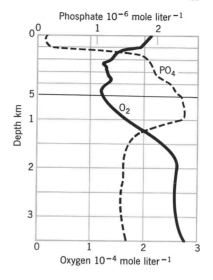

Atlantic is shown in Figure 26-9, with the phosphate data for comparison. The two curves are virtually mirror images of each other. In the surface, where phosphate is low, there is an excess of photosynthesis over respiration, and therefore the oxygen concentration is high. Below the photic zone, only respiration can take place, and therefore the oxygen concentration decreases.

The surface temperature at this location was 27°C, giving a solubility of oxygen of 2×10^{-4} moles liter^{-1} (Fig. 20-3, p. 330), slightly less than the actual surface concentration. As the amount of light decreases with depth, the rate of oxygen production by photosynthesis also decreases, while respiration continues at all depths. At some level within the photic zone, respiration must equal photosynthesis so that at this depth, the compensation depth, the amount of oxygen produced will equal the amount consumed.

Below the compensation depth, the oxygen concentration decreases, indicating active utilization of oxygen by animal respiration. Note that the minimum oxygen values roughly parallel the maximum concentrations in phosphate. Below 1 km the oxygen concentration increases. The higher concentration of oxygen in the bottom water is due to the fact that this water has a lower temperature (between 2 and 3°C), because it sank from the sea surface in high latitudes. But this water originally contained about 3.4×10^{-4} mole liter^{-1} of dissolved oxygen. Hence the bottom water has lost oxygen by respiration, since it sank from the sea surface in high latitudes.

If the oxygen became exhausted, the animals would not be able to survive. The presence of oxygen in the deep water is therefore essential for the maintenance of animal life in the depth of the ocean. To get a feeling for the oxygen problem, let us take 1 liter of seawater from the deep ocean and bring it to the surface. It will contain about 2×10^{-6} atom of phosphorous as phosphate. We then convert all the phosphate to organic plant matter by photosynthesis. In the plankton, the ratio of carbon to phosphorus atoms is about 100 to 1, so that the organic matter produced will contain 2×10^{-4} atom of carbon. When the water is equilibrated with air, it contains about 3×10^{-4} mole of O_2. This is more than enough oxygen to burn all the organic matter back to carbon dioxide and regenerate the nutrients. The exact amount of oxygen in the water at saturation depends on the temperature. If we had equilibrated the water in the tropics at 30°C, it would contain 2×10^{-4} mole of O_2, so that the oxygen would be depleted by respiration.

The Variation of Productivity in the Ocean

We must now ask how the fertility of the sea varies from place to place. On land, there are fertile regions where food can be grown in great abundance, and there are deserts which are able to support only a very sparse animal population. The productivity of the land depends primarily on the availability of water, the temperature, and the availability of mineral nutrients. If necessary, man can supply the latter by adding fertilizer to the land. In the ocean there obviously is no lack of water. At the same time, the variation of temperature in space and time is much more limited than on the land. Therefore we might expect the fertility of the ocean to be much more uniform than that of the land surface.

When we examine the distribution of life in the sea, we find that the ocean also contains biological deserts and areas teeming with life. Fishermen have traditionally concentrated their efforts in shallow coastal waters. Originally, this restriction was imposed by the primitive nature of boats and fishing gear. When technology overcame these limitations, however, it was found that much of the open ocean is a poor fishing ground. Since the fish ultimately depend on the phytoplankton for their food, let us survey the distribution of plant productivity, the *primary productivity,* in the world ocean.

TABLE **26-2** PRIMARY PRODUCTIVITY IN SELECTED AREAS
OF THE OCEAN

Location	Season	Productivity $(10^{-3}\ g\ C\ m^{-2}\ day^{-1})$
Arctic Ocean	Midsummer	0–24
Arctic Ocean	Late summer	0–6
North Atlantic near Spitsbergen	June	400–2400
Mediterranean	Midsummer	30–40
Sargasso Sea	Summer	100–200
North Atlantic Shelf	Spring	1900
Equatorial Atlantic	February	60–800

The development of the carbon-14 method (see pp. 378) has made
it possible to estimate the primary productivity in the open ocean. Because of the cyclic nature of the plant populations, many measurements
distributed in time and space are required to arrive at an accurate estimate of the annual productivity. Some determinations of the primary
productivity are listed in Table 26-2.

From the perspective of an orbiting satellite, the fertile land areas of
the earth can readily be distinguished from the deserts. The fertile areas
are colored green by vegetation, but the biological deserts display bare
sand or a permanent cover of ice and snow. The deserts and fertile portions of the ocean also have their characteristic colors. Clear ocean
water containing few plant cells is deep blue in color. The presence of
dense populations of diatoms imparts a brown-green color to the ocean.
Thus, by noting the color of the sea, one can get an idea of the areal
distribution of plant life. The colors observed are in general agreement
with the measurements of productivity and permit the construction of
an approximate chart of the primary productivity in the world ocean
(Fig. 26-10).

The Reasons for the Variation in Productivity

The fertility of the ocean varies widely. It is generally high in the
coastal regions and low in the centers of the subtropical gyres. How can
we account for this variation? The low productivity under the Arctic
sea ice is undoubtedly due to the attenuation of the light by the ice.
However, if light were the primary limiting factor throughout the ocean,
we would expect the productivity to be highest in the tropics and lowest

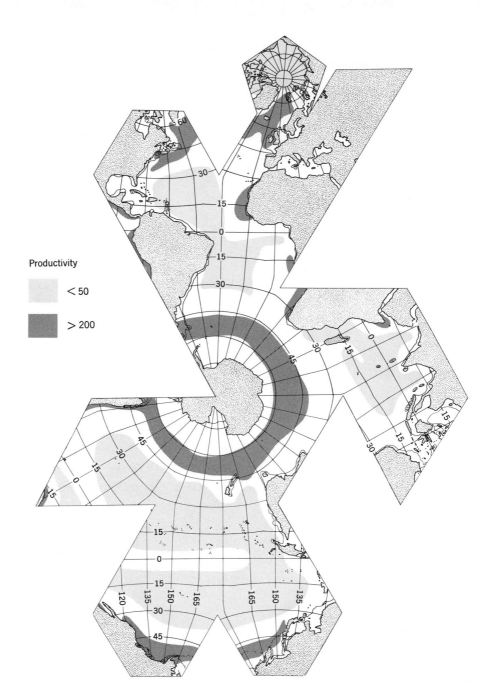

Figure **26-10** Approximate distribution of the primary productivity in the world ocean. Units are g carbon m^{-2} yr^{-1}.

in high latitudes. If anything, the situation seems to be reversed, with a high productivity in the subarctic seas.

In addition to light, matter is required for photosynthesis. The average rate of primary production of carbon by phytoplankton in the ocean is about 50 g m^{-2} yr^{-1}, or about 4 moles m^{-2} yr^{-1}. If we assume that this primary production is spread uniformly over a depth interval of 20 m, it will convert 0.2 millimole of carbon per year per liter of seawater to organic matter. The yearly requirement of the three elements C, N, and P and their abundances in deep ocean water in millimoles are as follows:

	Carbon	Nitrogen	Phosphorus
Needed per year	0.2	0.03	0.002
Concentration in deep water	2.3	0.03	0.002

We see that while deep ocean water contains a 10-year supply of carbon for the production of phytoplankton, phosphorus and nitrogen would be exhausted in one year. Therefore, the growth of plants will exhaust the supply of the nutrient elements long before it encounters a shortage of carbon.

Photosynthesis in the surface water of the ocean, followed by a downward transport of organic matter, depletes the surface water of nutrients. The nutrients are regenerated by respiration and returned to the surface by vertical mixing. When they are averaged over the annual cycle, a steady state is reached so that the downward transport of nutrients as organic matter equals the upward transport of nutrients by vertical mixing.

Consider a water column 1 m^2 in size. Let the nutrient concentration be x moles per liter in the upper layer and y moles per liter in the lower layer. By mixing, R liters per unit time are exchanged between the layers. R is thus the rate of vertical mixing per unit area per unit time (Fig. 26-11).

Per unit area per unit time, mixing transports Ry moles of nutrients from the deep layer to the upper layer and Rx moles from the upper layer to the lower layer. The net upward transport of nutrients, therefore, is equal to the mixing rate times the concentration difference:

$$\text{net upward transport} = Ry - Rx = R(y - x).$$

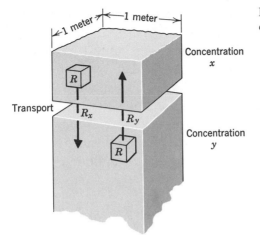

Figure **26-11** The vertical transport of dissolved nutrients.

The downward biological transport is roughly proportional to the primary productivity. Averaged over a yearly cycle, the upward and downward transports must be equal. Therefore, the primary productivity is approximately proportional to the product of the mixing rate times the concentration difference.

The data of Table 26-2 indicate that the primary productivity of different areas in the ocean varies by a factor of more than 100. The variation in primary productivity thus requires a similar variation in the vertical transport of nutrients $R(y - x)$. This can be the result of variation in R, in $y - x$, or in both parameters. The phosphate concentration at 500 m varies between 2 and 3×10^{-6} mole liter^{-1}. The surface concentration is always low and very seldom exceeds 10^{-6} mole liter^{-1}. The difference $y - x$ thus ranges at most from 3 to 1×10^{-6} mole liter^{-1}. Thus the much larger variation in the product $R(y - x)$ must be due almost exclusively to changes in R.

The primary productivity in the ocean can be limited either by grazing or by the availability of nutrients. If it were limited by grazing only, it should be independent of the rate of vertical mixing, R. The large observed variations in the primary productivity, hence in $R(y - x)$, would therefore require large variations in $y - x$. In other words, if the annual average primary productivity is not determined by the supply of nutrients, one would expect to find large variations in the difference in nutrient concentration between the surface water and the deep water. While

x does vary by large factors, the difference $y - x$ is relatively constant. We must therefore conclude that the primary productivity depends mainly on the rate of vertical mixing, which controls the transport of nutrients into the surface layer.

Vertical Mixing

The productivity of the ocean depends primarily on the rate at which nutrients are returned to the surface layer by vertical mixing. Upon what does the rate of vertical mixing depend? In Chapter 11, we studied the interaction between the ocean and the atmosphere. This interaction altered a thin mixed layer at the sea surface. During the period of warming, the thickness of the mixed layer decreased; with cooling, the mixed layer became deeper (Fig. 11-14, p. 172). The mixed layer is characterized by uniform temperature and salinity and, therefore, by uniform density. We see from Figure 26-8 that the nutrient concentration in the mixed layer is also uniform.

Beneath the mixed layer, the density of the water and the concentration of the nutrients increase rapidly with depth. The rate of primary production depends on how fast the nutrients can diffuse upward through the density gradient layer into the mixed layer. To get a feeling for the process of vertical mixing, let us compare two experiments. First we pour some water, colored by dye, into a glass. Next, we pour some clear water on top of it. It is difficult to add the clear water without a great deal of mixing. A very small amount of stirring will mix the color throughout the water. Next, we repeat the experiment, but this time we add a lot of salt to the colored water so that it is more dense than the clear water on top. Now, we have to stir more energetically in order to mix the waters: the density gradient at the interface inhibits mixing.

As the density difference between the waters is increased, the amount of vertical mixing produced by a constant expenditure of energy decreases. Thus one would expect an inverse correlation between the primary productivity in the ocean and the vertical density differences. In the typical data from the North Pacific, shown in Figure 26-12, note that the productivity is high where the vertical density contrast is small, and vice versa. The average annual productivity is controlled primarily by the supply of nutrients, which, in turn, depends on the vertical density gradient. Grazing can influence the seasonal variation in produc-

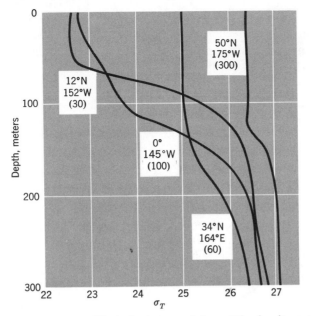

Figure **26-12**   Vertical winter variation of the density near the sea surface in the North Pacific. The numbers indicate the location and the primary productivity in g carbon m^{-2} yr^{-1}.

tivity; however, it does not appear to have a major influence on the annual average.

The Origin of the Invertebrate Phyla

In Chapter 3 we learned that the fossil record is discontinuous at the beginning of the Cambrian. The start of this period is marked by the sudden appearance of shelled representatives of most invertebrate animal phyla. Several scientists (Berkner and Marshall, 1965; Cloud, 1968) have suggested that this sudden diversification of life is the result of the geochemical evolution of the earth, that it marks the time when the oxygen content of the atmosphere became adequate for animal respiration (see Chap. 22, p. 360). The presence of sufficient oxygen then leads to the rapid evolution of multicellular animals.

The fossil record, however, gives no clues about the evolution of the phyla. They appear fully differentiated at the beginning of the Cambrian, and subsequent developments, except for the evolution of the vertebrates, are minor compared to the change from the Precambrian to the Cambrian. The author (Weyl, 1968a) has suggested a possible answer to this dilemma.

On page 422, we saw that the Langmuir circulation may have played an important role in the origin of life, by aggregating organic matter formed at the sea surface. This material is then swept downward and concentrates along density gradient surfaces. At the present time, the main density gradient surfaces (Fig. 26-12) penetrate the photic zone in low latitudes. Thus blue-green algae could have evolved in this layer from heterotrophic organisms that were supplied with organic matter by the Langmuir circulation. The plants would have to evolve at constant density, for they would be killed by the ultraviolet radiation if they floated to the surface. On the other hand, if they became too dense, they would sink out of the photic zone.

If the density gradient layer in the Precambrian tropical sea was indeed populated by blue-green algae, oxygen would have been generated in the layer by photosynthesis. Because of the low rate of diffusion across the layer, the oxygen concentration in the layer would be appreciable, even if the atmosphere and the deep ocean are kept devoid of oxygen by the oxidation of reduced minerals. Thus animals requiring oxygen for respiration could evolve in the layer long before the atmosphere contained sufficient oxygen. These animals would have to be planktonic within the layer, in order to remain within the oxygenated environment. The more rapid mixing processes over the continental shelves would probably make the benthic shelf environment inhospitable to animals.

Thus the postulated Precambrian environment is the direct antithesis of the present situation. Areas that now support only a low rate of productivity because of the low rate of diffusion would have been well oxygenated and so would have supported both plants and animals. Areas that are rich at present because of a high rate of vertical diffusion would have been low in oxygen and so could not have supported animals requiring respiration.

Once the oxygen level in the atmosphere became sufficiently high (about 1 percent of the present level), the animals that had evolved in the density gradient layer as planktonic forms could take on a benthic habitat on the continental shelves. As a result they could evolve a dense mineral skeleton and so be readily preserved in the fossil record. Their planktonic ancestors would have been soft-bodied, and they would have existed only in the density gradient layer over deep water. Thus their preservation in the fossil record would be very unlikely.

We see that the study of marine ecology of the present ocean suggests a possible answer to the sudden appearance of the invertebrate phyla at the beginning of the Cambrian. The phyla could have slowly evolved over at least a billion years as planktonic forms in the density gradient layer. The beginning of the Cambrian merely marks the time when the oxygen level in the atmosphere permitted the diverse plankton to inhabit the bottom of shallow seas and grow mineral skeletons. Further work is necessary to show if this hypothesis is valid or if the origin of the invertebrate phyla has a different explanation.

The Equatorial Currents

Let us now return to the present ocean. Surprisingly, Figure 26-12 indicates that there is high productivity along the equator, where the surface density is low. Although the vertical density difference is high, the rate of increase of density with depth is relatively low. To understand the higher productivity along the equator, we must consider the equatorial currents systems.

The circulation of the atmosphere and the oceans is affected by the rotation of the earth, which results in the Coriolis acceleration. At the equator, however, the Coriolis effect disappears. This disappearance gives rise to the fastest subsurface currents anywhere in the ocean. These currents, the equatorial undercurrents, are restricted to a very narrow belt at the equator and therefore were not clearly recognized until after 1950, when detailed surveys of this region were initiated.

The equatorial undercurrents arise as follows: The atmospheric circulation in low latitudes consists of the northeast trades north of the equator and the southeast trades south of the equator. These winds pile up water at the western margin of the ocean against the eastern shore of the bordering continent. The winds cause the surface water to move from east to west. Because of the pile-up at the western margin of the ocean, the water has a tendency to return from west to east. North and south of the equator, the eastward return is deflected by the Coriolis effect in such a way that these waters converge at the equator. Here the Coriolis effect disappears, giving rise to a rapid current (up to 150 cm

Figure **26-13** The winds, surface, and subsurface currents at the western margin of an ocean near the equator.

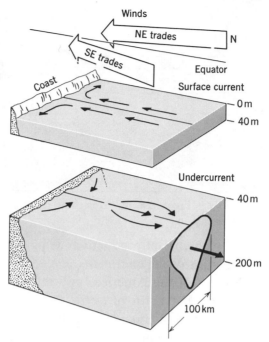

sec^{-1}) from west to east. Should the current stray off the equator, the Coriolis effect acts in such a way as to push it back toward the equator (Fig. 26-13).

The strong current shear between the west-flowing surface water and the east-flowing undercurrent gives rise to considerable vertical mixing. Thus vertical stirring at the equator is particularly strong. As a result, the vertical transport of nutrients is enhanced, and productivity is increased along the equator. A reduction in the surface winds may occasionally cause the equatorial undercurrent to become exposed at the sea surface.

Upwelling

Another peculiarity of the distribution of primary productivity is the relatively high productivity near the coast (Fig. 26-10). Where the depth of water is less than the depth of the winter mixed layer, the nutrients cannot sink through the density gradient. Thus the nutrients are readily brought back to the sea surface in winter.

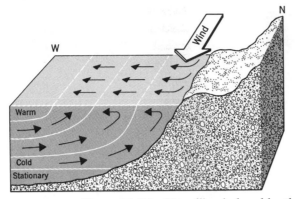

Figure **26-14**   Upwelling induced by the wind along the Oregon coast.

The high coastal productivity is particularly pronounced at the western margins of the continents. Excellent fishing grounds exist along the west coast of the Americas and off the west coast of Africa. In these locations, the return of nutrients to the surface water is increased in summer by a phenomenon called upwelling (Fig. 26-14). During this time of year, the predominant winds on the Oregon coast are from the north. If there were no Coriolis effect, they would cause the surface water to move southward along the shore. Due to the earth's rotation, however, the southward movement is deflected to the right, causing the surface water to be moved offshore. As a result, deeper, colder water, rich in nutrients, becomes exposed at the surface. When this nutrient-rich water is illuminated, it gives rise to a very high productivity. The upwelling extends vertically to a depth of about 200 m and the upwelled water extends from the shore to about 100 km offshore.

Upwelling results in good fishing and poor swimming. Owing to the upwelling, the surface water off the Oregon coast is colder in summer than in winter. Occasionally, if the winds stop for a few days, the upwelling is interrupted and the surface water is warmed, resulting in mild water temperatures. Considerable upwelling takes place along the Peru-Chile coast of South America. The cold waters off the Peru current, enriched by upwelling, are very productive. The abundance of fish supports a large population of sea birds, whose droppings have produced huge guano deposits along the coast. In some areas, guano deposits up to 300 m high have been laid down over the years.

El Niño

The tropical marine bird life of Peru not only is threatened by man but is also affected adversely by a natural phenomenon known as *El Niño,* the child. The Peru current swings west at 6°S off Punta Aguja and converges with the equatorial countercurrent which comes from the west (Fig. 12-5, p. 195). About every seven years, the intensity of the trade wind drops off and the equatorial waters spread further south, displacing the cool Peru current. At the same time, the reduction in the wind reduces the amount of upwelling so that the cold, nutrient-rich water of the Peru current is replaced by nutrient-poor tropical waters.

The change in productivity causes widespread destruction of plankton and fish. Dead fish accumulate on beaches, and the water becomes depleted in oxygen, leading to the generation of hydrogen sulphide by sulphate-reducing bacteria. The hydrogen-sulphide concentration may get sufficiently high to blacken the paint on ships. Because of this effect, the *El Niño* is also known as "The Painter." The lowered fish population leads to a mass mortality of sea birds during *El Niño* years, such as in 1925, 1930, 1941, 1951, 1953, 1957/58, and 1965.

Summary

The distribution of organisms in space and time is complex. The patchiness of the distributions and the vertical migrations of many species make it difficult to prepare an accurate census of populations. Communities of organisms comprising the same genera but differing at the species level inhabit similar environments that may be widely separated in space.

Evolution has resulted in complex adaptations of organisms to the varied marine environment. The food cycle in the sea starts with photosynthesis. The plant matter produced is then grazed by herbivores that, in turn, are eaten by carnivores. The bacteria and the benthic detritus-feeders ensure that the organic matter is recycled so that the nutrients are returned to the water, where they can once more be utilized for photosynthesis. Specific examples are the seasonal cycles in Long Island Sound and the flow of organic matter in the English Channel. The productivity of the ocean is controlled primarily by the vertical upward diffusion of nutrients into the photic zone. Of special interest is the high productivity along the equator and on the eastern margins of the ocean basins.

Study Questions

1. What are some of the possible adaptive advantages of vertical migration?
2. What difficulties do we encounter if we attempt to determine planktonic populations in a particular area? How would our sampling program have to be designed to overcome these difficulties?
3. Can Arctic whales easily migrate to the Antarctic? Explain.
4. Describe a terrestial analogue of the balanced aquarium.
5. Why does the nitrate concentration in Figure 26-5 decrease by a larger factor than the concentration of phosphate?
6. Discuss the variation of primary productivity in the world ocean.

Supplementary Reading

(Starred items require little or no scientific background.)

* Dietz, Robert S. (1963). "The Deep Scattering Layers," *Scientific American,* August.
Hedgpeth, J. W. (1957). *Treatise on Marine Ecology,* Vol. I, *Ecology.* Geological Society of America, Memoir 67.
* Lermond, J. W. (1966). "Peru Current," *Science and the Sea.* Washington, D. C.: U. S. Naval Oceanographic Office.
Raymont, John E. G. (1963). *Plankton and Productivity in the Oceans.* London: Pergamon Press.

Part VI
The Marine Environment

Physical, geological, chemical, and biological factors combine to shape the marine environment. The tropical coral reef illustrates the interaction between geology and biology, for the plants and animals produce solid rock and sediments while the geologic history controls the distribution of the biota. Estuaries where rivers enter the sea through drowned valleys are of particular importance to man. Here we find sharp salinity gradients which affect the currents and the flora and fauna. A study of estuaries leads naturally to a study of mediterranean seas. Then we must consider the deep circulation of the ocean, which brings oxygen into the deep and returns nutrients to the surface.

The deep circulation of the Atlantic Ocean differs markedly from that of the Pacific. This difference leads us to examine how the ocean contributes to controlling our climate. How are changes in the ocean related to variations of the climate, and what role did the ocean play during the ice ages? Perhaps a better understanding of the large-scale interaction between the ocean and the atmosphere will permit us to predict and even to purposefully modify the climate of the future. As a by-product of advanced technology, the marine environment is being altered by man in many ways, but the science of oceanography is not yet sufficiently advanced for us to predict the long-term consequences of our actions.

The Pacific Ocean as seen by the Applications Technology Satellite 1, on September 17, 1967. Five hurricanes are visible. Letters on photograph refer to the names of hurricanes. (NASA photo)

Skindivers examining the corals in the breaker zone of the reef. (T. F. Goreau)

27 Coral Reefs

In the immensity of the tropical Pacific and Indian oceans are found whole archipelagos of atolls, narrow islands of low elevation circling a shallow lagoon and themselves surrounded by a reef (Fig. 27-1). Outside the reef the depth of the water increases rapidly. Many a sailor must have wondered how these small islands originated within the vastness of the ocean. When we examine an atoll, we find that it consists entirely of calcium carbonate: the reef is a massive framework of wave-resistant, living corals, and the sediments that comprise the island are the broken-up remains of calcareous plants and animals.

The Coral-Reef Problem

The true stony corals belong to the order Scleractinia in the class Anthozoa of the phylum Cnidaria. They consist of solitary or colonial polyps that build a calcareous skeleton into which the polyps can withdraw. Figure 27-2a shows a small reef coral as it appears when the polyps are contracted, so that the calcareous skeleton can be seen clearly. The same coral is shown in Figure 27-2b with the tentacles fully expanded.

Corals exist at all depths within the ocean. However, reef-building corals occur only in shallow tropical waters that are brightly illuminated and whose temperature does not fall below 18°C.

Coral reefs cover some 2×10^8 km^2 of the tropical ocean. There are three types: fringing reefs, barrier reefs, and atolls. Fringing reefs develop in shallow water along the coast. Barrier reefs form an offshore breakwater parallel to the coasts of continents or isolated islands. The largest such reef, the Great Barrier Reef, extends some 2000 km parallel to the northeast coast of Queensland, forming the true eastern edge of the Australian continent. Others are found just east of the line of the Florida Keys in the Bahamas, and in coastal areas and around islands in the tropical western Atlantic. The third type of reef forms the atolls,

Figure **27-1**   Rongelap Atoll in the Marshall Islands as photographed from Gemini V Aug. 27, 1965. (NASA)

surrounding the small ring-shaped group of islands and protecting it from the ocean waves.

How did coral reefs originate? Since corals build reefs only in relatively shallow water, they require a stable, shallow substrate on which they can construct a wave-resistant framework. If coral reefs formed under static conditions, seamounts or guyots would have to extend up

Figure **27-2**   A small reef coral Meandrina *meandrites* that lives on sand bottom: (*a*) the colony is contracted and the skeletal detail is visible; (*b*) the same colony is fully expanded, showing the tentacles. The white spots on the tentacles are the stinging cells, nematocysts. (Photograph: T. F. Goreau)

(a)   (b)

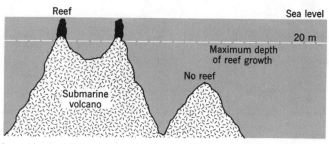

Figure **27-3**   The hypothetical formation of a coral atoll under static conditions.

to a depth of about 20 m below sea level to permit the establishment of a coral reef (Fig. 27-3). It is extremely unlikely that a large number of submarine volcanoes would just happen to reach that critical depth. The reefs of atolls would have to be constructed on the rim of an extinct submarine volcanic crater reaching almost to the sea surface.

A dynamic answer to the coral-reef problem was first proposed by Charles Darwin. Between 1831 and 1836, the young Darwin served as naturalist on H.M.S. *Beagle* during a three-year hydrographic mapping cruise around the world. In the course of his duties he examined a number of coral reefs and atolls.

Darwin argued as follows. Consider a submarine volcano gradually built up by lava until it forms a volcanic island. In the tropics, as the volcanic activity ceases, the margins of the cone at a water depth between 0 and 20 m will become populated by reef-building corals (Fig. 27-4a). Organic calcification and breakdown by waves will go on simultaneously, forming large quantities of coral and algal debris. Most of this debris will be swept down the steep side of the volcanic cone into deep water, causing a slight outward expansion of the reef. In addition, the debris will form a flat platform between the reef and the shore.

The ejected volcanic rock represents an excess weight on the ocean floor, and therefore the crust will be depressed. Once the volcanic activity ceases to add new material, subaerial and marine erosion will tend to level the island. In addition, the volcano slowly sinks as the crust comes back to equilibrium. Meanwhile the corals will continue to grow and maintain the reef near sea level. The carbonate sediments produced by the reef are redistributed by the waves to partially fill the gap between the reef and the volcanic cone, forming a shallow, ring-shaped lagoon. Much of the sediment is carried seaward to be deposited on the

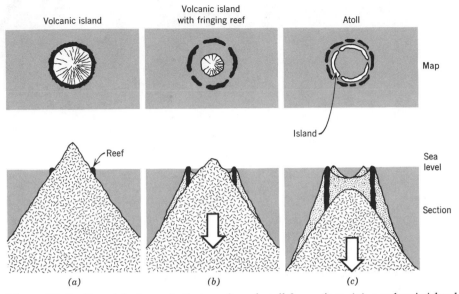

Figure **27-4** Darwin's dynamic theory of coral atoll formation: (*a*) a volcanic island with a fringing reef; (*b*) subsidence of the volcano and upward growth of the reef produce a small volcanic cone surrounded by a lagoon and a barrier reef; (*c*) further subsidence and upward growth of the reef produce an atoll.

outside slope of the reef. After some time we find a volcanic island surrounded by a lagoon and an outer reef (Fig. 27-4*b*).

As subsidence continues, the volcanic core of the island eventually sinks below the waves. By continued upward growth, the reef maintains itself near sea level while the sediments produced in the reef environment form sandbars inside the reef. The sandbars become piled up by storm waves and the wind to form a circlet of islands surrounding the lagoon (Fig. 27-4*c*).

Darwin's dynamic mechanism does not require that the volcanic islands have a specific elevation. All that is necessary is that at some time they reach a depth at which reef-building corals can become established. As the island subsides, the corals and associated organisms maintain the reef near sea level by the organic deposition of calcium carbonate.

If Darwin's theory of the origin of coral islands is correct, then all atolls must contain a volcanic core. To verify this hypothesis, the Royal Society of London in 1904 undertook to drill into Funafuti, one of the Ellice Islands group of the western Pacific. Recent to Pliocene limestones and dolomites of shallow-water origin were found, indicating a long history of reef building on a subsiding platform. Drilling was carried to a depth of 345 m before the hole had to be abandoned—unfortunately, not deep enough to encounter the supposed volcanic core of the island.

In 1952, the U. S. Geological Survey undertook a drilling program in the Marshall Islands (Schlanger et al., 1963). After penetrating exclusively through shallow-water limestones, two drill holes on Eniwetok Atoll reached the basaltic core of the island at depths of 1250 and 1400 m. Active reef building since Eocene times, some 60 million years ago, had kept the top of the subsiding volcano near sea level. Thus Darwin's theory was confirmed.

The Effect of Glacial Sea-Level Changes on Coral Reefs

Darwin published his first report on coral reefs in 1840. That same year, after a study of glaciers in the Alps, the Swiss-born naturalist Louis Jean Agassiz (1807–1873) suggested that there had been a Glacial Epoch. That the deposition of large ice sheets on the continents would have a global effect on the level of the ocean was first pointed out by Nathaniel Southgate Shaler in 1895. The glacial sea-level oscillations of 100 m amplitude (shown on Fig. 15-7, p. 249) are superposed on the much more gradual subsidence postulated by Darwin.

Sea level reached its present stand only about 6000 years ago; all modern coral reefs have therefore become established since that time. Some 15,000 years ago sea level was at least 100 m below its present stand, so that the coral reefs developed during the last glacial period are now well below the depth of active coral-reef formation. In many places, echo soundings reveal these drowned reefs.

At a depth of about 30 m, the drilling of Eniwetok encountered the top of a limestone zone that showed evidence of leaching by fresh water. The drilling program thus confirmed that the entire atoll had been exposed above sea level during glacial times, resulting in the alteration of the marine limestone by rainwater. The rate of sea-level rise during deglaciation (over 1 cm yr^{-1}) was apparently too rapid to maintain the reefs at their old location. Not until sea level became stabilized, some 6000 years ago, were the reef-building corals once more able to establish a wave-resistant framework.

The present reefs are thus built upon a former shoreline that was dry land during the last ice age. Since the reefs have had only a relatively short time in which to establish themselves, their shape is controlled largely by the old land surface. We cannot understand the distribution of the reefs without studying the glacial history of the platforms on which the reefs have developed. Gaps in the reef may represent glacial stream channels that cut into the formerly exposed limestone. In many areas there simply may not have been sufficient time for reefs to reestablish themselves.

A Traverse across a Coral Reef

To learn more about the coral-reef environment, let us examine the reefs that fringe the north shore of the island of Jamaica, in the Caribbean Sea. To get an over all view of the reef, let us first look at the shoreline from the air (Fig. 27-5). This photograph was taken on a very calm day from a helicopter, looking obliquely at the coast from an altitude of 50 m. The shore is just east of the town of Ochos Rios. In the foreground, the sea floor rises steeply at an angle of 45° or more. The light

Figure **27-5**   Aerial photograph of the reef at Ochos Rios on the north shore of Jamaica. (Goreau, 1959)

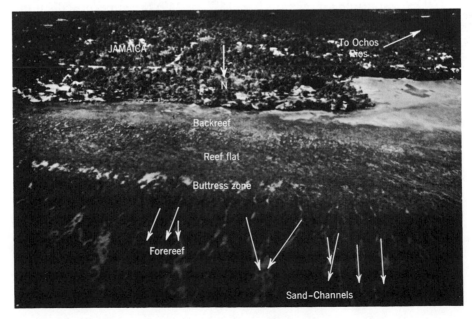

dendritic pattern in the foreground is produced by sand channels that drain sediments from the reef.

The front of the reef crest is marked by a number of dark stripes that run perpendicular to the coast. These are channels between massive corals arranged like buttresses that form the forward edge of the reef flat. If the sea were not calm, this buttress zone would take the brunt of the breaking waves. The line of breakers marking the reef front is often the mariner's first indication that he is approaching a reef-protected coast. Behind the buttress zone lies the reef flat. Here corals grow up to the low-tide level. Behind the reef flat is the slightly deeper back reef, or lagoon area, which terminates at the steep limestone cliffs of the shore. Except for small areas where there are sandy beaches, the cliffs have been etched in the tidal zone by marine organisms to form a sea nip (see Fig. 21-9).

Having examined the reef from the air, let us traverse the reef underwater to examine the various reef communities in more detail. The outer edge of the reef consists of very large treelike corals, the branches of which are oriented in line with the waves, so that they offer a minimum resistance to the surging water (Fig. 27-6). The massive branches, point-

Figure **27-6**   The elkhorn coral Acropora *palmata* in the breaker zone of the reef under very calm conditions. (Photograph: T. F. Goreau)

ing seaward at the reef front, break the energy of the surf and protect the rest of the reef. During strong storms some of the massive corals are broken off, only to be replaced by new growth. Thus the reef edge is a living breakwater where destruction is balanced by organic reconstruction.

As we move inward from the reef edge, the motion of the water is greatly reduced, and so more fragile corals are able to survive. This is the reef flat, rich in a variety of branching and massive spherical corals. The water is very shallow, and we must fend our way between the corals. Numerous fish swim about, and spiny sea urchins inhabit the nooks and crannies. Sponges and sea fans add variety, and calcareous red algae encrusting the coral debris cement it into a solid framework.

After we cross the reef flat, we come to the back reef, where the sea floor is covered by sand-size carbonate sediments of skeletal origin. Occasionally, isolated groups of corals form miniature "patch reefs" that rise almost to the sea surface.

Many calcareous green algae grow in the reef. When these algae die they break up and add their calcareous skeletons to the sediment. Some algae, for example Halimeda, (Fig. 27-10), contain broad, thin plates of calcium carbonate. In many areas, their remains comprise a large fraction of the calcareous sand. Mixed into the sand are shells of Foraminifera and mollusks and pieces of corals and calcareous red algae.

Some green algae, for example Penicillus, "Neptune's shaving brush" (Fig. 27-7), contain a skeleton made up of fine needles of calcium carbonate. When the algae die, they produce a lime mud which becomes suspended in the water by wave turbulence and so is carried out of the

Figure **27-7**   A grove of Penicillus at a depth of 25 m. (Photograph: E. A. Graham)

reef into deeper water. On broad shelves, however, the lime mud is able to accumulate. The sediments of Florida Bay and of large portions of the Bahama Banks are mostly lime mud produced by these algae and their relatives.

The Seaward Margin of the Forereef

With the aid of an aqualung, it is possible to examine the seaward margin of the reef to a depth of about 75 m. As we swim along the reef, we note that the reef front is interrupted by numerous channels that run perpendicular to the reef trend. The bottom of these seaward-sloping, steep-sided channels is covered by sediment that is slowly moving into deeper water (Fig. 27-5). These channels form "rivers of sand" that "drain" the reef of the sediment. Without such drainage, the reef would soon "drown" in its own sediment, and the corals would be buried in their own detritus and die. The moving sand prevents the establishment of new corals in the channels. In many places, however, the tops of the channels are bridged by the corals growing over them from both sides, forming tunnels and caves.

The seaward side of the reef front changes its character with depth (Fig. 27-8). The reef crest extends from the surface to a depth of 20 m. This zone is able to withstand the full force of the waves. Here the massive wave-resistant framework is composed primarily of branching corals of large size. The rate of calcification is very high and compensates for erosion.

At a depth of about 20 to 30 m, the wave energy is much reduced and the light intensity is only about one-fourth of its surface value. The corals are less massive but more diverse in population (Fig. 27-9). The water motion is still sufficiently strong to prevent the accumulation of fine sediment outside the reef framework.

Between about 25 and 70 m, the slope of the sea floor becomes very steep, being seldom less than 45° and often more than 70°. At a depth of 50 m the light intensity is reduced to 5 percent of the surface value. Because of the slower water motion, sediments begin to accumulate in areas where the slope is less than the angle of repose. These sediments consist of a mixture of sand and silt. In areas where the slope exceeds the angle of repose, the old rock surface and coral rubble debris are exposed and provide a base for the attachment of sessile organisms

Figure **27-8** Vertical zonation of the reef on the north shore of Jamaica. (*a*) Profile showing recent growth on old limestone. (After Goreau and Hartman, 1963) (*b*) Depth about 2 m. Front face of coral buttress. The large branching growth at the left are elkhorn coral (Acropora *palmata*) which occur only in the most turbulent regions of the reef. The lettuce-like coral in the center is Agaricia *tenuifolia,* a typical buttress builder. (*c*) Depth about 5 m. Fisheye lens view of a shallow reef buttress showing the seaward side. The narrow sand chute being examined by the diver drains sediment into deeper water. (*d*) Depth about 20 m. Fisheye lens view of deep coral spurs and chute. The environment here is less turbulent than in (*c*) and the elkhorn coral is replaced by the more fragile staghorn coral (Acropora *cervicornis*) living on top of the spur, while the sides of the chute are still formed by massive species. (*e*) Depth about 30 m. Vertical reef wall. Because of the steepness and relative lack of light, there are not so many reef corals. Their place is taken by whip gorgonia, whip antipatharia, sea fans, sponges and other benthic forms. A school of fish is swimming down the vertical face of the precipice, which is in reality an old wave-eroded shoreline drowned by post glacial rise of sea level. (Photographs: T. F. Goreau)

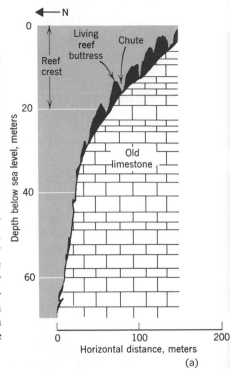

(a)

(b)

(c)

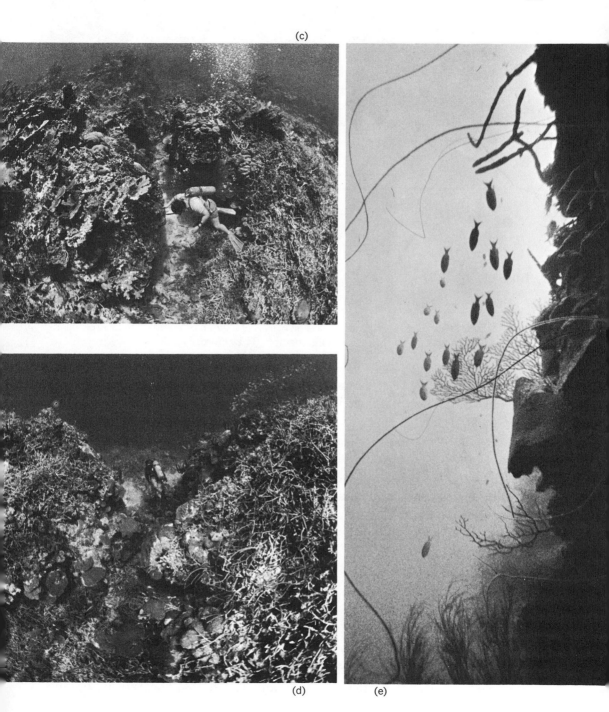

(d)

(e)

Figure **27-9**   Taking a coral-population census at a depth of about 20 m in a reef channel off Runaway Bay, Jamaica. (Photograph: T. F. Goreau)

(Fig. 27-10). The flora and fauna are varied; however, the variety and size of corals are reduced, and the colonies are less calcified. Species of coral that have massive forms in shallow water are reduced to thin, fragile sheets. As the light intensity decreases with depth, the amount of calcification is gradually diminished. Because of the low rate of calcification, the destructiveness of boring organisms becomes more significant. At shallower depth, the reef frame is stabilized by the cementing action of encrusting algae, but in deeper water, the light intensity is inadequate for their growth.

Photosynthesis

Corals are animals that feed on plankton and do not carry on photosynthesis, yet their high rate of calcification is dependent on the light intensity. Why, then, is rapid calcification in corals limited to the shallow tropical seas?

Using radioactive calcium, Goreau (1967) measured the rate of calcium uptake of particular corals and found that the rate is, on the

Figure **27-10** The reef off Runaway Bay, Jamaica at a depth of 28 m. The diver is collecting samples of Halimeda *copiosa,* a new species that is an important contributor of sand-size sediments. (Goreau, T. F., 1967; photograph by E. A. Graham)

average, about ten times faster during the day than during the night. Similar measurements of the calcification rates of calcareous algae indicate a light-to-dark ratio of only 1.4.

Using radioactive carbon-14, in the form of bicarbonate, it is possible to determine the photosynthetic fixation rate of carbon by coral colonies. Radioactive bicarbonate is added to the seawater in which the corals are growing. The subsequent appearance of carbon-14 in the organic matter of the coral can only be the result of photosynthetic conversion of CO_2 to organic matter. We find that, per hour, about 0.2 percent of the organic matter has been added to the reef-building coral by photosynthesis.

If we examine the coral polyps under the microscope, we find unicellular green algae, zooxanthella, living within the animal cells (Fig. 27-11). This intimate relationship between plant and animal cells is an example of symbiosis, the association of dissimilar organisms to their mutual advantage. All reef-building corals contain zooxanthella.

Symbiotic plants are not restricted to corals but also occur in other tropical marine invertebrates. For example, the giant clam, *Tridacna,* has zooxanthella in its brilliantly colored mantle. This clam, which can grow to a maximum length of about 1.7 m, is found in tropical reefs of

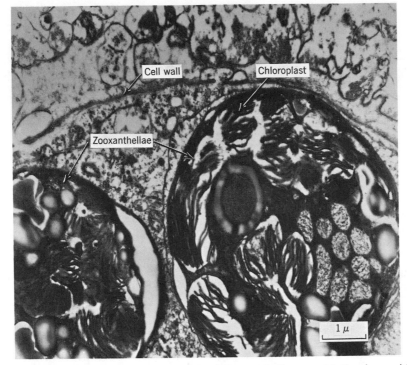

Figure **27-11** Electronmicrograph of Zooxanthellae, symbiotic algae within the tissues of a reef-building coral. Note that the algae are inside the cell of the coral polyp. (Goreau, 1961)

the Indian and Pacific oceans and lives with its mantle exposed to the light of the sun. It is not yet known whether zooxanthella play a role in the calcification of this giant Bivalvia.

Photosynthesis by zooxanthella can affect the calcification of corals in two ways. First, the extraction of CO_2 by the plant cells raises the pH of the animal's fluid and so increases the concentration of carbonate ion. Second, by photosynthesis the plant may produce organic compounds that facilitate the process of calcification.

Studies made with carbon-14 have shown that there is a transfer of organic matter between the animal host and the plant. Compounds photosynthesized by the plant cells are transferred to the animal host and contribute significantly to its nutrition. On the other hand, the plant cells are not able to synthesize certain essential organic compounds, hence must derive them from the animal host. Thus there is a two-way exchange between plant and animal. At the same time, the algae derive their CO_2 and mineral nutrients from the animal and contribute O_2 to the animal. Some soft corals apparently have completely lost the ability to feed on plankton and derive their nutrition almost exclusively from their symbiotic plants.

The symbiotic relationship between the coral and its plant cells reduces the organism's dependency on the environment. This is of great importance, since reef-building corals are restricted to tropical seas. We saw that the productivity of the tropical ocean is generally low, except in the immediate vicinity of the equator. The symbiotic relationship permits a more efficient utilization of the available nutrients. Thus the reef community is able to support a more abundant flora and fauna.

The symbiotic organisms form a virtually self-contained ecological system. If man had evolved such a relationship, we would be walking around with great green ears flapping in the sun. Photosynthesis in the ears would convert CO_2 from the bloodstream to oxygen and organic matter. At the same time, the waste products of the body would supply the necessary mineral nutrients for the symbiotic plant cells.

Summary

Coral reefs in tropical waters form wave barriers on the margins of continents, around islands and atolls. Darwin suggested that coral atolls result from continued reef growth over slowly subsiding volcanic islands. Gradual subsidence of the volcano is modulated by the more rapid glacial sea-level oscillations. All reefs existing today have been built up during the last 6000 years on a previously exposed weathered land surface.

A reef consists of the forereef zone, the reef flat, and the backreef region. The forward edge of the reef consists of massive corals arranged like buttresses which take the brunt of the waves. Numerous sand channels drain sediment from the reef into deep water. All reef-building corals contain symbiotic plant cells which contribute to the nutrition of the corals and permit more rapid calcification. As the light intensity decreases with depth, the rate of calcification by corals decreases, so that a species that is massive near the surface is only weakly calcified at depth.

Study Questions

1. How must Darwin's original theory of the formation of coral atolls be modified to take into account the effect of glacial sea-level changes?
2. How are the various sizes of sediment produced in the reef, and how are they redistributed?
3. Describe the symbiotic relationship between corals and zooxanthella. What advantages does this relationship have for the host and the algal cells?
4. How does the environment differ in the backreef region, the reef flat, the buttress zone, and the forereef zone?

Supplementary Reading

(Starred items require little or no scientific background.)

* Darwin, Charles. *The Structure and Distribution of Coral Reefs.* New York: D. Appleton. 1st ed., 1842; 3rd ed. 1898.
* De Beer, Sir Gavin (1965). *Charles Darwin, A Scientific Biography.* Garden City: Doubleday Anchor Books. Chap. 4.
Goreau, T. F., and Willard D. Hartman (1963). "Boring Sponges as Controlling Factors in the Formation and Maintenance of Coral Reefs." In *Mechanisms of Hard Tissue Destruction,* Publ. No. 75, *AAAS,* Washington, D. C., pp. 25–54.
Goreau, T. F. (1959). "The Ecology of Jamaican Coral Reefs. I. Species Composition and Zonation," *Ecology,* **40,** 1959, pp. 67–90.

28 Estuaries and Mediterranean Seas

The open ocean is relatively well mixed, and its characteristics change only gradually. Where portions of the ocean are surrounded by land so that they communicate with the world ocean only through narrow straits, mixing is inhibited. As a result, the salinity of these waters can be greatly modified by excess runoff from the land or by high evaporation to the dry continental air. The European Mediterranean was of great importance to the rise of Western civilization, and many of the largest cities of the world are located on estuaries. The marginal waters serve as harbors for commerce, as recreation areas, as a source of seafood, and as a receptacle for the liquid, solid, and thermal wastes of our "effluent" society.

Estuaries

An estuary is a semienclosed body of water that is connected with the ocean, whose water is measurably diluted by fresh water derived from the land. Estuaries are numerous in areas such as the east coast of the United States, where the drowned glacial river valleys have not yet been filled in by sediments. The salinity in an estuary varies from that of fresh river water to normal marine salinity, depending on the rate of freshwater discharge into the estuary from the land and the rate of exchange of water with the open ocean. This exchange may be greatly enhanced by tidal currents. The large salinity variation produces variations in density that dominate the vertical circulation in the estuary.

Consider a hypothetical estuary consisting of a drowned former river valley with a rather narrow connection to the open ocean (Fig. 28-1). Sea level outside the estuary will rise and fall with the tides. The water level within the estuary, however, will not necessarily be the same as that outside. Because of resonance effects between the natural period

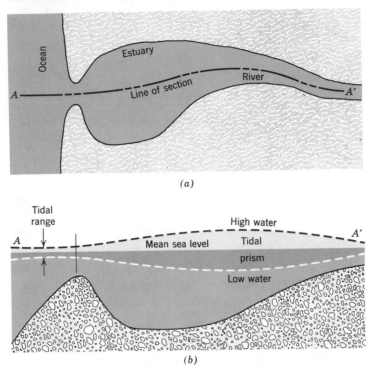

(a)

(b)

Figure **28-1**   A hypothetical estuary: (*a*) map; (*b*) vertical section showing the water levels at high and low tides (not to scale).

of water oscillation within the estuary and the frequency of the tide-generating forces (Chap. 15), the tidal range inside the estuary may be amplified. The volume of water between the high- and low-water levels is known as the *tidal prism*. During each tidal cycle, a volume of water equal to the volume of the tidal prism is exchanged with the ocean.

In addition to the outflow of river water, therefore, there is a back-and-forth exchange of water through the entrance of the estuary (Fig. 28-2). As the tide rises in the estuary, water flows in; as the tide falls, water is returned to the ocean. The tidal exchange is superposed on the net outflow of river water. The current is at the maximum at mid-tide, when the water level is changing most rapidly. The flow through the entrance reverses direction just before high tide and just after low tide, when the tidal currents cancel the flow of river water. The time of no net flow will depend on the relative volumes of river discharge and tidal exchange. If the river flow is large compared to the rate of tidal

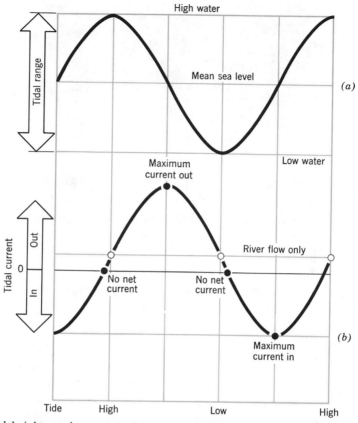

Figure **28-2** Tidal heights and currents: (*a*) sea-level variation in the estuary; (*b*) tidal current at the entrance to the estuary.

exchange, there will always be an outflow of water, modulated in intensity by the tidal exchange.

Water Circulation in the Estuary

The circulation of water within the estuary depends on the tidal range and the amount of vertical mixing between the fresh river water and the salt ocean water. First, let us consider the case in which the tidal prism is small with little vertical mixing (Fig. 28-3). The less dense river water floats on top of the denser seawater. There will be a small amount of mixing at the freshwater-seawater interface, and some seawater will be entrained and carried back out to sea, to be replaced by a slow bottom inflow of fresh seawater from the ocean. The salinity boundary between

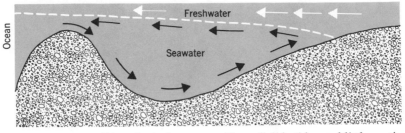

Figure **28-3**   Hypothetical estuary with negligible tides and little vertical mixing.

the two waters is relatively sharp; the freshwater is not significantly increased in salinity until the stronger wave action of the open ocean mixes the two waters and thus dissipates the freshwater tongue.

If the tidal exchange is still negligible but the vertical mixing within the estuary is increased, the situation shown in Figure 28-4 is obtained. Now there is considerably more exchange between the two waters, and therefore the vertical salinity gradient is less steep. More seawater has to flow into the estuary to replace the salt water that has been entrained by freshwater and carried out of the estuary by the surface current.

The vertical salinity structure in such an estuary will consist of three layers: a mixed surface layer of low salinity, whose salinity increases seaward; a bottom layer of close to marine salinity; and an intermediate salinity gradient layer, where salt is mixed upward. The motion of the surface layer is toward the sea, but the motion of the bottom seawater layer is toward the land. The dissolved and suspended load of the river water is carried seaward in the upper layer while coarser material is transported into the estuary from the ocean and carried along the bottom.

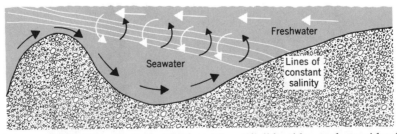

Figure **28-4**   Hypothetical estuary with negligible tides and considerable vertical mixing.

An example of an estuary in which the effect of the tide is negligible is the mouth of the Mississippi. Here the volume of river discharge is much greater than the small amount of tidal flow. The seaward-flowing river water is underlaid by a wedge of seawater. To replace the seawater entrained in the river, there is a slow upriver bottom flow of salt water. The distance the salt-water wedge extends upriver depends on the flow of the river. During flood stages the wedge will shrink; during low-river stages it will extend further upriver.

Finally, we must consider how the estuarine circulation is affected if there is significant tidal flow. As the water level in the estuary rises and falls with the tide, a volume of water equal to the tidal prism is exchanged between the estuary and the open ocean. Thus the tidal current is added to the river outflow and the inflow of bottom water. If freshwater runoff is small compared to the tidal exchange, the salinity in the embayment will not be diluted significantly. In that case, by definition, we are dealing with a bay rather than an estuary. There is, of course, no clear-cut division between the two; one type grades continuously into the other. While a bay of normal salinity will be populated by a normal near shore fauna, the fauna of an estuary is dominated by species that are able to tolerate low and variable salinity.

The effect of the tidal current is to increase the exchange of water between the ocean and the estuary. At the same time, the stronger currents caused by the tides will increase the amount of mixing. As a result, the vertical variation in salinity is considerably reduced. The dominant variation is now in the horizontal direction, from freshwater above the range of the tide to the salinity of the open ocean (Fig. 28-5). The salinity distribution varies from one estuary to another, and in the same estuary

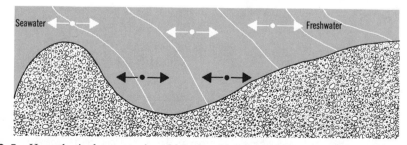

Figure **28-5** Hypothetical estuary in which the tidal flow is large compared to the freshwater discharge.

Figure **28-6** Tidal model of New York harbor, at the U.S. Army Engineer Waterways Experiment Station, Vicksburg, Mississippi. The model is used to study the transport of sediments in order to reduce shoaling in navigation channels. (U.S. Army photograph)

from flood time to drought. Scale models are often employed in order to study the complex circulation of estuaries (Fig. 28-6).

These are particularly useful if one wishes to predict the effect of proposed man-made alterations, such as deep channels, on the estuarine circulations.

The Effect of Man on Estuaries

Many of the most densely populated areas of the world are located on estuaries. In the process of urbanization, man has altered the geography by dredging channels and filling in marginal areas of the estuary. He uses the waters of the estuary for recreation and transportation and as a depository for sewage, industrial wastes, and heat from the condensers of steam electric-generating plants. These varied uses of the estuarine environment are not compatible with one another, and they alter the ecological system.

The flora and fauna of estuaries consist of communities that have evolved over geologic time and that are marked by high productivity and relatively low species diversity. In addition to its permanent population, an estuary also acts as a nursery for many animals that spend their adult life in the open ocean. When man destroys the estuarine environment, he not only reduces the living space for such estuarine species as oysters but also affects the abundance of certain fish offshore.

The effect of man's activity is often complex. For example, Ryther (1954) has investigated the abnormal algal growth found in some of the bays of Long Island. The bays were heavily polluted by the wastes of duck farms, which contribute large amounts of nutrients to the seawater. The phosphate-to-nitrate ratio in these wastes, however, is much higher than is normal for seawater. Imbalance in the nutrient ratios has caused certain algae to grow in enormous numbers. Such growth is detrimental to the normal plankton populations and so has led to a serious decrease in the oyster fishery.

The normal nutrient level in the sea is such that the organic matter produced by photosynthesis can be reoxidized by animals, using the oxygen dissolved in the water. When nutrients are added, the rate of photosynthesis in the surface layer is increased, so that more organic matter reaches the bottom of the estuary. The oxygen concentration, however, depends only on the temperature at which the water is equilibrated with the atmosphere. If the vertical circulation becomes reduced by thermal or salt stratification, there may not be enough oxygen present to enable animal respiration to consume the organic matter. As the oxygen becomes depleted, the animals can no longer survive, and bacteria that do not require oxygen take over. Although such oxygen depletion is very rare in the open ocean, it takes place in some estuaries where the exchange of bottom water with the ocean is reduced by shallow sills. The largest body of marine water devoid of dissolved oxygen is the Black Sea.

The Black Sea

We can consider the Black Sea as an estuary of the Mediterranean Sea (Fig. 28-7). For seawater to enter the Black Sea, it must first traverse the Mediterranean and head north between Greece and Turkey through the Aegean Sea. Then the water enters the narrow Dardanelles, crosses the Sea of Marmara, and finally flows through the narrow Bosporus into the Black Sea. At the northern end, the Black Sea connects with the Sea of Azov, the estuary of the River Don. Other major rivers that enter the Black Sea are the Danube and the Dnieper.

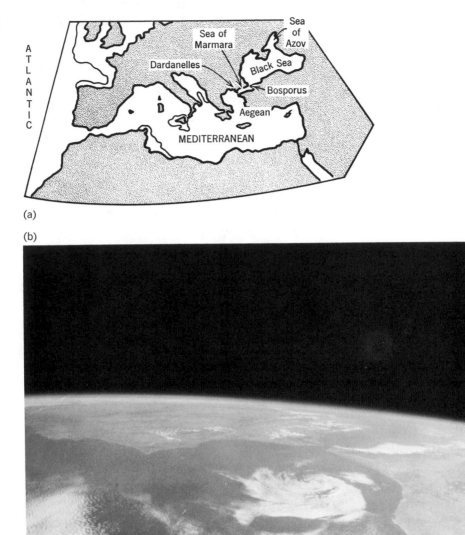

Figure **28-7** The Mediterranean area: (*a*) map; (*b*) Gemini X photo taken by Michael Collins looking from the Atlantic towards the Strait of Gibraltar and the Mediterranean. (Photograph: NASA)

The circulation of the Black Sea depends on the difference between evaporation, precipitation, and river addition. If more water were lost by evaporation than is added by precipitation and river flow, the net water flow would be from the Mediterranean, leading to an increase in salinity. If, on the other hand, the addition of water by rivers and precipitation exceeded the loss by evaporation, the Black Sea water would flow into the Mediterranean and would have a low salinity.

The approximate water fluxes of the Black Sea are as follows:

Evaporation	-15.6×10^9 cm^3 sec^{-1}
Precipitation	$+\ 7.6 \times 10^9$ cm^3 sec^{-1}
River run-off	$+12.6 \times 10^9$ cm^3 sec^{-1}
Total	$+\ 4.6 \times 10^9$ cm^3 sec^{-1}

Since there is an excess of freshwater addition, the Black Sea must lose water through the Bosporus to the Mediterranean; we might therefore expect the Black Sea to be a large freshwater lake. But when Count Marsigli measured the currents in the Bosporus in the eighteenth century, he found that there was not only an outflow of surface water from the Black Sea to the Mediterranean, as expected, but also a deeper current flowing in the opposite direction (Chap. 4, p. 36).

The inflow of saline bottom water through the Bosporus produces a salinity stratification in the Black Sea (Fig. 28-8). The surface salinity is low where the rivers enter and reaches a maximum of 18.2‰ in the center of the sea. The average surface salinity is slightly under 18‰. Below

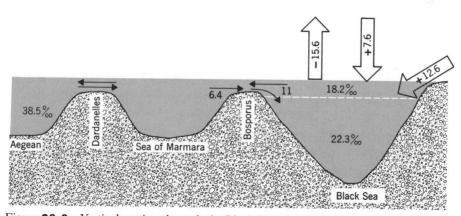

Figure **28-8**   Vertical section through the Black Sea, showing the salinity structure and the exchange of water. Fluxes are in 10^9 cm^3 sec^{-1} (not to scale).

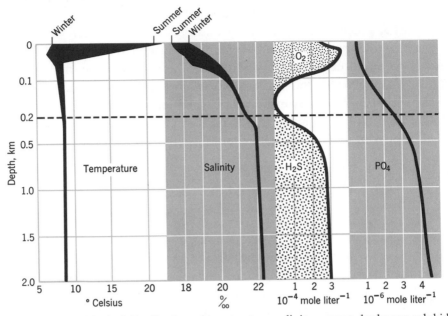

Figure **28-9**   Vertical distribution of temperature, salinity oxygen, hydrogen sulphide, and phosphate in the Black Sea. (Note change of vertical scale at 200 m.)

the seasonally mixed surface layer, the salinity increases rapidly to about 21.7‰ at 300 m. From there to the bottom, the salinity increases slightly to 22.3‰. The average temperature and salinity variations with depth are shown in Figure 28-9.

Since all the surface water in the Black Sea is less saline than the bottom water, the bottom water can be renewed only by the inflow of saline water from the Sea of Marmara. Let us make a rough estimate of the rate at which this bottom water is renewed. The Black Sea has an area 4.2×10^{15} cm² and a volume of 5.4×10^{20} cm³. Thus, below 100 m, the Black Sea contains 5.0×10^{20} cm³ of water. The average salinity of this deep water is about 22.2‰, hence its total salt content is 1.1×10^{19} g.

The water in the deeper parts of the Black Sea can be replaced only by the inflow of dense bottom water through the Bosporus. If the salinity is in a steady state, the amount of salt coming in must equal the amount of salt leaving, which is 11×10^9 cm³ sec⁻¹ at a salinity of 17.6‰, or 1.9×10^8 g salt sec⁻¹. Thus to replace all the salt below 100 m requires

$$\frac{1.1 \times 10^{19}}{1.9 \times 10^8} = 5.8 \times 10^{10} \text{ sec} = 1800 \text{ yr}$$

Oxygen in the Black Sea

The residence time of salt, hence of the water in the deeper parts of the Black Sea, is about 2000 years. The Black Sea is biologically highly productive, with a yearly total production of dry organic matter estimated at 10^{14} g. This amounts to a rate of production of about 100 g carbon m^{-2} yr^{-1}. The combination of high productivity and slow replacement of bottom water leads to a depletion of dissolved oxygen in the deep layers of the Black Sea. Once the oxygen is removed by respiration of animals near the surface, the deeper water is unable to support an animal population. One might therefore expect the organic debris, with its associated nutrients, to accumulate on the bottom.

The vertical variations of the properties of the water in the center of the Black Sea are shown in Figure 28-9. Near the surface there is a seasonal cycle, with the water being warmer and less salty in summer. The oxygen concentration decreases rapidly below 50 m and becomes zero near 175 m. Below this depth, the water contains increasing amounts of hydrogen sulphide and an increasing concentration of phosphate. If the oxidation of organic matter in the deep were prevented by a lack of oxygen, we would expect phosphate to accumulate in the sediments, so that its concentration in the deep water should be low.

In fact the data show an increase of phosphate toward the bottom, so that this nutrient must be regenerated. Normally regeneration is accomplished by the respiration of animals by the reaction:

$$CHOH + O_2 \xrightarrow{\text{animals}} CO_2 + H_2O$$

In the absence of free oxygen, the organic matter is consumed by sulphate-reducing bacteria. They are able to use the oxygen combined as sulphate to oxidize the organic matter and produce hydrogen sulphide by the following reactions:

$$2CHOH + SO_4{}^{--} \xrightarrow{\text{bacteria}} S^{--} + 2CO_2 + 2H_2O$$

$$S^{--} + 2H^+ \longrightarrow H_2S$$

The bacteria convert sulphate ion into sulphide ion and oxidize organic matter to CO_2 and water. The sulphide ion then combines with hydrogen ion to form hydrogen sulphide gas.

The sulphate-reducing bacteria thus oxidize the organic matter and keep the inorganic nutrients in circulation. The nutrients are gradually

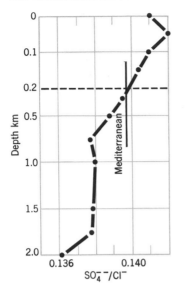

Figure **28-10**   The vertical variation of the sulphate to chloride ratio in the Black Sea. (After Skopintsev and Gubin, 1955)

mixed back into the surface layer, where they are utilized once more by phytoplankton. At the same time, the hydrogen sulphide diffuses into the oxygen-containing surface water, where the sulphide is reoxidized to sulphate:

$$S^{--} + 2O_2 \longrightarrow SO_4^{--}$$

As a result, the sulphate concentration is increased in the surface water of the Black Sea and depleted in the deep water (Fig. 28-10).

The Mediterranean

Because the Black Sea receives an excess of freshwater, its water is less saline than the ocean. The low rate of addition of saline bottom water through the Bosporus causes the oxygen in the Black Sea below 150 m to be depleted, so that the only living organisms are sulphate-reducing bacteria. In the Mediterranean, by contrast, evaporation exceeds precipitation and runoff. The approximate fluxes are as follows:

Evaporation	$- 115 \times 10^9$ cm³ sec⁻¹
Precipitation	$+ \ 32 \times 10^9$ cm³ sec⁻¹
Runoff	$+ \ 14 \times 10^9$ cm³ sec⁻¹
Total	$- \ 69 \times 10^9$ cm³ sec⁻¹

The excess evaporation of the Mediterranean produces an increase in salinity, relative to the Atlantic Ocean, outside the Strait of Gibraltar.

The ratio of the salinities is $1 + W/M$ (see Chap. 11, p. 164) where W/M is the ratio of the rate of water-vapor loss to the rate of mixing. The average surface salinity of the Mediterranean is about 38.5‰; that of the Atlantic outside the Strait is 36.6. The ratio, then, is $38.5/36.6 = 1.05$. Thus the rate of water exchange across the Strait must be 20 times the net evaporation rate, or about 1.5×10^{12} cm³ sec⁻¹.

The volume of the Mediterranean Sea is 3.7×10^{21} cm³. Therefore, the renewal time of the Mediterranean water is $3.7 \times 10^{21}/1.5 \times 10^{12} = 2.5 \times 10^9$ sec, or about 80 years. Thus the rate of exchange of the Mediterranean water is very much greater than that of the Black Sea. This is due to the fact that the Strait of Gibraltar is deeper and much wider than the Bosporus. The exchange of water in the two channels is reversed. Low-salinity surface water flows out of the Black Sea while saline dense water flows in. The surface water of the Strait of Gibraltar consists of Atlantic water that flows into the Mediterranean, while the denser, saltier Mediterranean water flows out over the bottom of the Strait.

The rapid inflow of Atlantic surface water does not ensure that the water in the bottom of the Mediterranean is renewed. Its renewal depends on the formation of bottom water within the Mediterranean. To understand the deep circulation in the Mediterranean, we must examine how the inflowing Atlantic water is modified. The surface circulation and salinity distribution are shown in Figure 28-11. After Atlantic water

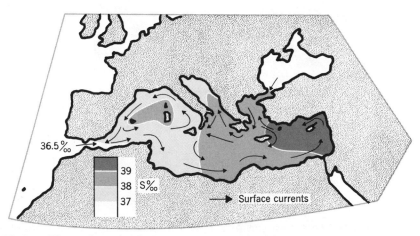

Figure **28-11**   The surface circulation and salinity distribution in the Mediterranean.

enters the Strait of Gibraltar, it flows along the coast of Africa to the east and then returns to the west along the European coast, forming a number of secondary gyres.

As the water flows from west to east, the excess of evaporation over precipitation causes the salinity to increase gradually. The distribution of surface salinity shows that the highest salinities, in excess of 39‰, are found in the northeastern portion of the Mediterranean. While the salinity primarily increases from west to east, the temperature in winter decreases from south to north (Fig. 7-1, p. 78). Thus the lowest winter water temperatures are found along the coast of France and in the northern Adriatic.

The Deep Circulation of the Mediterranean

The deep circulation of the Mediterranean depends on the relative density of the high-salinity eastern water versus the lower-salinity, colder water in the north-central portion of the sea. It turns out that the high-salinity warm water is slightly less dense than the colder water in the north. The high surface salinity in the eastern Mediterranean gives rise to an intermediate water which sinks in winter and flows from east to west at a depth of about 300 m (Wüst, 1960). As it flows westward, the intermediate water is diluted by mixing with the less saline water above and below and finally flows out into the Atlantic along the bottom of the Strait of Gibraltar.

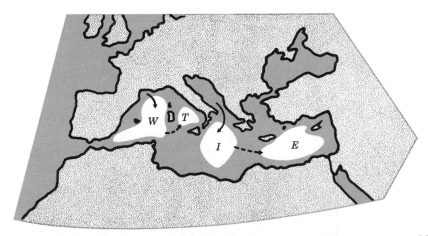

Figure **28-12**   The deep basins of the Mediterranean and the movement of bottom water. *W*, Western basin; *T*, Tyrrhenian basin; *I*, Ionian basin; *E*, Eastern basin.

The topography of the Black Sea is simple, since it consists of a single depression about 2.2 km deep. In contrast, the bottom topography of the Mediterranean is more complex. The Mediterranean contains four major depressions (Fig. 28-12)—the Western, the Tyrrhenian, the Ionian, and the Eastern basins.

Mediterranean bottom water forms in two areas (Wüst, 1961). The densest water is produced in the Adriatic with σ_T values of 29.25. This water flows southward along the bottom, mixes, and fills the bottom of the Ionian basin. From this basin, the water overflows into the Eastern basin. A second area of bottom-water formation is in the Gulf of Lyon. The water here attains σ_T values of 29.1 and fills the Western basin, from which it overflows into the Tyrrhenian basin. The characteristics of the bottom waters of the Mediterranean are as follows:

	Temperature	Salinity	σ_T
Eastern basin	13.2°C	38.6‰	29.25
Western basin	12.7°C	38.4‰	29.1

The productivity of the Mediterranean is low (Table 26-2, p. 435), and the rate of bottom-water formation in the Adriatic and the Gulf of Lyon is adequate to replenish the oxygen used by respiration in the deep. Apparently this has not always been so, for examination of cores of sediment from the bottom of the Mediterranean shows that they contain layers of sediment that were deposited in the absence of oxygen. Changes in climatic conditions in the past must periodically have reduced the rate of bottom-water formation to the point where the oxygen became exhausted.

The density of the surface water in the Mediterranean is much greater than that in the open ocean. In Figure 11-15 (p. 174) we saw that the surface density in the Atlantic and Pacific goes to σ_T values that are slightly above 27. In contrast, the σ_T of the Mediterranean exceeds 29. As the Mediterranean water flows out along the bottom of the Strait of Gibraltar, it mixes with the inflowing Atlantic water; however, its density after mixing is still well in excess of that of the Atlantic surface water. Thus the Mediterranean outflow runs along the bottom to a depth about 1.2 km, where it attains the same density as the Atlantic deep water. The Mediterranean outflow has a significant effect on the deep circulation of the Atlantic Ocean.

Summary

Estuaries are embayments of the ocean that are diluted by runoff from the land. The circulation in an estuary depends on the amount of fresh-water discharge, the strength of the tidal current, and the amount of vertical mixing.

Estuaries may be vertically stratified in salinity, or the main salinity gradient may be in the horizontal direction. Excess runoff into the Black Sea produces a salinity stratification which leads to a depletion of oxygen. As a result, sulphate-reducing bacteria convert organic matter to CO_2 by using sulphate as an oxygen source, and there is a consequent buildup of hydrogen sulphide.

Excess evaporation increases the salinity in the Mediterranean. The bottom water of the Mediterranean is formed where saline surface water is cooled in winter in the Gulf of Lyon and in the Adriatic. Saline Mediterranean water flows out at the bottom of the Strait of Gibraltar, mixes with the inflowing Atlantic surface water, and produces a saline tongue of water in the Atlantic at a depth of 1.2 km.

Study Questions

1. A river empties into an estuary. Where would you expect to find sediments carried by the river and marine sediments brought in by seawater?
2. Compare the water exchange between the Black Sea and the Mediterranean with that between an estuary and the open ocean.
3. Draw two sketches to contrast the flow through the Bosporus and that through the Strait of Gibraltar.
4. How do the Red Sea and the Persian Gulf exchange water with the Indian Ocean, and why?

Supplementary Reading

Lauff, George H., ed. (1967). *Estuaries.* Washington, D. C.: American Association for the Advancement of Science.

Zenkevitch, L. (1963). *Biology of the Seas of the U.S.S.R.* New York: Interscience Publishers. Chap. 9.

29 The Deep Circulation of the Ocean

Biological activity in the sea tends to deplete the surface water of nutrients and enrich the deep waters. At the same time, respiration in the deep ocean lowers the concentration of dissolved oxygen, which can be regenerated only in the photic zone or at the sea surface. Marine life, therefore, depends on an interchange of water between the surface and the deep sea. Without such interchange, the primary productivity would be low and the deep ocean would be devoid of animals. The circulation consists of two processes—advection and eddy diffusion.

Advection

Let us follow a blob of seawater marked with a dye as it moves through the sea (Fig. 29-1). On the surface, the blob will drift in the surface currents. At some stage in winter it may be cooled sufficiently to become denser than the adjacent surface water and sink. The blob will then continue to drift in the slower subsurface currents. Horizontal and vertical motion of water at the surface and within the sea is known as advection.

While the blob is still on the surface, its temperature and salinity can be altered by exchange of heat and water with the atmosphere (Chap. 11). Once it sinks below the surface, however, it can no longer interact with the atmosphere, and the salinity therefore remains constant. The temperature changes only as the depth of the blob changes as a result of the work done on the water by compression or expansion (p. 140). We can make allowances for the small temperature changes with depth by using the *potential temperature*. The potential temperature of a blob of water is the temperature it would have if it were raised to the surface without any exchange of heat. As the water is raised to the surface it expands and therefore does work against the rest of the ocean. As a result, the water cools slightly. Thus the potential temperature is

481

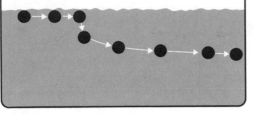

Figure **29-1**   Advection.

always less than the actual temperature. Figure 10-4 (p. 140) showed that the actual temperature of the water in the Philippine Trench increases with depth. The potential temperature of the water below 4 km, however, has a constant value of 1.2°C.

Advection below the sea surface alters neither the salinity nor the potential temperature. Below the photic zone the oxygen content of the water decreases with time because of respiration by animals. Therefore we call the salinity and the potential temperature *conservative* properties of the water. The oxygen concentration and the concentrations of nutrients, on the other hand, are examples of *nonconservative* properties.

Diffusion

Again we inject some dye into the ocean and follow the colored water as it moves through the sea. With time, the dye patch spreads out, and the color becomes diluted by mixing with fresh seawater (Fig. 29-2). Eventually, after thousands of years, the dye will be uniformly distributed throughout the ocean.

Diffusion in the sea results from two processes—molecular and eddy diffusion. In a stationary fluid, a dissolved substance gradually spreads through the fluid volume owing to the random molecular motions of the solvent and solute. On the time-space scale of the ocean, the unaided process of molecular diffusion is so slow as to be negligible. Even when we drop sugar into a cup of coffee, we do not wait for molecular dif-

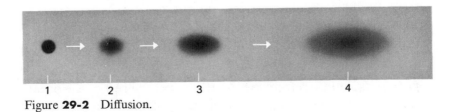

| 1 | 2 | 3 | 4 |

Figure **29-2**   Diffusion.

fusion to distribute the sugar uniformly throughout the cup but accelerate the process by stirring the coffee. Stirring brings coffee saturated with sugar into intimate contact with fresh coffee so that the molecular diffusion acts over much shorter distances to even out concentration gradients. Ultimately, however, we depend on molecular diffusion to produce a homogeneous solution.

Diffusion by stirring is called *eddy diffusion,* since it is produced by turbulent eddies. In the sea, the energy for stirring is provided by the atmosphere-ocean interaction and by the tides. Large, uniform motions, such as those of waves and ocean currents, are accompanied by turbulent motions which produce mixing. An example of this is the turbulence produced when waves break as they enter shallow water.

Advection and Diffusion

So far we have considered advection and diffusion separately. Actually the two occur simultaneously. Thus the advected dye spot in Figure 29-1 will spread out with time and become more diffuse (Fig. 29-3). In Figure 29-1, we could reverse the arrows and let the blob move from right to left and rise to the surface, since any direction of motion is physically possible. When we combine diffusion with advection, however, the sense of the motion is fixed. With time, the concentration of dye can only decrease and spread out, since there is no process within the sea by which the dye can be concentrated. At the sea surface, however, evaporation of water can increase the salinity, hence the dye concentration.

By following the motion of dye spots in the sea, we can study the circulation of the ocean. We can drop a dye into the sea from an airplane and take aerial photographs of the manner in which the colored patch moves and diffuses. To study subsurface currents, we would have to

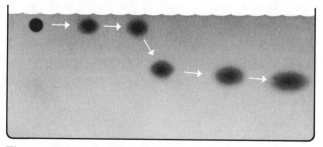

Figure **29-3** Advection and diffusion.

release a dye at depth and then collect water samples to determine the later distribution of the dye within the sea.

But we can form a rough idea of the ocean circulation without introducing a dye, by making use of the natural variations of conservative properties within the sea. For example, the outflow from the Mediterranean introduces a tongue of saline water into the North Atlantic. We can map this higher-salinity water and infer its motion from the fact that the water can only get less saline with time. Before we study the natural "dye markers" in the ocean, let us first look at other ways of measuring advective motions in the sea.

The Measurement of Ocean Currents

There are two methods by which ocean currents can be measured. The first method depends on measuring the rate at which the water flows by a fixed current meter. The second method uses a marker whose motion relative to the earth is determined when it is placed in the sea. To make measurements by the first method, the instrument must remain fixed relative to the floor of the ocean. It is not sufficient to stop a research vessel and lower a current meter over the side, for the ship will drift in the surface currents and in response to the wind. The research vessel must be anchored or the instrument must be held in place on a vertical line that is stretched between an anchor and a float (Fig. 29-4a).

So that the line will be minimally affected by surface waves and other motions, the float is kept well below the wave-tossed surface, and current meters are attached at various depths along the taut line. A current meter consists of a propeller whose rate of rotation measures the speed and a vane that measures the direction of the current relative to a magnetic compass. The meter includes a recorder so that data can be taken over a period of time. To recover the meter with its data, a release is placed above the anchor. After a fixed length of time, or on acoustic demand from the ship, the anchor is released and the float returns the instrument to the surface, where it can be retrieved by the research vessel. To aid in recovery, a light and a radio transmitter are enclosed in the float.

A simple but ingenious method for measuring currents near the floor of the ocean has been developed by J. N. Carruthers (1967). This technique employs a baby bottle, castor oil, and gelatin (Fig. 29-4b). If a float is anchored by a flexible line, it will stand straight only if there

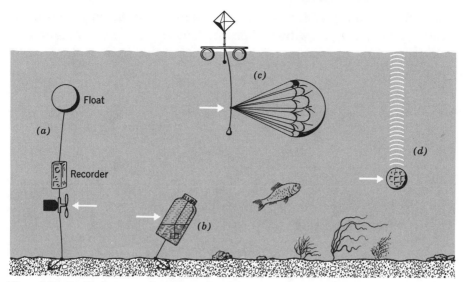

Figure **29-4** Various methods for measuring ocean currents: (*a*) current meter on tight anchored line; (*b*) Carruther's "Pisa"; (*c*) parachute drogue; (*d*) neutrally buoyant float.

are no currents. If a current is present, the float is deflected from the vertical, just as a tethered balloon will hang at an angle in the wind. There are various types of current meters, all of which measure the angle and its direction. The device must be calibrated to show how the deflection varies with the current.

Carruthers' device, which he calls the "Pisa," after the leaning tower, works as follows. A baby bottle is half-filled with a gelatin solution topped by castor oil. A small magnet is suspended from a stem attached to the bottle top. The bottle is heated so that the gelatin melts, and the bottle is then quickly lowered to the sea floor, where it floats above its anchor at an angle that depends on the bottom current. As the gelatin is cooled by the low temperature of the bottom water, it hardens, freezing the magnet in position. The bottle is then hauled back on board. The speed of the current can be determined from the angle of the interface between the gelatin and the castor oil, and the direction can be read from the magnet.

Other methods of measuring ocean currents depend on placing a marker within the ocean and observing its drift. One such method makes use of a parachute which is inflated in the water and drifts with the current. The parachute is carried to the depth at which the current is to be measured by a weight. A line of proper length connects the parachute to a surface float, consisting of an inflated truck-tire inner tube. A bamboo pole is fastened to the tube with a weight at one end and a radar reflector, light, and flag at the other end. The bamboo pole acts as a mast so that the motion of the float can be monitored from the research vessel by radar. The light and flag assist in recovering the float at night or during the day (Fig. 29-4c).

The large water resistance of the parachute will cause the float to drift with the currents at the depth of the chute. We now determine the drift of the float over a period of a day. In order to provide a fixed reference point, another float with a radar reflector is anchored to the bottom. By determining the relative positions of the two floats on the radar screen, it is possible to determine the drift as a function of time. At any one time, one can set out a number of parachutes at various depths and thus simultaneously determine the currents at different depths.

Another method of following deep currents has been developed by J. Swallow (1955). This method depends on a free float within the sea, the density of which is adjusted so that it will float at the desired depth after being dropped overboard. So that the ship can follow its horizontal motions, the float is equipped with a sound source (Fig. 29-4d).

The results of a typical series of current measurements with parachute drogues are shown in Figure 29-5. Two drogues each were placed at depths of 10 and 75 m, and one drogue was placed at a depth of 500 m. The drogues were followed for 23.8 hours. Although the detailed trajectories are quite irregular, the average drifts of duplicate drogues for the one-day period agree quite well. The trajectories show that instantaneous single-current readings are meaningless. An accurate idea of the long-term drift of the water requires measurements carried out over a period of time, so that irregular and tidal motions average out. For the data shown, the drift velocities were relatively large. In many areas of the sea, however, the irregular motions are much more rapid than the steady drift, so that it becomes extremely difficult and time-consuming to determine the average drift velocities.

Ocean currents can also be determined indirectly by measuring the distribution of density in the sea. From the density variations, we can infer the distribution of pressure within the ocean, relative to a refer-

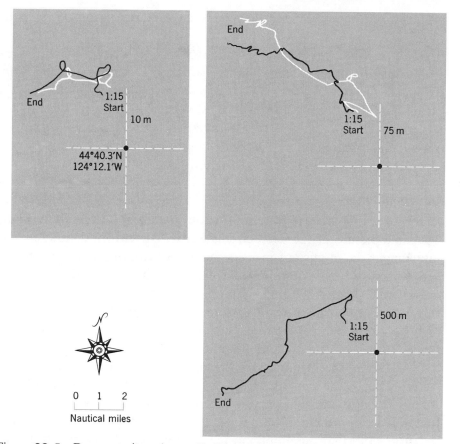

Figure **29-5** Drogue trajectories at 10, 75 and 500 m depth; duration of drift 23.8 hours. (Wyatt et al., 1967)

ence surface. If the ocean is in a steady state, the velocities relative to the reference surface can be obtained by equating the horizontal pressure gradients to the Coriolis acceleration. The method is similar to that employed in Chapter 9 (Fig. 9-8, p. 121) to determine the wind velocity from the surface pressure gradient. It works well for rapid surface currents, where horizontal density gradients are relatively large and the motion along the reference surface can be assumed to be zero. In the deep ocean, however, the horizontal variations of density are generally small, and the motion may extend all the way to the sea floor. If direct current measurements are made at one level, observations of the horizontal variations of density then permit the computation of velocities at other levels.

The direct current measurements available at the present time are insufficient in number and quality to describe the detailed motions within the sea. Much work remains to be done to understand the surface currents of the oceans and how they vary with time in response to changes in the atmosphere. To describe the deeper motions within the sea we must, therefore, depend largely on the natural dye tracers provided in the ocean by extremes of salinity and temperature.

Extremum Surfaces

Because of the irregular motions within the sea, it is difficult to obtain a reliable picture of the deep circulation by current measurements. To supplement direct current measurements, we can make use of the dispersal with time of natural tracers within the sea. Assume that, at a particular depth, water of higher salinity is introduced into the ocean. The vertical salinity structure will then look like Figure 29-6a. Injection of high-salinity water, for example, results from the overflow of Mediterranean water into the Atlantic Ocean.

With depth, the salinity goes through a maximum value. After the water in the salinity-maximum layer has moved horizontally and been subject to eddy diffusion, the vertical salinity structure will have a less intense maximum (Fig. 29-6b). By studying the salinity structure, we can therefore infer that the motion at the level of the salinity maximum has been from a to b.

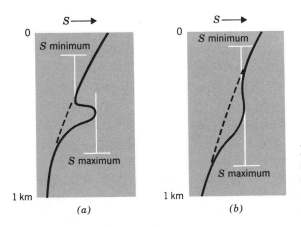

Figure **29-6** Vertical salinity structure at two points in the ocean. The motion along the salinity maximum is from a to b.

Just as the salinity at a salinity maximum can only decrease with time, so the salinity of a salinity minimum must increase with time, as a result of mixing with the saltier water above and below. The motion along a salinity-minimum surface is therefore from lower to higher salinities. In the same manner, the temperature at a temperature minimum within the water column can only increase with time, while a temperature maximum must decrease with time.

Because of the time irreversibility of the diffusion process, it is possible to infer the motion of seawater along extremum surfaces. Since the temperature of seawater generally decreases from the surface to the bottom, only rarely do we find temperature extremum surfaces. The vertical salinity structure, on the other hand, usually shows a number of maxima and minima, which permit us to infer the deep motion within the ocean. To do this, one must plot the value of the salinity along the extremum surfaces on a map. The currents along these surfaces then flow from more extreme to less extreme values. Let us examine the salinity extremum surfaces in the world ocean, starting near the sea surface.

The Near-Surface Salinity Maximum

When we study the vertical salinity distribution in the ocean, we find a near-surface salinity maximum in the lower latitudes. This maximum surface is usually at a depth of about 100 m. The distribution of salinity along the salinity maximum is shown in Figure 29-7. The salinities along the maximum surface range from above 37‰ in the Atlantic to below 34.5‰ in parts of the North Pacific.

In the Atlantic and the Pacific oceans, the highest salinities occur symmetrically about 20° north and south of the equator. Note that the salinities are significantly higher in the Atlantic than in the Pacific. The difference between the North Atlantic and the North Pacific is most pronounced. The salinities decrease toward the equator and toward higher latitudes. The salinity-maximum surface is formed by the sinking of the surface water in winter in the regions of high surface salinity. This water then spreads out toward the north and south. Along the equator, the salinity-maximum surface is perturbed by the equatorial undercurrents (Fig. 26-13, p. 443) which entrain the salinity-maximum water. It is not possible to show this effect on the scale of Figure 29-7. Thus, in effect, we have four discrete salinity-maximum surfaces in the Atlantic

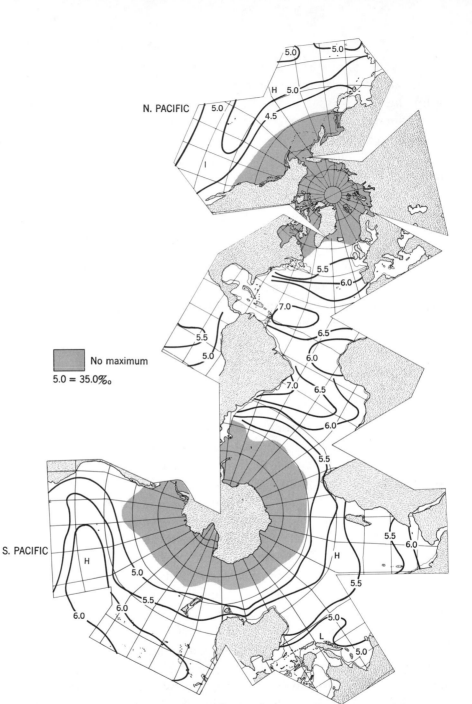

Figure **29-7** Salinity distribution of the near-surface salinity maximum surface. Numbers are salinity in ‰–30, e.g. 6.0 = 36.0‰. (After Weyl, 1968b)

and Pacific oceans. South of 45°S, the salinity maximum disappears. It is also absent in the extreme north.

The salinity-maximum surface in the South Indian Ocean is similar to that in the South Pacific. In the North Indian Ocean, the salinity distribution is altered by the effect of the monsoons on the surface circulation (shown in Figure 12-7, p. 198). In winter the surface current, driven by the northeast monsoon, is generally from the east. During this time, cold, dry air flows out from the continent of Asia. As it warms over the ocean, the air withdraws moisture from the sea and produces high salinities, particularly in the northeast part of the Indian Ocean (Fig. 12-4, p. 192). During the summer, the monsoon winds blow from the southwest onto the Asian continent. This reverses the ocean circulation so that the surface currents flow from west to east. As the moist oceanic air moves over India and southeast Asia, it produces some of the heaviest rains on earth. These rains reduce the surface salinity, primarily in the east.

During the period of evaporation, the surface currents flow from east to west; during the period of excess precipitation, they flow from west to east. As a result, we obtain a large gradient in salinity between the west and east. The highest surface salinities, in excess of 40‰, occur in the Red Sea and the Persian Gulf. These high salinities are produced by dry winds that blow over these narrow areas. Figure 29-7 shows that the North Indian salinity-maximum surface is most intense in the west and decreases as the water spreads out toward the east and toward the equator.

The Intermediate-Depth Salinity Minimum

The extremum surfaces must alternate. Thus, below the near-surface salinity maximum, we find a surface along which the salinity is a minimum. This surface generally occurs at a depth of about 500 to 1000 m. The salinity along this surface, shown in Figure 29-8, ranges from over 35‰ to below 34‰.

Let us start our examination of the intermediate-depth salinity-minimum surface in the Southern Ocean. South of about 45°S the minimum surface is absent. When it first appears, the minimum has its lowest values, near 34.2‰. As we move northward into the three major oceans, the salinity increases, indicating that the general flow is from south to north.

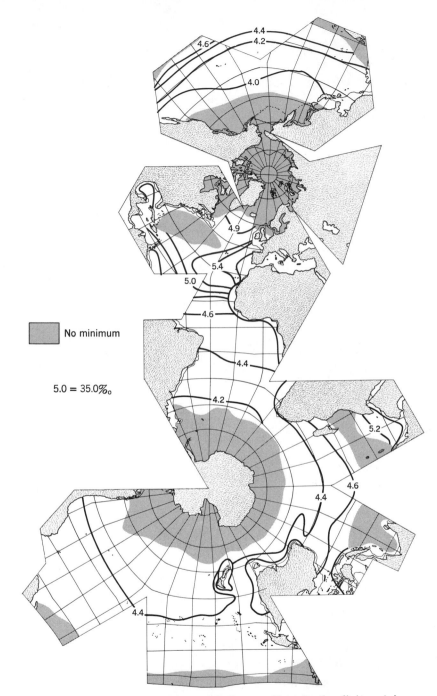

Figure **29-8** Salinity distribution of the intermediate-depth salinity-minimum surface. Numbers are salinity in ‰–30, e.g. 4.4 = 34.4‰. (After Weyl, 1968b)

The salinity-minimum surface that originates in the Southern Ocean extends northward to different latitudes in the three oceans. In the northeast Atlantic it extends to about 45°N, while in the Pacific it extends only to about 15°N. In the Indian Ocean, on the other hand, the minimum does not extend north of 10°S. The highest salinities in the minimum layer are found in the North Atlantic near the Strait of Gibraltar. The high salinity results from mixing with the outflow from the Mediterranean which underlies the salinity-minimum water.

In the North Pacific there is a separate minimum surface which has its lowest salinity near 40°N, with increasing salinity toward the equator. This surface can be followed to the equator, where it overlays the salinity-minimum surface that extends northward from the Southern Ocean. The two minimum surfaces are separated by a weak salinity maximum. A minor minimum surface exists in the extreme North Atlantic and in the northwest Indian Ocean.

The Deep Salinity-Maximum Surface

Below the intermediate salinity-minimum surface the salinity increases once more. In the Pacific proper, the salinity increases all the way to the ocean floor, so that no distinct maximum surface is formed. In most of the Atlantic, on the other hand, the salinity increases to a maximum and then decreases again as we approach the bottom of the ocean. A salinity-maximum surface is formed at a depth of between 1 and 2 km. The salinity along this surface is shown in Figure 29-9.

The highest salinities along the salinity-maximum surface occur outside the Strait of Gibraltar. Here, the water that flows out of the Mediterranean enters the deep Atlantic circulation at a depth of about 1.2 km. A distinct maximum is absent in the western North Atlantic, where the salinity increases all the way to the sea floor. The decrease in salinity along the maximum layer indicates that the water is moving southward in the Atlantic and then turns eastward around the tip of Africa to flow around the continent of Antarctica.

In the Indian Ocean, the deep salinity-maximum surface extends to about 15°S, but it extends only to about 55°S in the Pacific. Separate deep salinity-maximum surfaces of relatively small areal extent exist in the northwest Indian and Atlantic oceans. The deep salinity-maximum surface is the lowest salinity extremum. Below it, the salinity decreases to the sea floor.

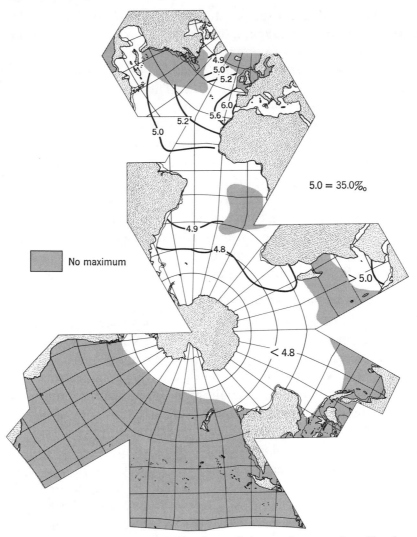

Figure **29-9** Salinity distribution of the deep salinity-maximum surface. Numbers are salinity in ‰−30, e.g. 4.8 = 34.8‰. (After Weyl, 1968b)

Bottom Potential-Temperature Distribution

In order to study the bottom currents of the deep ocean, we must consider the distribution of the potential temperature. With depth, the potential temperature decreases almost everywhere in the ocean. Thus upward mixing of bottom water can lead only to an increase in potential temperature with time. In addition, the bottom water is warmed by heat that is conducted upward from the interior of the earth. This heat amounts, on the average, to 50 cal cm^{-2} yr^{-1}. Thus, per year, the geothermal effect, if spread over a 1-km-high column of seawater, increases the average temperature of the water by 5 $\times$ 10^{-4}°C.

Figure 29-10 shows the distribution of the bottom potential-temperature in the major basins of the world ocean that extend below 5 km. Because mixing and geothermal heating can lead only to an increase in potential temperature, the sense of the bottom-water motion must be toward increasing potential temperature. The lowest temperature, -0.8°C., occurs just off the Weddell Sea in the Antarctic. The temperature then increases northward in the Atlantic. The Mid-Atlantic Ridge (Fig. 17-11, p. 286) separates the deep Atlantic into an eastern and a western basin. The bottom water from the Antarctic enters the western basin and then spills over into the eastern basin through the Romanche trench near the equator.

The bottom water in the Pacific also flows from south to north. The southeastern basin derives its bottom water from the Ross Sea area, while the other basins are filled by water from the west. The water enters the main western Pacific basin at a potential temperature of 0.3°C, reaches 0.85°C at the equator, and is warmed to 1.1°C by the time it reaches the bottom of the northeast Pacific. The basins of the Indian Ocean are also filled from the south, with cold bottom water flowing in from both the west and the east.

The distribution of bottom potential temperature shows that bottom water is injected into the world ocean from near the Antarctic continent and flows to the north. The Arctic Ocean furnishes only a small amount of bottom water, which sinks in the Norwegian Sea and then spills into the Arctic Ocean with a bottom potential temperature of -0.4°C. This water is isolated from the North Atlantic by a relatively high ridge between Greenland-Iceland and the Faeroe-Shetland Islands. Some Arctic bottom water, however, does enter the northwest Atlantic basin, with a potential temperature of 1.5°C.

The Deep Circulation in the Atlantic and Pacific Oceans

We have examined the three salinity-extremum surfaces and the distribution of potential temperature along the bottom to obtain a picture of the deep circulation of the world ocean. Now let us combine this in-

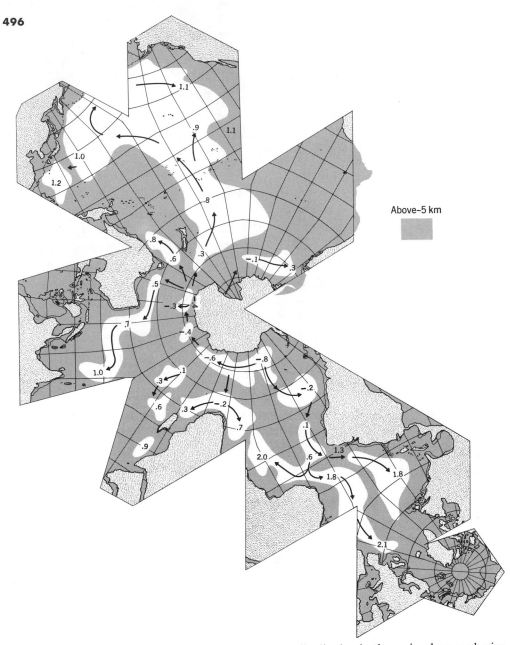

Figure **29-10** Bottom potential temperature distribution in the major deep sea basins of the world ocean that extend below 5 km.

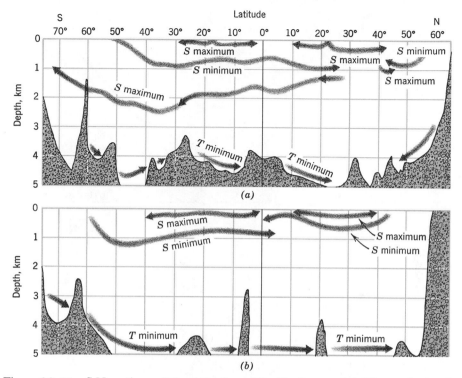

Figure **29-11** S-N sections of the Atlantic and Pacific Oceans. (*a*) Western basin of the Atlantic Ocean. (After Wüst, 1935) (*b*) Pacific Ocean at 160°W. (After Reid, 1965)

formation and look at south-north sections through the two major oceans (Fig. 29-11).

The bottom waters of both oceans are derived from the Antarctic; however, there is some advection of Arctic bottom water into the northern part of the west Atlantic basin. The southward-flowing, deep salinity maximum is found only in the Atlantic, and in this ocean the salinity-minimum surfaces are much less symmetrical about the equator. Both oceans have shallow salinity maxima on either side of the equator.

The bottom water that fills the deeper parts of the world ocean forms on the shelves of the Antarctic continent. The process is illustrated in Figure 29-12. As one moves south in the sub-Antarctic region, the surface temperature and salinity decrease rapidly. Meanwhile, the more saline water of the deep salinity maximum is moving south out of the Atlantic Ocean and circles the Southern Ocean. In winter, seawater freezes (Fig. 11-24, p. 182) around Antarctica. As the ice forms, it rejects

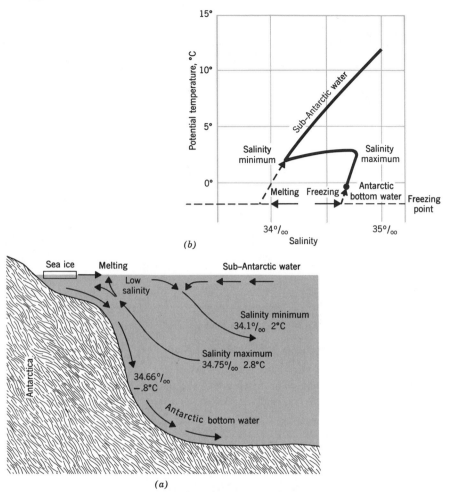

Figure **29-12** Bottom-water formation near Antarctica: (*a*) schematic cross section; (*b*) potential temperature-salinity diagram.

most of the salt from the water. Thus the water under the ice increases in salinity and so becomes more dense. As a result, particularly over the Antarctic shelves, the denser surface water mixes with the warmer and saltier water underneath to produce Antarctic bottom water. The sea ice drifts northward, impelled by the winds, before melting. As a result of the ice motion, a salinity minimum forms in summer in the surface water some distance away from Antarctica. The mixing of the low-salinity, cold surface water with the warmer, more saline sub-Antarctic water leads to the sinking of the salinity-minimum intermediate water, which then flows northward. Thus the freezing of Antarctic sea ice, particularly in the Weddell Sea, plays an important role in the formation of bottom water.

The sinking of water in the Antarctic region renews the bottom water in the ocean. Without such renewal, respiration by animals in the deep would deplete the dissolved oxygen in the water. The cool Antarctic surface water has a high concentration of dissolved oxygen which is mixed downward during the formation of Antarctic bottom water. As the water moves toward the north, the oxygen concentration is reduced by animal respiration. In the extreme South Pacific, the bottom water contains about 2.3×10^{-4} mole liter^{-1} of dissolved oxygen. This is reduced to about 1.5×10^{-4} mole liter^{-1} by the time the bottom water reaches the north Pacific.

The behavior of the extremum surfaces and the reduction in oxygen by respiration give an indication of the direction of the movement of the deep water; however, they tell us nothing about the rates at which these motions take place. To get an idea of the time scale of the deep circulation, we can make use of natural radioactive carbon-14. This isotope, produced in the atmosphere by the interaction of cosmic rays with nitrogen gas, enters the sea in the form of CO_2. Once the water sinks below the surface, no further carbon-14 is added and the radioactivity gradually decays. From the amount of carbon-14 that remains in the deep water, one can estimate how long this water has been out of contact with the atmosphere.

The data show that the waters of the Atlantic have been in the deep for about 400 years. The deep Pacific waters are considerably older, about 1500 years. Unfortunately, the carbon-14 data do not yet permit us to deduce a detailed picture of the deep circulation. They demonstrate, however, that the deep circulation of the Atlantic Ocean is considerably more vigorous than that of the Pacific. This is a result of the greater surface-density contrast in the Atlantic (Fig. 11-15, p. 174). Also, the Atlantic is more favorably located to receive Antarctic bottom water from the Weddell Sea (Fig. 29-10).

The Water Characteristics of the World Ocean

The cross sections in Figure 29-11 show that most of the water of the Atlantic and Pacific is below the salinity-minimum surface. The deep water is therefore relatively uniform in temperature and salinity. Mont-

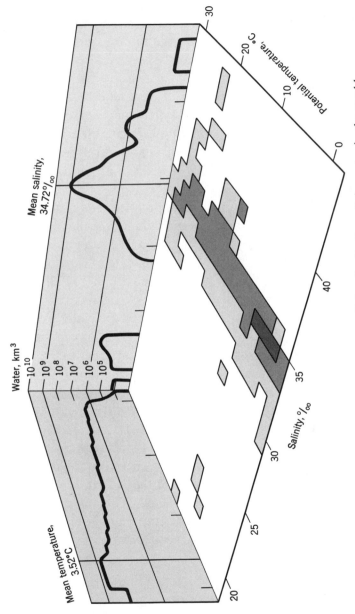

Figure **29-13** Potential temperature-salinity distribution of the water in the world ocean. 75% of the water of the ocean falls into the region of darkest shading and 99% falls into the dark area. (After Montgomery, 1958)

gomery (1958) has summarized the characteristics of seawater in the world ocean by determining the volumes of water that fall into given intervals of potential temperature and salinity. His findings are summarized in Figure 29-13. About 75 percent of all the water in the ocean falls between a salinity of 34 and 35‰ and a potential temperature between 0 and 6°C. About 99 percent of the water falls inside the dark shading in Figure 29-13, with the remaining 1 percent spread over a much broader area. The figure also shows the volumetric distribution of potential temperature and salinity on a logarithmic scale.

The total volume of the ocean is 1.37×10^9 km³. Of this, 1.26×10^9 km³ of water have a salinity between 34 and 35‰, and 9.80×10^8 km³ have a temperature between 0 and 4°C. The average potential temperatures and salinities of the major oceans are as follows:

	Potential Temperature (°C)	Salinity (‰)
Atlantic Ocean	3.73	34.90
Pacific Ocean	3.36	34.62
Indian Ocean	3.72	34.76
World Ocean	3.52	34.72

Summary

The circulation of the ocean is a combination of advection and eddy diffusion. The advective motion displaces the water; diffusion mixes it with adjacent waters. With time, an extremum within the water column becomes less intense because of mixing. Velocities in the ocean can be determined by measuring the drift of water relative to a fixed current meter, or by following the motion of a marker that drifts with the water. Velocities can also be computed from the distribution of density, provided that the velocities along a reference surface are known and that the motions are in steady state. To determine the circulation of the deeper water within the ocean, we must make use of the variations along salinity and temperature extremum surfaces. The motion is such that, with time, a salinity maximum becomes less salty, a salinity minimum becomes saltier, and the bottom potential temperature increases.

Bottom water is formed in the Antarctic region and flows toward the north in all oceans. A deep salinity-maximum surface is formed in the North Atlantic and flows into the Southern Ocean. A salinity-minimum surface is formed in the Southern Ocean, and the water flows northward into the three oceans. A separate major minimum surface forms in the North Pacific and flows toward the equator. A shallow salinity-maximum surface forms at either side of the equator and spreads out laterally.

The injection of Antarctic bottom water renews the supply of oxygen in the deep for animal respiration. We can obtain a rough idea of the rate of this renewal by measuring the radioactivity of natural carbon-14 in the deep ocean. The average potential temperature of the ocean is 3.52°C, and its average salinity is 34.72‰.

Study Questions

1. Suppose that you are approaching an unknown coast and need fresh water. How could you determine in which direction to turn to find the nearest major river? Would this method always work?
2. Describe two methods that might be used to measure the winds in the atmosphere, similar to the methods used for measuring ocean currents. Why is it simpler to measure the wind?
3. Determine the approximate average speed and direction of the drogues from the trajectories in Figure 29-5. A nautical mile per hour (1 knot) = 51.5 cm sec^{-1}.
4. How do the salinity distributions of the Atlantic and Pacific oceans differ?
5. The volume of the Atlantic is 3.24×10^8 km^3. That of the Pacific is 7.08×10^8 km^3. What volume of fresh water would have to be transferred from the Pacific to the Atlantic, to be replaced by an equal volume of sea-water, in order that the average salinities of the two oceans be equal?

Supplementary Reading

(Starred items require little or no scientific background.)

* Kort, V. G. (1962). "The Antarctic Ocean," *Scientific American*, September.
Neumann, G., and W. J. Pierson, Jr. (1966). *Principles of Physical Oceanography*. Englewood Cliffs, N. J.: Prentice-Hall. Chap. 14.
* Stommel, H. (1955). "The Anatomy of the Atlantic," *Scientific American*, January.
Von Arx, W. S. (1962). *An Introduction to Physical Oceanography*. Reading, Mass.: Addison-Wesley. Chaps. 8, 9.

30 The Ocean and Climatic Change

The physical environment on the surface of our planet has three components—geography, chemistry, and climate. We have seen that the geography of the earth changes slowly through geologic time. The chemical evolution of the ocean has been strongly affected by the presence of life. Life processes enter into many of the cycles that help to stabilize the chemical composition of seawater. Finally, we must consider the role the ocean plays in climatic stability and change.

Climate is the statistical aggregate of weather from which the daily, seasonal, and year-to-year variations can be determined. The mean monthly climate of the earth in January and July was shown in part in Figure 7-1 (p. 78) and Figure 10-11 (p. 151). The surface distributions of temperature, pressure, and wind are strongly affected by the oceans. We shall now examine the long-time variations of climate. Specifically, we wish to investigate the role the ocean might play in climatic change.

Climatic Changes

Everyone complains about the weather, and we often hear that the weather is not what it used to be. Our feelings about the weather are subjective; to discover climatic trends, however, we cannot resort to an opinion poll. During recent times, huge amounts of data have been gathered by the weather bureaus of the world. When we look at these data, we find that the average changes are remarkably small. Mitchell (1963) has analyzed the change in the average annual temperature between two 30-year periods, 1890–1919 and 1920–1949, and finds a change of only $+0.22 \pm 0.04°C$ between 60°N and 60°S. Although this change appears to be significant, it could not have been deduced from casual examination of the data.

In more localized areas, however, the changes in climate have been much more dramatic. For example, between 1962 and 1965, the northeastern United States suffered an unusual drought, which had grave economic consequences. Unfortunately, the United States Weather Bureau had no Joseph on its staff to predict the duration of the dry spell. Since Freud, the interpretation of dreams has been taken over by another science, and meteorology has not yet developed methods for long-term weather prediction.

The climatic variations during the last 1000 years have been most dramatic in the extreme North Atlantic, and these variations have had a profound effect on world history. Although we do not have instrumental records that cover this period, it is possible to obtain an objective picture by examining written records of physical events.

In Iceland, agriculture is seriously affected by climatic variations. Because of the impact of sea ice on farming, the diaries of Icelandic farmers from earliest times note the presence of ice on the coast. From such diaries, Koch (1945) has compiled a record of sea-ice conditions on the coast of Iceland starting with A.D. 860, ten years after the island was discovered by the Scandinavians (Fig. 30-1).

Until the end of the twelfth century, ice conditions were relatively mild. As a result, when Eric the Red was outlawed from Iceland in 981, he was able to sail due west and colonize Greenland. This led to the exploration of the east coast of North America by his son, Leif Ericson, at the end of the tenth century (Fig. 4-6, p. 33).

During the thirteenth and fourteenth centuries, the ice conditions deteriorated and blocked the direct route between Iceland and Greenland (Fig. 30-2). The colder weather enabled the Eskimos to wipe out

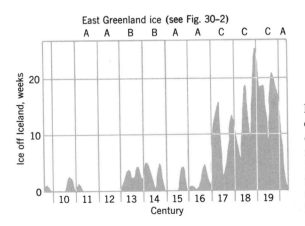

Figure **30-1** Twenty-year average of the number of weeks of ice near Iceland since the ninth century. (After Koch, 1945). The letters *A*, *B*, *C* refer to the probable ice limit in Greenland waters (see Fig. 30-2).

Figure **30-2** Probable limit of sea ice around Greenland. (After Koch, 1945) (A) Eleventh, twelfth, fifteenth, sixteenth, and twentieth centuries; (B) thirteenth and fourteenth centuries; (C) seventeenth, eighteenth, and nineteenth centuries.

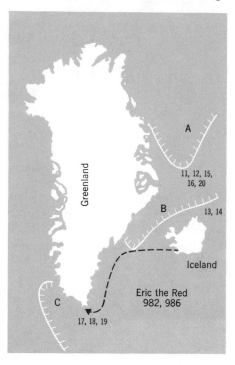

most of the Norse settlements in Greeland. As a result, the Vikings did not settle in America. The climate improved during the fifteenth and sixteenth centuries; however, by that time the Inugsuk Eskimos had taken over Greenland, and the last of the Vikings died at the beginning of the sixteenth century. During the next three centuries, the climate deteriorated drastically. The Eskimo settlements in Greenland became isolated, and survival in east Greenland became impossible. Agriculture in Iceland deteriorated to such an extent that evacuation to Europe was seriously considered. At the beginning of the twentieth century, the climate improved once more, and the Danes were able to establish administrative headquarters along most of the Greenland coast.

H. H. Lamb (1964) has investigated European records for evidence of mild and severe winters. Using such data as the dates of freezing and melting of rivers and lakes, he has developed a winter severity index. His findings are summarized in Figure 30-3. The upper curve shows the results for the London-Paris region, while the lower curve is for the region around Kiev, near 30°E, 50°N.

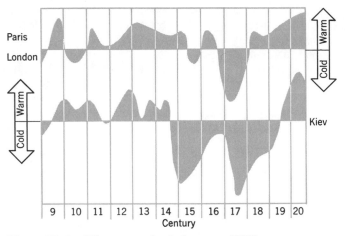

Figure **30-3**   Winter severity index near 50°N; upper curve near 0° longitude; lower curve near 30°E. (After Lamb, 1964)

When we compare the climatic curves in Figures 30-1 and 30-3, we find that the climate was relatively mild during the first half of the second millenium. Conditions then became much colder in a period often referred to as the "little ice age." They improved once more near the end of the nineteenth century.

Although the gross climatic features in western Europe were similar, the timing of the changes differ from one place to the other. Therefore we are not dealing with simultaneous changes in the hemispheric climate but, rather, with a more complex pattern of change.

The most dramatic climatic changes that have affected man have been the ice ages during the Quaternary. The advances and retreats of the continental ice sheets in North America and northwestern Eurasia have altered sea level by 100 m (Fig. 15-7, p. 249). There is no indication that this succession of glacial and interglacial periods is over. Another glacial period can probably be anticipated within the next 10,000 years. If man does not destroy himself by nuclear war or by poisoning the environment first, will we understand our environment sufficiently by then to anticipate and possibly control the next ice age? If we survive the cold war, our descendants will have to prepare for a war against the cold.

The Causes of Climatic Change

The Sun. We have seen that the climate has undergone significant changes. We must now inquire into the causes of climatic change and particularly into the role the ocean may play in controlling the climate. Before we can do this, however, we must once more start with the prime source of energy, the sun. In Chapter 6 we saw that the motions of the earth give rise to a seasonal variation in the amount of solar radiation received. In addition to the strong annual cycle, there are also very much weaker, longer cycles, which result from the variation in the ellipticity of the earth's orbit (92,000 years), the variation of the major axis of the earth's orbit (21,000 years), and the variation in tilt of the earth's axis relative to the plane of its orbit (40,000 years).

The combined effect of these cycles is small. Milankovitch (1938) has calculated that the variation in insolation is equivalent at latitude 65°N to about a 5° shift in latitude. It is interesting, however, that the frequencies of these variations are of the same order as the glacial cycles. If the surface of our planet acts so as to amplify this weak thermal signal, astronomical variations may become significant.

Milankovitch's calculations of the changes in incoming solar radiation are based on the changes in geometry only and assume that the energy output of the sun remains constant. We do know that the sun undergoes short time fluctuations; for example, the number of sunspots varies with an 11-year period. Unfortunately, we do not yet have instrumental data to show how constant is the total energy radiated by the sun. Measurements at the earth's surface determine only the amount of energy that has passed through the atmosphere. To measure true variations in the solar-energy output, we must make measurements outside the atmosphere over long periods of time. It is to be hoped that satellite data or a solar-monitoring station on the moon will soon tell us about the constancy of the solar-energy output. However, it will be a long time before we can establish the possible existence of long-term cycles.

At the present time, the question of variability in solar output cannot be answered. If no other causes can be found, one can always assume that climatic change is due to variability of the sun. Meanwhile, however, let us investigate whether climatic changes could result from processes on earth.

The Atmosphere. The energy received from the sun interacts with the atmosphere (Fig. 7-6, p. 85); however, the atmosphere has a very short time constant. The weather changes from day to day, and meteorologists are not able to predict the weather more than a few days in advance. The water content of the atmosphere is exchanged with that of the ocean about once a week (Fig. 10-10, p. 150). The mass of the atmosphere is equivalent to 10 m of water, while its heat capacity is equivalent to only 2.4 m of water. Thus, while the atmosphere is responsible for the day-to-day variations in the weather, its memory is too short to account for long-term climatic variations.

An interesting confirmation of this statement has been provided by Mintz (1968). He simulated the atmosphere by using a large computer which solves the equations of motion of the atmosphere. Mintz starts out with an atmosphere at rest and then allows the latitudinal varia-tion of solar energy to put his model atmosphere into motion. After only about one month, the model displays a normal weather pattern. In this short span of time, the model atmosphere has forgotten its static past. The thermal time constant of the atmosphere is much shorter than the annual cycles. Therefore we must look elsewhere for causes of climatic change.

The chemical composition of the atmosphere has a long time constant. The transparency of the atmosphere for infrared radiation depends on the concentration of carbon dioxide. Even though the atmosphere con-tains only 0.03 percent CO_2, this small amount of gas significantly affects the heat balance. If we calculate the temperature of the earth for a CO_2-free atmosphere, we find that the climate would be about 10° cooler. The actual cooling would be even greater, since the change in tempera-ture would reduce the water content of the atmosphere, which, in turn, would lower the temperature still more. Small variations in the CO_2 content of the atmosphere can thus affect the climate.

In Chapter 21 (Fig. 21-10, p. 345) we saw how man's industrial activity has increased the amount of CO_2 in the atmosphere. From 1890 to 1940 the mean temperature of the world increased by about 0.5°C. It was therefore only natural to assume that this increase resulted from the addition of CO_2 by fossil-fuel burning. Because of the ever-increasing rate of fuel burning, one would expect the upward trend in world tem-peratures to become ever steeper. Instead, the climatic record shows that there has been a downward trend in world temperatures since 1940. Thus, while the CO_2 content of the atmosphere undoubtedly plays a role, it alone cannot be responsible for the present climatic trends.

The Surface of the Earth. The energy from the sun, after it traverses the atmosphere, is intercepted by the land and ocean surface of the earth. In Chapter 7, we saw that the seasonal variation of the temperature is much larger over the continents than over the ocean. Seawater has a much larger storage capacity for heat than the land, and therefore the ocean tends to even out the annual temperature fluctuations. The winter mixed layer is about 100 m deep, so that the heat capacity of the ocean surface is 40 times that of the atmosphere. The annual cycle in ocean surface temperature, however, indicates that the thermal time constant of the sea surface is less than one year. Since the annual variations on the land are even larger, its memory appears to be even shorter.

That the land can have a long-time climatic effect, however, is illustrated by Greenland. Figure 7-1 (p. 78) shows that the temperatures over Greenland are abnormally low. This is a direct consequence of the Greenland ice cap. The presence of ice adds elevation to the land, and ice is a good reflector of solar radiation. Both effects lead to surface cooling. Thus the presence of ice lowers the surface temperature and helps to preserve the ice cover.

If we used giant bulldozers to scrape the ice cover off Greenland, we would obviously lower its elevation. At the same time, the bare ground would absorb much of the solar radiation now reflected by the ice. As a result, once the ice were removed, the surface temperature would probably rise sufficiently high to prevent the reestablishment of the permanent ice cover. The present Greenland ice sheet is a remnant of the last glaciation.

Sea ice has a similar effect. Although it cannot significantly alter the elevation of the sea surface, it is a good reflector for solar radiation, while open water is a good absorber. If we were to melt all the ice in the Arctic Ocean, the increased absorption of solar radiation would probably prevent the ocean from freezing over again. In this way climate depends on history and tends to resist change. A glaciated area tends to remain covered by ice, and an ice-free surface, by absorbing more solar energy, tends to remain warm.

A Theory of the Ice Ages

In Chapter 29, we saw that the circulation of the Atlantic Ocean differs significantly from that of the Pacific. To a large extent this difference is a result of the higher salinity of the Atlantic. Figure 11-22 (p. 180) shows that sea ice in the Pacific extends to 60°N, but areas of the Atlantic are ice-free to 75°N. The low surface salinity in the Pacific limits the downward convection of heat, and therefore sea ice forms readily. The uniform vertical salinity structure in the eastern North Atlantic (Fig. 11-23, p. 181), on the other hand, permits convection to the ocean bottom, and so the sea does not freeze over.

What would happen if the North Atlantic had a surface salinity as low as that of the Pacific? In that case, we would expect sea ice to extend as far south as in the Pacific, to 60°N. Thus the north coast of Great Britain would be locked in ice (Fig. 30-4). The distribution of sea ice would now be symmetrical about the North Pole, and southward extension of the sea ice would cool northern Europe to temperatures like those of Alaska (Fig. 7-1, p. 78). At the same time, the polar high-pressure system would become more symmetrical and would displace the low-pressure region over Iceland further south (Fig. 10-11, p. 151).

The resulting change in the winds would alter the surface currents in the North Atlantic to a system similar to that of the Pacific. Instead of a warm, salty current flowing north along the coasts of Scandinavia, we would have a separate North Atlantic subarctic gyre (Fig. 30-4). This gyre would prevent the waters from the warm, salty subtropical gyre from carrying heat and salt to the edge of the ice. Instead, the excess of precipitation over evaporation would keep the surface salinity low in the north and thus preserve the sea ice. Once we reduced the salinity of the surface water in the North Atlantic, the salinity would stay low, and the sea ice would be preserved.

Let us now look at the maximum extent of the glacial ice during the ice ages, as indicated in Figure 30-4. We note that the ice sheets are concentrated primarily around the North Atlantic. In the North Pacific the situation differs less from present conditions. It does not appear unreasonable to suppose that the cooling of the North Atlantic, due to the additional sea ice postulated, could have led to the formation of glacial ice sheets.

So far, our hypothetical reconstruction is a piece of science fiction. If by magic we could make the Atlantic behave like the Pacific, we would probably find ourselves in another ice age. To see if such a transformation would have been possible, we must ask why the Atlantic differs from the Pacific at the present time. If this difference is due to the differences

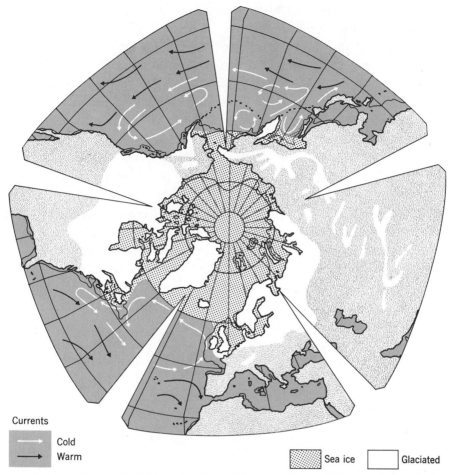

Currents

Cold

Warm

Sea ice   Glaciated

Figure **30-4** Hypothetical reconstruction of the oceanic conditions during a glacial period. The extent of sea ice and of the glaciated land surface are indicated. (After Weyl, 1968b)

in geography, then the reconstruction has no scientific merit since the changes in geography since the last glacial have been very minor. If, on the other hand, the difference could be the result of a minor but persistent change in the atmospheric circulation, then Figure 30-4 can be taken more seriously.

The difference between the present North Atlantic and North Pacific is related to their differences in surface salinity (Figure 12-4, p. 192). If we could reduce the surface salinity of the North Atlantic, we would be closer to establishing the hypothetical glacial conditions. The required change in salinity could be brought about by reducing the transfer of water vapor from the Atlantic to the Pacific. A large amount of water vapor is currently transported across the Isthmus of Panama by the trade winds in winter (Fig. 10-11a, p. 151). In summer, on the other hand, the winds are diverted in a more northerly direction, and the water is shed against the backbone of the Americas, the Cordilleran, and flows back into the Atlantic (Fig. 10-11b, p. 151).

If the atmospheric circulation in the Caribbean region became more summer-like over a long period of time, the Atlantic salinity would gradually be reduced. If the water-vapor flux across the Isthmus were completely blocked for 600 years, and if other conditions remained the same, the average Atlantic salinity would equal that of the Pacific. Such a shift might result from the slight changes in relative insolation brought about by the long-cycle changes in the geometry of the earth-sun system.

Thus small changes in the atmospheric circulation, if they are sufficiently persistent, could produce major changes in the ocean circulation, resulting in a major change in climate. It is possible to check the reconstruction of the glacial ocean by looking at the record of marine organisms as preserved in the deep-sea sediments. There should have been large changes in the eastern North Atlantic, north of 40°N, while the conditions in the North Pacific should have been relatively unchanged. The data we have so far seem to confirm this; however, more work needs to be done. This will undoubtedly show that the situation is more complex.

Summary

The world average climate has been stable during the first half of the twentieth century. Climatic conditions in the North Atlantic, however, have undergone major changes since the year 900, and these changes have affected history. The largest climatic variations since the appearance of man are the ice ages of the Quaternary. Climatic change can result from variations of the sun, changes in the atmosphere, and changes on the surface of the earth.

Variations in the ocean circulation could be brought about by minor long-term changes in the average behavior of the atmosphere. Such changes in the ocean may be responsible for short-time climatic variations as well as for the major variations in climate during the Quaternary.

We have much to learn about the ocean and how it interacts with the atmosphere and with life on earth. Such knowledge is essential if we are to be able to predict the future evolution of our environment. Perhaps this knowledge will enable man to live in better harmony with nature and to control the climate of the future. In the past, man retreated before the advancing ice of the glacial periods. When the glaciers melted and sea level became established at its present level, irrigation agriculture gave rise to civilization. Now, 6000 years later, man has the capability to poison his environment. Assuming that he does not do so, will he be able to prevent the next ice age?

Study Questions

1. Compare the timing of the "little ice age" as revealed in the four areas shown in Figures 30-1, 30-2, and 30-3.
2. What effect does a snow cover have on climate?
3. Compare Figure 30-4 with Figure 12-5. How does the suggested glacial circulation differ from the present?

Supplementary Reading

(Starred items require little or no scientific background)

Mitchell, J. Murray, Jr. (1965). "Theoretical Paleoclimatology." In H. E. Wright, Jr., and D. G. Frey, eds., *The Quaternary of the United States.* Princeton: Princeton University Press. Pp. 881–901.
*Mowat, Farley (1965). *Westviking.* Boston: Little, Brown. (A discussion of the Viking explorations of North America.)
Namias, Jerome (1965). "Short-Period Climatic Fluctuations," *Science,* **147,** February, pp. 696–706.
——— (1967). "Long-Range Weather Forecasting—History, Current Status and Outlook," *Bulletin of the American Meteorological Society,* **49,** pp. 438–470.
Weyl, Peter K. (1968). "The Role of the Oceans in Climatic Change: A Theory of the Ice Ages," *Meteorological Monographs,* **8,** February, pp. 37–62.

References

BAGNOLD, R. A. (1962). "The Sea," *Interscience,* **3,** pp. 507–528.

BARGHOORN, E. S. and SCHOPF, J. W. (1966). *Science* **152,** pp. 758⁻763.

—— and TYLER, A. S. (1965). *Science* **147,** p. 563.

BASCOM, W. (1964). *Waves and Beaches.* New York: Anchor Science Study Series, Doubleday.

BERKNER, L. V., and MARSHALL, L. C. (1965). *J. Atmos. Sci.,* **22,** p. 225.

—— (1966). *J. Atmos. Sci.,* **23** (2), p. 133.

BROWN, C. W., and KEELING, C. D. (1965). *J. Geophys. Res.,* **70,** p. 6077.

BROWN, H. (1952). *The Atmospheres of the Earth and Planets.* Chicago: University of Chicago Press. Pp. 258–266.

BULLARD, E. C., et al. (1965). *Phil. Trans. Roy. Soc. London, A,* **258,** pp. 41–51.

CARRUTHERS, J. N. (1967). *Fishing News,* June 23, p. 29.

CLOUD, P. E. (1968). In *Evolution and Ecology.* New Haven: Yale University Press.

CONALLY, J. R., and EWING, M. (1965). *Science,* **150,** pp. 1822–1824.

CUSHING, D. H. (1959). "On the Nature of Production in the Sea," Ministry of Agriculture, Fisheries and Food, *Fishery Investigations,* Ser. II, **XXII,** No. 6, London, HMSO.

DEFFEYES, K. S. (1965). *Limnol. Oceanogr.,* **10,** p. 412.

DEUTSCH, E. R. (1966). Royal Society of Canada, Spec. Publ. No. 9, p. 34.

DODIMEAD, A. J., FAVORITE, F., and HIRANO, T. (1962). *Review of the Oceanography of the Subarctic Pacific,* Fisheries Research Board, Canada, POG, Nanaimo.

DOELL, R. R., and DALRYMPLE, G. B. (1966). *Science,* **152,** p. 1060.

EMERY, K. O. (1960). *The Sea off Southern California.* New York: John Wiley and Sons.

FUGLISTER, F. C. (1960). *Atlantic Ocean Atlas,* Woods Hole: Woods Hole Oceanographic Institution.

FULLER, B. (1943). *Life,* March 1.

FULTZ, D. (1961). *Advan. Geophys.,* **7.**

GOREAU, T. F. (1959). *Ecology,* **40,** pp. 67–90.

—— (1961). *Endeavour,* **20,** pp. 32–39.

—— (1967). *Bull. Marine Sci.,* **17,** pp. 432–441.

——, and HARTMAN, W. D. (1963). *AAAS* Publ. 75, p. 36.

HARDY, SIR A. (1967). *Great Waters.* New York: Harper and Row.

HARVEY, H. W. (1950). *J. Marine Biol. Assoc.,* **25,** pp. 97–137.

HEDGPETH, J. W. (1957). *Geol. Soc. Am. Memo.* 67, **1,** p. 17.

HEEZEN, B. C., and EWING, M. (1952). *Am. J. Sci.,* **250,** pp. 849–873.

———, THARP, M., and EWING, M. (1959). "The Floors of the Oceans, I," *Geol. Soc. Am., Spec. Paper 65.*

HESS, H. H. (1946). *Am. J. Sci.,* **244,** pp. 772–791.

HJULSTROEM, F. (1939). In Trask, *Recent Marine Sediments.* American Association of Petroleum Geology.

HOLLAND, H. D. (1965). *Proc. Nat. Acad. Sci. U. S.,* **53** (6), p. 1173.

HURLEY, P. M., et al. (1967). *Science,* **157,** p. 495.

ISACKS, B., OLIVER, J., and SYKES, L. R. (1968). *J. Geophys. Res.,* **73,** pp. 5855–5899.

JUDSON, S. (1968). *Science,* **160,** pp. 1444–1446.

———, and RITTER, D. F. (1964). *J. Geophys. Res.,* **69,** p. 3395.

KOCH, L. (1945). *Meddelelser om Gronland,* **130,** No. 3, pp. 1–373.

LAMB, H. H. (1964). *Geol. Rundschau,* **54,** pp. 486–504.

LANGMUIR, I. (1938). *Science,* **87,** p. 119.

LA VIOLETTE, P. E., and CHABOT, P. (1967). *Deep-Sea Res.* **14,** pp. 485–486.

LE PICHON, X. (1968). *J. Geophys. Res.,* **73,** pp. 3661–3697.

LIVINGSTON, D. A. (1963). *U. S. Geol. Surv., Profess. Paper* 440 G.

MENARD, H. W., and SMITH, S. M. (1966). *J. Geophys. Res.,* **71,** p. 4305.

MILANKOVITCH, M. (1938). *Handbuch der Geophysik,* **9,** pp. 593–698.

MILLER, S. L. (1959). In *The Origin of Life on the Earth,* **1.** New York: Pergamon.

MINTZ, Y. (1968). *Meteorol. Monographs,* **8,** pp. 20–36.

MITCHELL, J. M., JR. (1963). In *Changes of Climate,* p. 1161, Arid zone Research XX, UNESCO, Paris.

MONTGOMERY, R. B. (1958). *Deep-Sea Res.,* **5,** pp. 134–148.

MUNK, W. H., and RILEY, G. A. (1952). *J. Marine Res.,* **11,** p. 215.

NEWELL, N. P., et al. (1959). *Bull. Am. Museum Nat. Hist.,* **117,** Article 4.

PALES, J. C., and KEELING, C. D. (1965). *J. Geophys. Res.,* **70,** p. 6053.

RASMUSSON, E. M. (1967). *Monthly Weather Rev.,* July, p. 403.

REID, J. L., JR. (1965). *Intermediate Waters of the Pacific Ocean.* Baltimore: The Johns Hopkins Press.

REVELLE, R., and SUESS, H. (1957). *Tellus,* **9,** pp. 18–27.

RILEY, G. A., et al. (1956). *Bull. Bingham Ocean Coll.,* **15.**

———, STOMMEL, H., and BUMPUS, D. F. B. (1949). *Bull. Bingham Ocean Coll.,* **12,** No. 3.

ROBINSON, R. A. (1965). *J. Paleontol.,* **39,** pp. 355–364.

RUBEY, W. W. (1951). *Bull. Geol. Soc. Am.,* **62,** pp. 1111–1148.

RYTHER, J. H. (1954a). *Ecology,* **35,** p. 522.

——— (1954b). *Biol. Bull.,* **106,** pp. 198–209.

SCHLANGER, S. O., et al. (1963). *U. S. Geol. Surv., Profess. Paper* 260 BB.

SCHOPF, J. W. (1967). *McGraw-Hill Yearbook of Science and Technology.*

SELLERS, W. D. (1965). *Physical Climatology.* Chicago: University of Chicago Press.

SHAW, E. (1962). "The Schooling of Fishes," *Sci. Am.,* June.

SHEPARD, F. P. (1963). *Submarine Geology* 2nd ed. New York: Harper and Row.

SKOPINTZEV, B., and GUBIN, F. (1955). *Trans. Hydrophys. Inst. Sea,* AN, USSR, **5,** (R).

SOROKIN, C., and KRAUSS, R. W. (1958). *Plant Physiol., 33,* pp. 109–113.

STARR, V. P., PEIXOTO, J. P., and CRISI, A. R. (1965). *Tellus, 17,* pp. 463–472.

STEMANN, N. E. (1956). In *The Galathea Deep Sea Expedition.* London: Allen and Unwin. pp. 55–64.

STOVER, C. W. (1968). *J. Geophys. Res., 73,* p. 3817.

SUTCLIFFE, W. H., JR., BAYLOR, E. R., and MENZEL, D. W. (1963). *Deep-Sea Res., 10,* p. 233.

SWALLOW, J. C. (1955). *Deep-Sea Res., 3,* pp. 74–81.

THORSEN, G. (1957). In *Geol. Soc. Am. Memo.* 67, **1,** pp. 505–507.

VINE, F. J. (1966). *Science, 154,* p. 1405.

———, and MATTHEWS, D. H. (1963). *Nature, 199,* p. 947.

WEYL, P. K. (1967). *Studies Tropi. Oceanogr.,* Miami, **5,** p. 178.

——— (1968a). *Science, 161,* pp. 158–160.

——— (1968b). *Meteorol. Monographs, 8,* pp. 37–62.

WILSON, J. T. (1965). *Science, 150,* pp. 482–485.

WIMPENNY, R. S. (1966). *Plankton of the Sea.* New York: American Elsevier Publishing Co., p. 140.

WORZEL, J. L., and SHURBET, G. L. (1955). *Proc. Nat. Acad. Sci., 41,* pp. 458–469.

WÜST, G. (1935). *Meteor Reports, VI,* 1st part. Berlin: Walter de Gruyter.

——— (1954). *Kiel. Meeresforsch., 10.*

——— (1960). *Deut. Hydrograph. Zeit., 13,* pp. 105–131.

——— (1961). *Deut. Hydrograph. Zeit., 14,* pp. 81–92.

WYATT, B., et al. (1967). Department of Oceanography, Oregon State University, Data Report No. 26, Ref. 67–20.

ZUBOV, N. N. (1943). *Arctic Sea Ice,* English transl. San Diego: U. S. Navy Electronics Laboratory.

Appendix 1 Exponential Numbers

To simplify the writing of very large and very small numbers, scientists use powers of ten:

$$1\ 000\ 000\ 000 = 10^9$$
$$1\ 000\ 000 = 10^6$$
$$1\ 000 = 10^3$$
$$100 = 10^2$$
$$10 = 10^1$$
$$1 = 10^0$$
$$0.1 = 10^{-1}$$
$$0.01 = 10^{-2}$$
$$0.001 = 10^{-3}$$
$$0.000\ 001 = 10^{-6}$$
$$0.000\ 000\ 001 = 10^{-9}$$

To multiply two powers of ten, the exponents are added:

$$100 \times 1\ 000 = 100\ 000 \text{ or } 10^2 \times 10^3 = 10^5$$

In general form: $\qquad 10^a \times 10^b = 10^{a+b}$

To divide a power of ten by another, the second exponent is subtracted from the first:

$$1\ 000 : 100 = 10 \text{ or } 10^3 : 10^2 = 10^1$$

In general form: $\qquad 10^a : 10^b = 10^{a-b}$

To raise a power of ten to a power, the exponent is multiplied by the power.

$$(100)^3 = 100 \times 100 \times 100 = 1\ 000\ 000 \text{ or } (10^2)^3 = 10^6$$

In general form: $\qquad (10^a)^b = 10^{ab}$

To take a square or cube root of a power of ten, the exponent is divided by 2 or 3 respectively:

$$\sqrt{10^6} = 10^3, \quad \sqrt[3]{10^6} = 10^2, \quad \text{and} \quad \sqrt{10^5}$$
$$= 10^2\sqrt{10} = 10^{2.5}, \quad \text{since} \quad \sqrt{10} = 10^{\frac{1}{2}}$$

To add or subtract numbers expressed as powers of ten, they must be converted to the same power:

$$3 \times 10^5 + 2 \times 10^4 = 3 \times 10^5 + 0.2 \times 10^5 = 3.2 \times 10^5$$
$$6 \times 10^{-4} - 4 \times 10^{-5} = 6 \times 10^{-4} - 0.4 \times 10^{-4} = 5.6 \times 10^{-4}$$

Problems

1. $2 \times 10^5 \times 4 \times 10^3 = 8 \times 10^8$
2. $3 \times 10^4 \times 5 \times 10^{-6} = \quad *$
3. $5 \times 10^{-6} \times 2 \times 10^{-3} =$
4. $8 \times 10^{-3} : 4 \times 10^6 = 2 \times 10^{-9}$
5. $3 \times 10^9 : 2 \times 10^3 =$
6. $6 \times 10^{-3} : 3 \times 10^{-8} =$
7. $(2 \times 10^3)^3 = 2^3 \times 10^9 = 8 \times 10^9$
8. $(3 \times 10^{-4})^2 =$
9. $(5 \times 10^2)^3 =$
10. $\sqrt{4 \times 10^{-6}} = 2 \times 10^{-3}$
11. $\sqrt[3]{27 \times 10^9} =$
12. $\sqrt{6.4 \times 10^{-5}} =$
13. $5 \times 10^{-3} + 10^{-2} =$
14. $7 \times 10^9 + 3 \times 10^6 =$

*The answers to these problems on page 522.

Appendix 2 The Metric System and Other Units

Metric Units	Other Units

Length

1 micron, $1\mu = 10^{-6}$ m	1 inch = 2.54 cm
1 millimeter, 1 mm = 10^{-3} m	1 foot = 30.3 cm
1 centimeter, 1 cm = 10^{-2} m	1 fathom = 6 feet = 1.83 m
1 meter, 1 m	1 mile = 1.61 km
1 kilometer, 1 km = 10^3 m	1 nautical mile = 1.85 km
	= 1 minute of latitude

Volume

1 liter = 10^3 cm³

Time

1 year = 3.16 × 10⁷ seconds
1 day = 8.64 × 10⁴ seconds

Mass

1 kilogram, 1 kg = 10^3 gram, g	1 pound = 0.454 kg

Speed

1 km hr⁻¹ = 27.8 cm sec⁻¹	1 knot = 51.5 cm sec⁻¹
	= 1 nautical mile hr⁻¹

Acceleration

1 gal = 1 cm sec⁻²
acceleration of gravity a_g = 980 gal

Pressure

1 bar = 10⁶ dynes cm⁻²	1 atmosphere = 1.013 bar

| Metric Units | | | Other Units |

Energy

| 1 joule $= 10^7$ ergs | 1 calorie $= 4.19$ joules |
| | $= 4.31 \times 10^{-2}$ liter atmospheres |

Illumination

721 foot candles $= 1$ cal cm^{-2} hr^{-1}

Temperature

°C	°K	°F	
0	273.2	32	Freezing point of water
100	373.2	212	Boiling point of water
10	283.2	50	
20	293.2	68	
30	303.2	86	
40	313.2	104	
−40		−40	

Conversion formula:

$$9 \times (°C + 40) = 5 \times (°F + 40)$$

Example: convert 10°C to °F
$9 \times (10 + 40) = 450 = 5 \times (°F + 40)$
°F $+ 40 = 90$, therefore 10°C $= 50$°F

Answers to Problems in Appendix 1

2. $1.5 \times 10^{-1} = .15$
3. 10^{-2}
5. 1.5×10^6
6. 2×10^{-11}
8. 9×10^{-8}
9. 1.25×10^8
11. 3×10^3
12. 8×10^{-3}
13. 1.5×10^{-2}
14. 7.003×10^9

INDEX

Numbers of pages on which terms are defined are in **boldface**. Asterisks indicate illustrations.

Iron, 359
Isacks, B., 299, 516
Isostasy, 210, 217, 223, 224
Isotherm, **174**

Jack-sail-by-the-wind, 397*
Jackson-Harmsworth expedition, 50
Jamaica, 454
Japan, 242
Jawless fish, 373, 409
Jellyfish, 370*, 373, 398*
Jigsaw Puzzle, 295-296
Johnson, M. W., 6
Judson, S., 220*, 221, 233, 234, 516

Kanamori, H., 307
Keeling, C. D., 345-347, 514, 516
Kelp, 389*, 391
Kelvin, Lord, 4
Kelvin Temperature Scale, 69, 135, 522
Kinsman, B., 185
Knot, 36, 521
Koch, L., 504, 505*, 516
Kort, V. G., 502
Krakatoa, 242
Krauss, R. W., 376, 516
Krill, 403, 403*, 420, 423
Krypton, 312, 353, 353
Kuenen, Ph. H., 234, 289
Kummel, B., 26
Kuroshio Current, 130, 131*, 197, 199
Kylstra, J. A., 416

Labrador Current, 181, 182
Lamb, H. H., 505, 506*, 516
Laminar flow, 226
Lamp shells, 405
Lamprey, 409*
Land, and water, distribution of, 64*, 65
  plants, 21, 367
Langmuir, I., 421, 516
Langmuir circulation, 421*, 422, 441
Latitude, 31, 37, 55*, 201-294
  length of degree of, 59
Lauff, G. H., 480
La Violette, P. E., 195*, 516
Le Pichon, X., 299, 301*, 304, 516
Lermond, J. W., 446
Lewed, George Henry, 3
Life, basis of, 365-374
  in deep ocean, 40, 41
  diversity of, 371-374

history of, 15-26
  origin of, 359, 360, 422, 441
  spontaneous creation of, 365, 366
  zones, 367, 368*
Light, *see* Radiation
Limestone, 282, 355, 453, 458*
Limey, 38
Limpet, 343, 370*, 406
Linnaeus, 371
Liquid-vapor equilibrium, 141
Lithosphere, **301**
  blocks of, 299-301
"Little" ice age, 506
Livingston, D. A., 324, 516
Lobster, 371, 372*, 402, 404, 424
Lodestone, 33
Logarithmic scale, 80*, 81
Logbook, 36, 39
London Bridge, 238
Long Island Sound, 426-429
Longitude, 31, 37, 38, 55*, 291, 292
Long-shore drift of sediments, 256-260
Lunar tide, 239-242
Lung, 411-415
Lyon, Gulf of, 479

Macoma community, 422, 423*
McAlester, A. L., 26
Magellan, Ferdinand, 35
Magellan, Strait of, 35
Magnesium, 316, 324, 343
Magnetic, compass, 33
  field of earth, 292-294
  growth stripes, 297-299
  polarity reversals, 297*
Magnetite, 293
Malacostraca, 371, 402, 403
Malthus, T. R., 374
Mammalia (Mammals), 21, 22*, 23*, 373,
    411, 412-415
Man, evolution of, 27
  in the sea, 412-415
Manganese nodules, 281*
Mantle of the earth, **216**, 217, 353
Map, 53-65
Marmara, Sea of, 471, 472*, 473
Marshall, L. C., 359, 440, 514
Marshall Islands, 28*, 450*, 453
Marsigli, Ferdinando, 36, 473
Martin, O. L., Jr., 185
Matthews, D. H., 297, 517
Maury, Matthew Fontaine, 39